Stochastic Processes in DEMOGRAPHY *and* APPLICATIONS

SUDDHENDU BISWAS
G. L. SRIWASTAV

CENTRAL

Stochastic Processes in DEMOGRAPHY *and* APPLICATIONS

Suddhendu Biswas
Formerly Professor and Head, Department of Statistics
University of Delhi

and

G. L. Sriwastav
Professor, Department in Statistics
Dibrugarh University, Assam

New Central Book Agency (P) Ltd
LONDON
DELHI KOLKATA PUNE HYDERABAD ERNAKULAM

REGD OFFICE

8/1 Chintamoni Das Lane, Kolkata 700 009, India
email: ncbapvtltd@eth.net

OVERSEAS

NCBA (UK) Ltd, 149 Park Avenue North, Northampton, NN3 2HY, UK

EXPORT

NCBA Exports Pvt. Ltd, 212 Shahpur Jat, New Delhi 110 049
email: ncbaexp@ncbapvtltd.com

BRANCHES

208B Peacock Lane, Shahpur Jat
New Delhi 110 049
email: ncbadel@ncbapvtltd.com

Shop Nos. 3 & 4, Vinayak Towers
681/B Budhwar Peth, Appa Balwant Chowk
Pune 411 002
email: ncbapune@ncbapvtltd.com

House No. 3-1-315, 1st Floor
Nimboliadda, Kachiguda, **Hyderabad** 500 027
email: ncbahydb@ncbapvtltd.com

GSS Shopping Complex, 1st Floor
Opposite College Post Office, Convent Road
Ernakulam 682 035
email: ncbaernk@ncbapvtltd.com

STOCHASTIC PROCESSES IN DEMOGRAPHY AND APPLICATIONS

First Published: July 2006
Reprinted: January 2011

PUBLISHER

New Central Book Agency (P) Ltd
8/1 Chintamoni Das Lane, Kolkata 700 009

PRINTER

New Central Book Agency (P) Ltd
Web-Offset Division, Dhulagarh, Sankrail, Howrah

TYPESETTER

Anin, Kolkata

COVER DESIGNER

Soumen Paul

COVER PRINTER

Liba Graphics, Kolkata

PROJECT TEAM

Joydeep Ghosh and Soumen Paul

ISBN: 81-7381-517-8

Price: ₹ 250.00

DEDICATED TO

The living memory of my parents
Late Amalendu Biswas
Late Bhakti Biswas

[Suddhendu Biswas]

and

The living memory of my mother
Late Sharda Devi

[G. L. Sriwastav]

Contents

Preface to the First Central Edition

The book was out of stock since a long time. In view of the great demand of this text, now, a completely revised version of the text is presented to make the coverage more uptodate and the presentation more comprehensive. Many topics which may not be exactly pertinent to the average readers in mathematical Demography—as Martingales Theory, Cox's proportional hazard model and other diffusion models have been excluded which are available in the first Author's first and second reprinted edition of the book Applied Stochastic Processes—A Biostatistical and Population oriented Approach (1993–first edition–New Age International and the John Wiley & Sons) (2004–second reprinted edition of New Central Book Agency).

To make the reading more comprehensive, we have attempted to elaborate the appendix of the book while providing outlines of Mathematics and Statistics needed to understand the methodologies in the Text. The last chapter of the book—on some Miscellaneous items in Indian Demography—have been considerably elaborated by inclusion of details of Sample Registration System (SRS) which provides the major source of vital statistics in India. The authors are grateful to Mr. Amitabha Sen and Mr. Soumen Paul for their best efforts to bring the edition to see the light of the day in such a elegant form.

It will be a matter of gratification of the Authors if the new edition also serves the purpose of students of Mathematical Demography. Suggestions from Teachers and Students towards the improvement of the text and pointing out errors due to oversight would greatly be appreciated.

1st June 2006

SUDDHENDU BISWAS
GOPAL L. SRIWASTAV

Preface to the First Edition

The book is a revision of my lecture notes delivered to students of the Indian Statistical Institute, Calcutta, University of Poona, University of Roorkee and the University of Delhi during the last twenty-five years or so. It is needless to add that during this period the subject has undergone significant changes—both in respect of inputs as well as the methods of treatments. New inputs as Stochastic modelling in Demography, Competing risk theory and survival theory, problems of incomplete demographic data, (Censored sampling) and sampling and non-sampling problems in the estimation of demographic parameters have completely reoriented the traditional structure of pedagogy through deterministic-cum-actuarial approach. Further, this has motivated a modern stochastic process-oriented treatment to the subject. While the traditional textbooks following actuarial approaches have become little out of place to the statisticians at large, who would prefer a model oriented approach and appropriate estimational techniques amenable to experimental data based on small samples (because of the actuarial methods being generally based on the tacit assumption of the availability of a large body of the data which often can be handled by deterministic models), very few textbooks are available with the modern approaches. Even if a few are existent, almost no book is available which dives deep into the core of the subject. An attempt has, therefore, been made in this book to present the techniques of population studies and applications using a Stochastic process-oriented approach.

Thus the text beings with chapter one, where motivation is primarily to provide a survey of the techniques of Stochastic processes and renewal theory which may be pertinent to the study of Demographic models. Techniques developed recently, as Cox's regression model, followed by a new estimational technique as the method of partial likelihood, (the validity of which depends on Martingales Central limit theorem), have also been attempted to be covered within the purview of the book. A majority of the models so developed are utilised for the analysis of fertility, mortality and population growth, under various set-ups. Thus a variety of models depending on the age-sex factors as well as the density of the population etc. have all been critically examined. A chapter on Competing risk theory, survival theory and censored sampling is also a new feature of the book; which possibly bridges a gap between Demography and life-testing or Reliability. At places, attempts have been made to analyse Demographic models with the help of established theories of Stochastic processes (ex. Geiger-Muller Counter theory, point processes and Martingales theory etc.) to which the original contribution of the author is also linked. Finally, the last chapter of the book comprises of the techniques of Demography, especially suited to cover the deficiencies of the existing vital statistics in India; as 'Reverse Survival' and 'Differencing method' for estimating the birth and death rates from Census data and Chandrasekharan's & Deming's methodology estimating vital rates from registration data. This, again, to the best of the knowledge of the author, has not found its place in a textbook of Demography in its present form and motivation.

The book is expected to be helpful to the Postgraduate and advanced level of undergraduate students taking one or two semester course(s) in Mathematical Demography or Applied Stochastic processes with major emphasis in Population studies or Biostatistics. A knowledge of Calculus and Linear Algebra up to degree level of Indian universities is, however, a prerequisite. Knowledge of Probability and Statistics at the elementary level is also essential in understanding the development and application of the subject.

I have endeavoured to make the presentation lucid as far as practicable by sprinkling quite a large number of examples. However, an important omission in the book is the 'Stochastic analysis on Migration' which, indeed, constitutes a very important item of consideration in the analysis of population growth. Nevertheless, the above gap or omission can be filled up in a subsequent edition while noting the demand of the same. I further solicit the co-operation of the users of the book which may need revision and correction, particularly the errors which might have crept in inadvertently while completing the book.

I express my sincere regards and gratitudes to the eminent academicians who have profoundly influenced my thinking and encouraged me in preparing the text; especially Prof. P.K. Bose (the former pro-Vice Chancellor and the Head of the Department of Statistics, University of Calcutta) and professor V.S. Huzurbazar (Department of Statistics and Operations Research, University of Denver, Denver, Colorado, USA, and formerly the professor and the Head of the Department of Mathematics and Statistics, University of Poona). I should also be failing in my duties if I do not convey my regards to Mr. S.J. Poti, formerly the Head of the Department of Demography, Indian Statistical Institute, whose stimulating discussion and excellent guidance during my tenure there made the learning of the subject highly interesting. Finally I convey my heartiest thanks to all the colleagues of mine, especially to Dr. P.K. Kapur, Department of O.R., University of Delhi, whose valuable suggestions have gone a long way to improve the presentation of the book. My special thanks to Mrs. Romi Srivastava, a Ph.D. student of mine for undertaking the pain of correcting the proof very meticulously. I also thank my other students Mr. Tapan Kumar Pachal, Dr. (Miss) Gurprit Nair, Dr. (Mrs) Ganga Shrestha, Md. Abul Basher Mian, Mr. Nather Abas Ebraheem, Mr. Hameed Saad Noor, Mrs. Mariamma Thomas, Mr. A.K. Prabhakar and Mr. Monthiar Abid Abdallah for their sincere cooperation. Last, but not the least, my remembrances go to many unnamed colleagues and students of mine, my interactions with whom have led to new insights into the subject; whose needs and requirements have become the parameters of the book. I am thankful to M/s Wiley Eastern Limited, publishers, for all their efficient arrangement leading to publication of the book with excellent printing and elegant get-up.

April 1988 SUDDHENDU BISWAS
Department of Statistics
University of Delhi, Delhi

1 Elements of Stochastic Processes and Renewal Theory

1.0 Introduction

'Demography' or 'Population Studies' literally means the quantitative and qualitative study of the Population. 'Demo' means the people or Population at large and 'graphy' means a technique of projecting the picture of the same. Thus literally speaking 'Demography' is a science or technique leading to the analysis of a Population. If the features of a population (e.g. sex ratio, survival etc.) are looked from biological perspective, the subject becomes a branch of Biometry. On the other hand, if some characteristics of a population are considered as the outcome of a socio-economic process, then the study of the population should, of course be based on socio-economic consideration; and as such it may be viewed as a branch of 'Sociometry'. But strictly speaking, the scope of the subject being analysis of a population, which being qualitative as well as quantitative in nature, acquires the status of an independent discipline, covering the areas of Biology, Sociology and Economics while using Statistical methodologies. Here we look upon the subject from 'Statistical' point of view. That is we envisage to analyze the subject by employing appropriate Statistical or Probabilistic techniques. The application of the Probabilistic techniques in 'Biology' leads to the formation of 'Biometry', the 'Sociology' into 'Sociometry' and 'Demography' to 'Demometry'. However, the scope of the subject, which we envisage to study here, being not completed fully by probabilistic techniques only, while there exists a host of coverage with descriptive or qualitative elements, the name 'Demometry' may not be appropriate for the subject whose study constitutes the objective of this book. As such we retain the name 'Demography'.

Of late, the subject has grown considerable importance because of the problems arising out of high growth rates in certain specific populations leading to several challenges in the national economy. Because the solution necessitates a balancing in the level of resources with rapidly growing population.

The movement of a population as a result of various interacting factors like mortality, fertility and migration as well as the status of a population indicated by several biological and socio-economic indicators such as morbidity (or sickness), expectation of life, labour force, work force, national income, marriage patterns and related customs, sex ratio, family limitational measures, all can be considered as time dependent random variables or Stochastic Processes. As such, the techniques of stochastic processes become relevant in pursuing quantitative studies in Demography; especially while handling experimental data of demographic surveys based on small samples. The appropriate stochastic modeling techniques not only enables to obtain reliable estimates of certain useful demographic parameters (which are otherwise not estimable directly, often being non observable) but also provides indispensable tools for estimating parameters which are often associated with a high degree of non sampling or response errors. With this motivation we propose to introduce a study of demography using the techniques of Stochastic processes in the text.

This book assumes that reader has some basic knowledge of Calculus, Matrices, Probability theory and Statistical Methodology. Of course, some important results on these topics, which are frequently used in this book are given in the Appendix at the end of the book.

1.1 Stochastic Process

Definition: An arbitrary finite or infinite family of random variables $X(t)$ (or X_t), $t \in T$ where T is a parametric space, is called a Stochastic Process and t is called an index parameter. A stochastic process is completely specified by the joint probability distribution of $X_t \forall s,\ t \in T$. The values assumed by the process are called states and the set of possible values is known as 'State space'. The set of possible values of the indexing parameter is called the 'Parameter space', which can be either discrete or continuous. Even though in most of the physical problems, time is the natural index parameter, other kinds of parameters such as space, length etc., may also be used. A realization of a Stochastic Process $X(t)$ where $t \in T$, is the set of assigned values of $X(t)$ corresponding to $\forall t \in T$. The realization may be described by plotting $X(t)$ against t.

Point process: A point process or a **Counting process** is a stochastic process with integer valued state space concerning random collection of point occurrences. If we impose certain emphasis on the points on which events occur rather than measuring the quantitative aspects of events then the domain of study becomes 'Point Processes' which falls within the purview of Stochastic Processes.

1.1.1 Some Characteristics of Stochastic Processes

The basic characteristics which are satisfied by some special kind of stochastic processes are as follows:

(i) Stationarity, (ii) Orderliness and (iii) Absence of after effect.

Stationarity implies that for $\forall t \geq 0$ and an integer $k \geq 0$, the probability of k number of events in an interval say $(\alpha, \alpha+t)$ denoted by $P_k(\alpha, \alpha+t)$ is independent of $\alpha \forall\ \alpha \geq 0$. In other words the probability of k occurrences will depend only on k and t and is independent of α or the probability of a given number of events will be independent of the translation of the origin.

Orderliness: We have $\sum_{k=0}^{\infty} P_k(\alpha, \alpha+t) = 1, \forall \alpha \geq 0$ and k be any non-negative integer.

Orderliness implies the probability of more than one event in a small interval of time $(0, t)$ as $t \to 0$ is of the order zero.

$$\Rightarrow \quad \lim_{t \to 0} \frac{1 - P_0(\alpha, \alpha+t) - P_1(\alpha, \alpha+t)}{t} = 0. \tag{1.1.1}$$

In case the process is stationary, the orderliness implies

$$\lim \frac{1 - P_0(\alpha, \alpha + t) - P_1(\alpha, \alpha + t)}{t} = \lim \frac{1 - P_0(t) - P_1(t)}{t} = 0, \text{ as } t \to 0. \quad (1.1.2)$$

Absence of after effect: If $P_k(\alpha, \alpha + t)$ is independent of the sequence of events in $(0, \alpha)$ then we call the Stochastic process to be independent from the effect of 'after effect'. The absence of 'after effect' thus implies the mutual independence of subsets of a point process taken over different points of time which do not mutually overlap.

1.1.2 Classification of Stochastic Processes

Let us take the example of a life table function l_x, the number of survivors at each age $x(x = 0, 1, 2, \ldots)$. $\{l_x\}$ may be regarded as a Stochastic Process, the index parameter x belonging to a discrete non-negative integer valued parametric space. The values of $l_x \, \forall \, x$ are also integer valued non negative quantities which constitute the state space. Hence the state space is also discrete valued and non negative. However, in general a Stochastic process may be classified on the basis of Parametric space and State space as follows:

(i) Discrete State space and discrete Parametric space.
(ii) Discrete State space but continuous Parametric space.
(iii) Continuous State space but discrete Parametric space.
(iv) Continuous State space and continuous Parametric space.

Transition Matrix and Transition Probability

We illustrate the concept by a one dimensional random walk. Suppose, a gambler in each step of gambling may increase his pay off by one and decrease also his pay off by one with probabilities p and $q = (1 - p)$, respectively. Suppose that each step is taken independently of every other step. The process thus described is known as unrestricted random walk. We raise the question viz., given the gambler is in position i, initially what is the probability it is in state j in n steps, this is precisely denoted as $p_{ij}^{(n)}$, $i, j = -\infty, \ldots, -1, 0, 1, \ldots \infty$, $n = 1, 2, \ldots$ This defines n step transition probability. Similarly, when we raise the question viz, given the position of the gambler initially as i, what is the probability that it becomes j in one step, the answer is p_{ij}. Obviously, if $j = i + 1$ then $p_{ij} = p$ if $j = i - 1$ then $p_{ij} = q$ and $p_{ij} = 0$ otherwise. Theoretically, $[p_{ij}]$ could be imagined as a transition matrix of infinite order. However, in general, it can be imagined as a finite transition matrix of order m as

$$P = \begin{bmatrix} p_{11} \; p_{12} \; p_{13} & \cdots & p_{1m} \\ p_{21} \; p_{22} & \cdots & p_{2m} \\ \cdots \; \cdots & \cdots & \cdots \\ p_{m1} \; p_{m2} & \cdots & p_{mm} \end{bmatrix} \quad (1.1.3)$$

subject to the condition $p_{ij} \geq 0 \; \sum_j p_{ij} = 1 \, \forall \, i = 1, 2, \ldots, m$. P is called a Stochastic matrix. Also if $\sum_j p_{ij} = 1$ and $\sum_i p_{ij} = 1$ holds simultaneously then P is called a Doubly Stochastic Matrix. Some authors classify the Stochastic Process over discrete parameter space as 'Stochastic Sequence' and 'Stochastic Process' over continuous parameter space as 'Stochastic Processes' in the proper sense.

1.1.3 Markov Process

A discrete valued Stochastic Process $X(t)$ is called a Markov Process if for $t_0 < t_1 < \cdots < t_j \in T$ and for any integers $k_0, k_1, \ldots, k_j$

$$P\{X(t_j) = k_j | X(t_0) = k_0, X(t_1) = k_1, \ldots X(t_{j-1}) = k_{j-1}\}$$
$$= P\{X(t_j) = k_j | X(t_{j-1}) = k_{j-1}\} \qquad (1.1.4)$$

Thus in a Markov process given $X(t_i)$ (present or the current known value), the conditional probability distribution of $X(t_j)$ is independent of $X(t_0), X(t_1), \ldots,$ $X(t_{i-1})$ (i.e. the past known values). However, it is not necessary that a Markov process should be discrete valued. A continuous valued process will also be called a Markov Process if

$$P\{k'_j \le X(t_j) \le k_j | X(t_0) = k_0, X(t_1) = k_1, \ldots X(t_i) = k_i\}$$
$$= P\{k'_j \le X(t_j) \le k_j | X(t_i) = k_i\},$$

for any real number $k'_j \le k_j$.

1.2 Markov Chain

If the parameter space of a Markov Process is discrete then the Markov Process is called A 'Markov chain'. A definition of Markov chain may be given from life-table. If $\{l_x\}$ be the number of survivors at age $x = 0, 1, 2, \ldots$ then $\{l_x\}$ conforms to a simple Markov chain, since

$$P\{l_x = k_x \mid l_0 = k_0, l_1 = k_1, \ldots l_{x-1} = k_{x-1}\} = P\{l_x = k_x \mid l_{x-1} = k_{x-1}\}$$
$$\forall \text{ positive integer } k_i,\ i = 0, 1, 2, \ldots, x. \qquad (1.2.1)$$

A Markov chain is said to be homogeneous with respect to the parameter space if the transition probability $p_{ij}(t, \tau) = P[X(t) = j \mid X(\tau) = i]$ depends only on $(t - \tau)$ and not on t and τ separately.

1.2.1 Classification of the State of a Markov Chain

The state 'j' of a Markov chain ($j = 0, 1, 2, \ldots$) is said to be periodic with period 'π' (> 1) if

$$p_{jj}^{(m\pi)} > 0 \quad \forall m = 1, 2, \ldots, \qquad (1.2.2)$$

$$p_{jj}^{(n)} = 0 \quad \forall n \ne m\pi \ldots$$

If $\pi = 1$, then the state 'j' is called 'aperiodic'.

Let the r.v. T_{jj} be time at which the particle returns to the state 'j' for the first time. In this case we say that the state 'j' is 'recurrent' if $P[T_{jj} < \infty] = 1$ and 'transient' if $P[T_{jj} < \infty] < 1$.

Again a state 'i' is called 'null recurrent' if i is recurrent and $E(T_{ii}) = \infty$ and 'positive recurrent' and $E(T_{ii}) < \infty$.

Now, a state 'i' leads to a state 'j' (which we write $i \to j$) if for some integer $k \ge 1$, $p_{ij}^{(k)} > 0$. If $p_{ii}^{(m)} = 1$ and $p_{ij}^{(m)} = 0 \, \forall m$, for $j \ne i$, then state 'i' is called an 'absorbing state'.

We say that the states 'i' and 'j' communicate if $i \to j$ and $j \to i$. Denoting '$\sim$' as an equivalence relation, which implies

(i) $i \to i$

(ii) $i \to j \Rightarrow j \to i$

(iii) $i \to j,\ j \to k \Rightarrow i \to k$

which are the conditions of reflexivity, symmetry and transitivity, respectively. This defines an equivalent class among the states in the Markov chain. In other words, the entire set of the states in the Markov chain can be partitioned into equivalent classes called 'irreducible classes' of the Markov chain. Thus a Markov chain of just one class is called irreducible. It follows that while it is possible for a state belonging to an irreducible class to visit another state in another class, but it can never return. Otherwise two irreducible classes would communicate and form a single class. It can be seen that both periodicity and recurrence (or transience) are class properties.

For a Markov chain with finitely many states, it follows that at least one state must be 'recurrent' and 'recurrent state' must necessarily be 'positive recurrent'.

For a further discussion see Fellar (1951).

1.2.2 Kolmogorov's Theorem

If a Markov chain (M.C) given by its transition probability matrix $P = [p_{ij}]$ is irreducible and aperiodic then $\exists$ one $\pi \ni$

$$\pi' = \pi' P \Rightarrow \pi = P'\pi \tag{1.2.3}$$

then π is called the equilibrium distribution of the states. If the M.C. has an equilibrium distribution then it is called 'Ergodic'. In this class we have a single irreducible class and every state is recurrent and aperiodic.

1.3 Some Equations Associated With Transition Probabilities

Here we shall discuss some equations which will give important relationships among transition probabilities.

1.3.1 Chapman–Kolmogorov Equation

Markov property implies important relations among the transition probabilities $p_{ij}(t,\tau)$. Let ξ be a fixed point in (τ,t) i.e., $\tau < \xi < t$ and let $X(\tau)$, $X(\xi)$ and $X(t)$ be the corresponding random variables. Because of the Markovity

$$P\{X(t) = k|X(\tau) = i,\ X(\xi) = j\} = P\{X(t) = k|X(\xi) = j\} = p_{jk}(\xi,t) \tag{1.3.1}$$

We therefore have,

$$\begin{aligned} P\{X(\xi) = j,\ X(t) = k|X(\tau) = i\} &= P\{X(t) = k|X(\xi) = j\}\{X(\xi) = j|X(\tau) = i\} \\ &= p_{jk}(\xi,t)p_{ij}(\tau,\xi) = p_{ij}(\tau,\xi)p_{jk}(\xi,t) \end{aligned} \tag{1.3.2}$$

The above represents a probability of transition from $X(\tau) = i$ to $X(t) = k$ through $X(\xi) = j$. Since passage from i to k can take place through all the values of $\xi \in (\tau,t)$ therefore, we may write

$$p_{ik}(\tau,t) = \sum_{\xi\in(\tau,t)} p_{ij}(\tau,\xi)p_{jk}(\xi,t). \tag{1.3.3}$$

Equation (1.3.3) is known as Chapman-Kolgomorov equation.

A special case: In case of a Homogeneous Markov Process

$$\begin{aligned} p_{ik}(\tau,t) &= p_{ik}(t-\tau) \\ p_{ij}(\tau,\xi) &= p_{ij}(\xi-\tau) \\ p_{jk}(\xi,t) &= p_{jk}(t-\xi) \end{aligned} \tag{1.3.4}$$

Thus (1.3.3) is reduced to

$$p_{ik}(t-\tau) = \sum_{i\in(\tau,t)} p_{ij}(\xi-\tau)p_{jk}(t-\xi) \tag{1.3.5}$$

which is Chapman-Kolmogorov equation of a homogeneous Markov process.

1.3.2 Kolmogorov Differential Equation

One may derive Kolmogorov differential equation from Chapman—Kolmogorov equation under the following regularity assumptions:

(i) $\forall\, i\,\exists$ a continuous function $v_{ij}(\tau) \ni$

$$\lim_{\delta\to 0}\left(\frac{1-p_{ii}(\tau,\tau+\delta)}{\delta}\right) = -v_{ii}(\tau) \quad \text{where} \quad v_{ii}\le 0 \tag{1.3.6}$$

(ii) $\forall$ pair of integers $i,\ j, i\neq j\exists$ a continuous function $v_{ij}(\tau) \ni$

$$\lim_{\delta\to 0}\frac{p_{ij}(\tau,\tau+\delta)}{\delta} = v_{ij}(\tau) \tag{1.3.7}$$

Further, for fixed j the passage from i to j in $(\tau,\tau+\delta)$ in (1.3.7) is uniform with respect to i. The function $v_{ij}(\tau)$ is called the *intensity function* of the process.

If we put $p_{ij}(\tau,\tau)=\delta_{ij}$, where δ_{ij} is Kroneckar delta,

then (1.3.6) $\quad \Rightarrow \lim_{\delta\to 0}\dfrac{p_{ii}(\tau,\tau)-p_{ii}(\tau,\tau+\delta)}{\delta} = -v_{ii}(\tau)$

$$\Rightarrow -\frac{d}{dt}(p_{ii}(\tau,t))|_{t=\tau} = -v_{ii}(\tau)$$

$$\Rightarrow \frac{d}{dt}(p_{ii}(\tau,t))|_{t=\tau} = v_{ii}(\tau) \tag{1.3.8}$$

Similarly, (1.3.7) can be written as $\lim_{\delta\to 0}\dfrac{p_{ij}(\tau,\tau+\delta)-p_{ij}(\tau,\tau)}{\delta} = v_{ij}(\tau)$

$$\Rightarrow \frac{d}{dt}p_{ij}(\tau,t)|_{t=\tau} = v_{ij}(\tau) \tag{1.3.9}$$

Further, (1.3.6) $\Rightarrow 1-p_{ii}(\tau,\tau+\delta) = -v_{ii}(\tau)\cdot\delta+0(\delta)$

$$\Rightarrow 1+v_{ii}(\tau)\delta+0(\delta) = p_{ii}(\tau,\tau+\delta) \tag{1.3.10}$$

and (1.3.7) $\quad \Rightarrow p_{ij}(\tau,\tau+\delta) = v_{ij}(\tau)\delta+0(\delta). \qquad (1.3.11)$

Noting $\quad \sum_j p_{ij}(\tau,t) = 1\,\forall\,\tau<t, \qquad (1.3.12)$

and putting $t=\tau+\delta$ we get

$$\sum_j p_{ij}(\tau,t) = \sum_j p_{ij}(\tau,\tau+\delta) = p_{ii}(\tau,\tau+\delta)+\sum_{j\neq i}p_{ij}(\tau,\tau+\delta)$$

$$= 1+v_{ii}(\tau)\delta+0(\delta)+\sum v_{ij}(\tau)\cdot\delta+0(\delta) = 1$$

$$\Rightarrow \delta \cdot v_{ii}(\tau) + 0(\delta) = -\delta \sum_{j \neq i} v_{ij}(\tau) + 0(\delta)$$

$$\Rightarrow v_{ii}(\tau) = -\sum_{j \neq i} v_{ij}(\tau) \tag{1.3.13}$$

1.3.3 Kolmogorov's Forward equation

Let $\tau < t < t + \delta$. We have from Chapman-Kolmogorov equation (1.3.3)

$$p_{ik}(\tau, t+\delta) = p_{ik}(\tau, t)p_{kk}(t, t+\delta) + \sum_{j \neq k} p_{ij}(\tau, t)p_{jk}(t, t+\delta),$$

where $p_{kk}(\tau, t+\delta) = 1 + v_{kk}(t) \cdot \delta + 0(\delta)$ from (1.3.10)

and $p_{jk}(t, t+\delta) = v_{jk}(t) \cdot \delta + 0(\delta)$ from (1.3.11)

$$\Rightarrow p_{ik}(\tau, t+\delta) = p_{ik}(\tau, t)(1 + v_{kk}(t) \cdot \delta + 0(\delta)) + \sum_{j \neq k} p_{ij}(\tau, t)[v_{jk}(t)\delta + 0(\delta)]$$

$$\Rightarrow \frac{p_{ik}(\tau, t+\delta) - p_{ik}(\tau, t)}{\delta} = p_{ik}(\tau, t)v_{kk}(t) + \frac{0(\delta)}{\delta} + \sum_{j \neq k} p_{ij}(\tau, t)\left[v_{jk}(t) + \frac{0(\delta)}{\delta}\right].$$

Taking limit as $\delta \to 0$ on both sides under the regularity condition

$$\Rightarrow \frac{\partial p_{ik}(\tau, t)}{\partial t} = p_{ik}(\tau, t)v_{kk}(t) + \sum_{j \neq k} p_{ij}(\tau, t)v_{jk}(t)$$

$$\Rightarrow \frac{\partial p_{ik}(\tau, t)}{\partial t} = \sum_{j} p_{ij}(\tau, t)v_{jk}(t). \tag{1.3.14}$$

Equation (1.3.14) represents Kolmogorov's forward equation with initial condition

$$p_{ij}(\tau, \tau) = \delta_{ij}. \tag{1.3.15}$$

1.3.4 Kolmogorov's Backward Equation

Again from Chapman-Kolmogorov's equation (1.3.3)

$$P_{ik}(\tau - \delta, t) = \sum_{j} p_{ij}(\tau - \delta, \tau)p_{jk}(\tau, t); \; \delta > 0$$

$$= p_{ii}(\tau - \delta, \tau)p_{ik}(\tau, t) + \sum_{j \neq i} p_{ij}(\tau - \delta, \tau)p_{jk}(\tau, t)$$

Also $p_{ii}(\tau - \delta, \tau) = 1 + v_{ii}(\tau - \delta) \cdot \delta + 0(\delta)$

and $p_{ij}(\tau - \delta, \tau) = v_{ij}(\tau - \delta) \cdot \delta + 0(\delta)$

$$\Rightarrow p_{ik}(\tau - \delta, t) = [1 + v_{ii}(\tau - \delta) \cdot \delta + 0(\delta)]p_{ik}(\tau, t) + \sum_{j \neq i} [v_{ij}(\tau - \delta) \cdot \delta + 0(\delta)]p_{jk}(\tau, t)$$

$$\Rightarrow \frac{p_{ik}(\tau - \delta, t) - p_{ik}(\tau, t)}{-\delta} = \left[-v_{ii}(\tau - \delta) + \frac{0(\delta)}{-\delta}\right] p_{ik}(\tau, t) - \sum_{j \neq i} v_{ij}(\tau - \delta)p_{jk}(\tau, t) - \frac{0(\delta)}{\delta} \sum_{j \neq i} p_{jk}(\tau, t)$$

Taking limit as $\delta \to 0$ on both sides,

$$\Rightarrow \quad \frac{\partial p_{ik}(\tau,t)}{\partial \tau} = -v_{ii}(\tau)p_{ik}(\tau,t) - \sum_{j\neq i} v_{ij}(\tau)p_{jk}(\tau,t)$$

$$\Rightarrow \quad \frac{\partial p_{ik}(\tau,t)}{\partial \tau} = -\sum_{j} v_{ij}(\tau)p_{jk}(\tau,t) \tag{1.3.16}$$

This is Kolmogorov's backward equation with initial condition

$$p_{ik}(t,t) = \delta_{ik} \tag{1.3.17}$$

Given that $\sum_k p_{ik}(\tau,t) = 1$, there exists a unique solution of $p_{ik}(\tau,t)$ satisfying both the forward and back ward as well as Chapman Kolmogorov equation. In general, $p_{ik}(\tau,t)$ is non-homogeneous with respect to time; however, it may be shown that if the intensity functions $v_{ij}(t)$ are equal to constant i.e. $v_{ij}(t)$ is independent of time, then the process is independent of time. In this case,

$$\left.\begin{aligned} \frac{dp_{ik}}{dt} &= \sum_j p_{ij}(t)v_{jk} \\ \frac{dp_{ik}}{dt} &= \sum_j v_{ij}p_{jk}(t) \end{aligned}\right\}, \tag{1.3.18}$$

where $p_{ik}(0) = \delta_{ik}$.

The intensity functions $v_{ij},\ i,j = 1,2,\ldots$ may be represented in the form of a matrix as $V = [v_{ij}]$. V is called the intensity matrix.

Example 1.1: Intensity Matrix for a Poisson process is

$$V = \begin{pmatrix} -\lambda & \lambda & 0 & \cdots & \cdots & 0 \\ 0 & -\lambda & \lambda & \cdots & \cdots & 0 \\ 0 & 0 & -\lambda & \lambda & \cdots & 0 \\ \vdots & \vdots & \vdots & \vdots & & \vdots \\ 0 & 0 & 0 & 0 & \cdots & -\lambda \end{pmatrix}$$

Kolmogorov forward and backward equations in respect of V are given as

$$\frac{\partial p_{ij}(\tau,t)}{\partial t} = p_{ij}(\tau,t)(-\lambda) + p_{j-1}(\tau,t)\cdot\lambda$$

$$\frac{\partial p_{ij}(\tau,t)}{\partial t} = \lambda p_{ij}(\tau,t) - \lambda p_{i+1j}(\tau,t),$$

respectively.

Example 1.2: Application of Kolmogorov's Differential Equation in Population Models.

Consider a Markov process $X(t),\ t \in [0,\infty)$, where $X(t)$ represents the size of the population at time t.

If $X(t) = r(r = 0,1,2,\ldots)$ then during an infinitesimal time interval $(t, t+\delta t)$, $X(t)$ increases by unity (either by birth or by immigration) with probability $\lambda_r\delta t +$

$0(\delta t)$ decreases by one (on account of death or emigration) with probability $\mu_r\delta t + 0(\delta t)$ or does not change (remains stationary) with probability $1-(\lambda_r+\mu_r)\delta t+0(\delta t)$.

Obtain the intensity matrix V of the process given that $\mu_0 = 0$. Also obtain the Kolmogorov forward equation and determine the equilibrium solution if it exists.

Solution: The state space of the Markov Process is $\{0, 1, 2, 3, \ldots\}$. We have

$$\left.\begin{aligned} &P\{X(t+\delta) = r+1|X(t) = r\} = \lambda_r\delta + 0(\delta) \\ &P\{X(t+\delta) = r-1|X(t) = r\} = \mu_r\delta + 0(\delta) \\ &P\{X(t+\delta) = r|X(t) = r\} = 1-(\lambda_r+\mu_r)\cdot\delta + 0(\delta) \\ &P\{X(t+\delta) = r'|X(t) = r\} = 0(\delta) \quad \text{for} \quad r' \neq r-1, r, r+1. \end{aligned}\right\} \tag{1.3.20}$$

The intensity matrix is

$$V = \begin{array}{c|cccccc} & 0 & 1 & 2 & 3 & 4 & \cdots \\ \hline 0 & -\lambda_0 & \lambda_0 & 0 & 0 & 0 & \cdots \\ 1 & \mu_1 & -\mu_1-\lambda_1 & \lambda_1 & 0 & 0 & \cdots \\ 2 & 0 & \mu_2 & -\mu_2-\lambda_2 & \lambda_2 & 0 & \cdots \\ 3 & 0 & 0 & \mu_3 & -\mu_3-\lambda_3 & \lambda_3 & \cdots \\ \vdots & \cdots & \cdots & \cdots & \cdots & \cdots & \cdots \end{array} \tag{1.3.21}$$

Then the Kolmogorov's differential equations are

$$\frac{dp_{ro}}{dt} = \lambda_0 p_{r0} + \mu_1 p_{r1}$$

$$\frac{dp_{rl}}{dt} = \lambda_{l-1}p_{r\,l-1} - (\mu_l+\lambda_l)p_{r\,l} + \mu_{l+1}p_{r\,l+1}, \quad (l = 1, 2, 3, \ldots).$$

Let us write the equilibrium or steady state solution of $p_{rl}(t) \to \pi_l$ as $t \to \infty$.

Then $\quad 0 = \lambda_0\pi_0 + \mu_1\pi_1$

$$0 = (\lambda_{l-1}\pi_{l-1} - \mu_l\pi_l) - (\lambda_l\pi_l - \mu_{l+1}\pi_{l+1}), \quad (l = 1,\ 2, \ldots).$$

Denoting $I_l = \lambda_l\pi_l - \mu_{l+1}\pi_{l+1}$, $l = 0, 1, 2, \ldots$

$$0 = -I_0$$

$$0 = I_{l-1} - I_l,$$

so that $I_l = 0, \quad (l = 0,\ 1,\ 2, \ldots)$.

Therefore, $\dfrac{\pi_{l+1}}{\pi_l} = \dfrac{\lambda_l}{\mu_{l+1}}$

$$\Rightarrow \qquad \pi_l = \left(\frac{\pi_l}{\pi_{l-1}}\frac{\pi_{l-1}}{\pi_{l-2}}\cdots\frac{\pi_1}{\pi_0}\right)\cdot\pi_0 = v_i\pi_0, \tag{1.3.23}$$

where $\quad v_l = \dfrac{\lambda_{l-1}}{\mu_l}\cdot\dfrac{\lambda_{l-2}}{\mu_{l-1}}\cdots\dfrac{\lambda_0}{\mu_1}$ and $v_0 = 1$.

Then $$\sum_l \pi_l = 1 \Rightarrow 1 = \pi_0\sum_l v_l \Rightarrow \pi_0 = \frac{1}{\sum\limits_{l=0}^{\infty} v_l}$$

$$\Rightarrow \quad \pi_l = \frac{v_l}{\sum_{l=0}^{\infty} v_l}, \quad (l = 0, 1, 2, 3, \ldots).$$

This will be a probability distribution if $\sum_{l=0}^{\infty} v_l$ converges.

Example 1.3: A useful result in two state Markov chain in continuous time.

If an event (say sickness) occurs with intensity λ and the complimentary event (say recovery) with intensity μ, then show that

$$(p_{ij}(t)) = \begin{pmatrix} \frac{\mu}{\lambda+\mu} + \frac{\lambda}{\lambda+\mu} e^{-(\lambda+\mu)t} & \frac{\lambda}{\lambda+\mu} - \frac{\lambda}{\lambda+\mu} e^{-(\lambda+\mu)t} \\ \frac{\mu}{\lambda+\mu} - \frac{\mu}{\lambda+\mu} e^{-(\lambda+\mu)t} & \frac{\lambda}{\lambda+\mu} + \frac{\mu}{\lambda+\mu} e^{-(\lambda+\mu)t} \end{pmatrix} \tag{1.3.25}$$

where $p_{ij}(t) = P[X(t) = j | X(0) = i] \; i, \; j = 0, 1.$ (1.3.26)

$X(t)$ is the continuous parameter Markov chain

$$\text{and} \quad (\mu_{ij}(t)) = \begin{pmatrix} \frac{\mu}{\lambda+\mu} t + \frac{\lambda}{(\lambda+\mu)^2}(1 - e^{-(\lambda+\mu)t}) & \frac{\lambda}{(\lambda+\mu)} t - \frac{\lambda}{(\lambda+\mu)^2}(1 - e^{-(\lambda+\mu)t}) \\ \frac{\mu_t}{\lambda+\mu} t - \frac{\mu}{(\lambda+\mu)^2}(1 - e^{-(\lambda+\mu)t}) & \frac{\lambda}{\lambda+\mu} t - \frac{\mu}{(\lambda+\mu)^2}(1 - e^{-(\lambda+\mu)t}) \end{pmatrix} \tag{1.3.27}$$

$$\mu_{ij}(t) = E[X(t) = j | X(0) = i] = \int_0^t p_{ij}(\tau) d\tau; \quad i, \; j = 0, 1, 2, \tag{1.3.28}$$

i.e. expected amount of time in $(0, t)$ the system is in the state 'j' given that it is initially in 'i'.

Proof: We have the Kolmogorov's equation

$$p_{00}(t + \delta t) = p_{00}(t)(1 - (\lambda \delta t + 0(\delta t)) + p_{01}(t)(\mu \delta t + 0(\delta t))$$

$$\Rightarrow \quad p'_{00}(t) = -\lambda p_{00}(t) + \mu(1 - p_{00}(t)) \text{ as } \delta t \to 0 \quad (\because p_{01}(t) = 1 - p_{00}(t)).$$

$$\text{Again,} \quad L(p'_{00}(t)) = \int_0^\infty e^{-st} p'_{00} dt = e^{-st} p'_{00}(t) \Big|_0^\infty + s \int_0^\infty e^{-st} p_{00}(t) dt,$$

(integrating by parts).

$$= -1 + sL(p_{00}(t)) = -\lambda L(p_{00}(t)) + \mu L(1) - \mu L(p_{00}(t))$$

where 'L' stands for Laplace transform

$$\Rightarrow \quad -1 + sL(p_{00}(t)) = \mu L(1) - (\lambda + \mu) L(p_{00}(t))$$

$$\Rightarrow \quad -1(s + \lambda + \mu) L(p_{00}(t)) = \mu L(1) = \frac{\mu}{s}$$

$$\Rightarrow \qquad L(p_{00}(t)) = \frac{\frac{\mu}{s}+1}{(s+\lambda+\mu)} = \frac{\mu+s}{s(s+\lambda+\mu)}$$

$$= \frac{\mu}{\lambda+\mu}\left(\frac{1}{s} - \frac{1}{s+\lambda+\mu}\right) + \frac{1}{s+\lambda+\mu}$$

$$\Rightarrow \qquad p_{00}(t) = \frac{\mu}{\lambda+\mu}(1-e^{-(\lambda+\mu)t}) + e^{-(\lambda+\mu)t}$$

$$= \frac{\mu}{\lambda+\mu} + e^{-(\lambda+\mu)t}\frac{\lambda}{\lambda+\mu} \tag{1.3.29}$$

$$\Rightarrow \qquad p_{01}(t) = 1 - p_{00}(t) = \frac{\lambda}{\lambda+\mu} - \frac{\lambda}{\lambda+\mu}e^{-(\lambda+\mu)t} \tag{1.3.30}$$

While interchanging λ with μ in $p_{01}(t)$, we have,

$$p_{10}(t) = \frac{\mu}{\lambda+\mu} - \frac{\mu}{\lambda+\mu}e^{-(\lambda+\mu)t} \tag{1.3.31}$$

Finally, $p_{11}(t) = 1 - p_{10}(t) = \dfrac{\lambda}{\lambda+\mu} + \dfrac{\mu}{\lambda+\mu}e^{-(\lambda+\mu)t}$ (1.3.32)

From (1.3.29) to (1.3.32) we get (1.3.25).

Further, from (1.3.28) we have,

$$\mu_{00}(t) = \int_0^t p_{00}(\tau)d\tau = \frac{\mu t}{\lambda+\mu} + \frac{\lambda}{(\lambda+\mu)^2}(1-e^{-(\lambda+\mu)t}), \text{ from (1.3.29).}$$

Precisely, in the same way from (1.3.30), (1.3.31) (1.3.32), respectively, we have

$$\mu_{01}(t) = \frac{\lambda t}{\lambda+\mu} - \frac{\lambda}{(\lambda+\mu)^2}(1-e^{-(\lambda+\mu)t}),$$

$$\mu_{10}(t) = \frac{\mu t}{\lambda+\mu} - \frac{\mu}{(\lambda+\mu)^2}(1-e^{-(\lambda+\mu)t})$$

and $\quad \mu_{11}(t) = \dfrac{\lambda t}{\lambda+\mu} + \dfrac{\mu}{(\lambda+\mu)^2}(1-e^{-(\lambda+\mu)t}).$

which proves (1.3.26).

Note that as $t \to \infty$

$$\lim p_{00}(t) = p_{00} = \frac{\mu}{\lambda+\mu},$$

$$\lim p_{01}(t) = P_{01} = \frac{\lambda}{\lambda+\mu}$$

$$\lim p_{10}(t) = p_{10} = \frac{\mu}{\lambda+\mu}$$

and $\quad \lim p_{11}(t) = p_{11} = \dfrac{\lambda}{\lambda+\mu},$

are the stationary distributions of the two state Markov chain. These are also known as the steady state distribution of the Markov chain.

1.4 Stochastic Population Models by Markov Processes

1.4.1 Poisson Processes

Denoting by $p_k(t) = P[X(t) = k|(0,t)]$, the Poisson process is given by Kolmogorov's equations

$$\left.\begin{aligned} p_k(t+\delta) &= p_k(t)(1-\lambda\delta) + p_{k-1}(t)\lambda\delta + 0(\delta); \quad k = 1,\ 2, \ldots \\ p_0(t+\delta) &= p_0(t)(1-\lambda\delta) + 0(\delta). \end{aligned}\right\} \tag{1.4.1}$$

Here the probability of an event occurring in $(t, t+\delta) = \lambda\delta + 0(\delta)$, where $\lambda(0 \leq \lambda < \infty)$ is a constant independent of t or δ (condition of Stationarity) and the probability of more than one event in $(t, t+\delta)$ is of the order zero (condition of orderliness).

$$\left.\begin{aligned} \frac{dp_k(t)}{dt} &= -\lambda p_k(t) + \lambda p_{k-1}(t), \quad k = 1, 2, \ldots \\ \frac{dp_0(t)}{dt} &= -\lambda p_0(t) \end{aligned}\right\} \tag{1.4.2}$$

The solution of (1.4.2) is given by

$$p_k(t) = e^{-\lambda t}\frac{(\lambda t)^k}{k!} \tag{1.4.3}$$

i.e. the number of occurrences in $(0, t)$ follows the Poisson distribution with parameter λt. Further, it can be shown that the time between successive occurrences follows exponential distribution with mean $\frac{1}{\lambda}$ (vide Example 1.4)

1.4.2 Time Dependent Poisson Process

This is obtained by Kolmogorov's forward equation

$$\left.\begin{aligned} &p_n(t+\delta t) = p_{n-1}(t)[\lambda\delta t + 0(\delta t)] + p_n(t)[1 - (\lambda\delta t + 0(\delta t))] \\ &\qquad\qquad\qquad\qquad\qquad\qquad\qquad\qquad n = 1, 2, \ldots \\ &\text{and} \quad p_0(t+\delta t) = p_0(t) \quad [1 - (\lambda\delta t + 0(\delta t))] \end{aligned}\right\} \tag{1.4.4}$$

The solution is given by

$$p_n(t) = \exp\left(-\int_0^t \lambda(\tau)d\tau\right)\frac{\left[\int_0^t \lambda(\tau)\right]^n}{n!}. \tag{1.4.5}$$

Obviously we have,

$$\lim_{\delta\to 0}\frac{p_0(t+\delta t) - p_0(t)}{\delta t} = -\lambda(t)p_0(t) + \lim_{\delta\to 0}\frac{0(\delta t)}{\delta t}$$

$$\Rightarrow \quad \int_0^t \frac{p_0'(\tau)d\tau}{p_0(\tau)} = -\int_0^t \lambda(\tau)d\tau = -\Lambda(t), \text{ say,}$$

$$\Rightarrow \quad p_0(t) = \exp\left(\int_0^t \lambda(\tau)d\tau\right) = \exp(-\Lambda(t)) \quad (\because p_0(0) = 1)$$

Similarly, one can show

$$\lim_{\delta\to 0}\frac{p_1(t+\delta t)-p_1(t)}{\delta t} = -\lambda(t)(p_1(t)-p_0(t)) + \lim_{\delta\to 0}\frac{0(\delta t)}{\delta t}$$

$$\Rightarrow \quad p_1'(t)+\lambda(t)p_1(t)=\lambda(t)p_0(t),$$

which is a linear differential equation of order one.

Multiplying both sides of the differential equation by the integrating factor

$$\exp\left[\int_0^t \lambda(\tau)d\tau\right]$$

$$\Rightarrow \quad \exp\left[\int_0^t \lambda(\tau)d\tau\right]\frac{dp_1(t)}{dt} + \lambda(t)p_1(t)\exp\left[\int_0^t \lambda(\tau)d\tau\right]$$

$$= \lambda(t)p_0(t)\exp\left[\int_0^t \lambda(\tau)d\tau\right] = \lambda(t) \quad \left(\because p_0(t)=\exp\left[-\int_0^t \lambda(\tau)d\tau\right]\right)$$

$$\Rightarrow \quad \frac{d}{dt}\left\{p_1(t)\exp\left[\int_0^t \lambda(\tau)d\tau\right]\right\} = \lambda(t)$$

$$\Rightarrow \quad p_1(t)=\exp\left[-\int_0^t \lambda(\tau)d\tau\right]\left[\int_0^t \lambda(\tau)d\tau\right] = \exp\left[-\int_0^t \lambda(\tau)d\tau\right]\cdot\Lambda(t)$$

Thus we see structure of $p_n(t)$ may be given by

$$p_n(t)=\exp\left(-\int_0^t \lambda(\tau)d\tau\right)\frac{(f(t))^n}{n!},$$

where $f(t)$ is an unknown function whose structure is to be determined.

We can write,

$$p_n(t)=\exp[-\Lambda(t)]\frac{(f(t))^n}{n!} \text{ in (1.4.4).}$$

$$\Rightarrow \quad \frac{f'(t)}{f(t)} = -\lambda(t) \Rightarrow \frac{d}{dt}(\log f(t)) = -\lambda(t)$$

$$\Rightarrow \quad f(t)=\exp\left(-\int_0^t \lambda(\tau)d\tau\right) = \exp[-\Lambda(t)]$$

$$\Rightarrow \quad p_n(t)=\exp[-\Lambda(t)]\frac{(\Lambda(t))^n}{n!}, \quad n=0,\ 1,\ 2\ldots, \text{ which proves (1.4.5).}$$

$$\therefore \quad E[X(t)]=\sum_{n=0}^{\infty} np_n(t)=\sum_{n=0}^{\infty} n\frac{[\Lambda(t)]^n}{n!}e^{-\Lambda(t)}=\Lambda(t). \tag{1.4.7}$$

1.4.3 Weighted Poisson Process

If $p_{n|\lambda}(t)$ is a given Poisson process with parameter λt and λ is a r.v. with a p.d.f. $f(\lambda)$ i.e. $f(\lambda)$ is the prior distribution of λ then we define

$$p_n(t)=\int_0^\infty p_{n/\lambda}f(\lambda)d\lambda \text{ as the weighted Poisson process} \tag{1.4.8}$$

Now, $$f(\lambda) = \frac{e^{-a\lambda}\lambda^{k-1}a^k}{\Gamma(\lambda)}; \ \lambda > 0 \text{ is the best prior distribution of } \lambda \text{ in the Bayesian sense} \tag{1.4.9}$$

where $$p_{n|\lambda}(t) = e^{-\lambda t}\frac{(\lambda t)^n}{n!} \tag{1.4.10}$$

$$\Rightarrow \quad p_n(t) = \binom{n+k-1}{k}\left(\frac{t}{a+t}\right)^a\left(\frac{a}{a+t}\right)^k \tag{1.4.11}$$

which is a negative binomial distribution for given t with parameters k and $p\left(=\frac{a}{a+t}\right)$.

Further, $$E(X(t)) = \frac{kt}{a} \quad \text{and} \quad \operatorname{Var}(X(t)) = \frac{kt}{a}\left(1+\frac{kt}{a}\right) \tag{1.4.12}$$

1.4.4 Pure Birth Process

It is defined as, given $X(t) = k$ and $X(0) = k_0$,

(i) the conditional probability that a new event will occur during $(t, t+\Delta)$ is $\lambda_k\Delta + 0(\Delta)$

(ii) the conditional probability that more than one event will occur is $o(\Delta)$.

The Kolmogorov equations are

$$\left.\begin{aligned} \frac{dp_k(t)}{dt} &= -\lambda p_{k0}(t) \quad \text{for} \quad k = k_0 \\ \frac{dp(t)}{dt} &= -\lambda_k p_k(t) - \lambda_{k-1}p_{k-1}(t) \quad \text{for} \quad k > k_0 \end{aligned}\right\} \tag{1.4.13}$$

with initial conditions as $p_{k_0}(0) = 1$, $p_k(0) = 0$ for $k \neq k_0$.

The solution is given by

$$p_k(t) = (-1)^{k-k_0}\lambda_{k_0}\lambda_{k_1}\ldots\lambda_{k-1}\left[\sum_{i=k_0}^{k}\frac{\exp(-\lambda_i t)}{\prod\limits_{\substack{j=k_0 \\ j\neq i}}^{k}(\lambda_j-\lambda_i)}\right] \tag{1.4.14}$$

(A) Yule's Process

It is a particular case of Pure Birth process by putting $\lambda_k = k\lambda$.

The Kolmogorov equations are

$$\left.\begin{aligned} \frac{d}{dt}p_{k_0}(t) &= -k_0\lambda p_{k_0}(t) \\ \frac{dp_k(t)}{dt} &= -k\lambda p_k(\lambda) + (k-1)\lambda p_{k-1}(t) \end{aligned}\right\} \tag{1.4.15}$$

The solution is

$$p_k(t) = \binom{k-1}{k-k_0}\exp(-k_0\lambda t)[1-\exp(\lambda t)]^{k-k_0}$$

$$= (-1)^{k-k_0}(k_0\lambda)\ldots((k-1)\lambda)\left[\sum_{i=k_0}^{k}\frac{e^{-i\lambda t}}{\prod\limits_{\substack{i=k_0\\ i\neq j}}^{k}(i\lambda-j\lambda)}\right] \tag{1.4.16}$$

where $(k_0\lambda)\ldots((k-1)\lambda) = \lambda^{k-k_0}\begin{pmatrix} k-1 \\ k-k_0 \end{pmatrix}(k-k_0)!$

and $\prod\limits_{\substack{i=k_0\\ i\neq j}}^{k}(i\lambda - j\lambda) = \lambda^{k-k_0}(-1)^{k-i}\begin{pmatrix} k-k_0 \\ i-k_0 \end{pmatrix}^{-1}(k-k_0)!$

Here,
$$\left.\begin{aligned} E(X(t)) &= k_0 e^{\lambda t} \\ \text{Var}(X(t)) &= k_0 e^{\lambda t}(e^{\lambda t}-1) \end{aligned}\right\} \tag{1.4.17}$$

(B) Polya Process

Polya process is a generalization of Pure birth process in the sense that the Probability of birth in time $(t, t+\Delta)$ is $\lambda_k\Delta + 0(\Delta)$ and λ_k depends on k and t.

We set, $\lambda_k(t) = \dfrac{\lambda + \lambda ak}{1+\lambda at}$ where λ and a are non-negative constants.

We have the initial condition $X(0) = k_0$ Kolmogorov equations are

$$\left.\begin{aligned} \frac{d}{dt}p_{k0}(t) &= \frac{\lambda+\lambda ak_0}{1+\lambda at}p_{k0}(t) \\ \frac{d}{dt}p_k(t) &= -\frac{\lambda+\lambda ak}{1+\lambda at}p_k(t) + \frac{\lambda+\lambda a(k-1)}{1+\lambda at}p_{k-1}(t), \text{ for } k > k_0. \end{aligned}\right\} \tag{1.4.18}$$

The p.g.f. of $X(t)$ is

$$\phi_X(s;t) = s^{k_0}\left[\frac{1/(1+\lambda at)}{1-s(\lambda at/(1+\lambda at))}\right]^{k_0+1/a} \tag{1.4.19}$$

Here, $X(t)$, excepting for the additive constant k_0, behaves like a negative binomial variable for a given t, with parameters p and r where

$$p = \frac{1}{1+\lambda at} \quad \text{and} \quad r = k_0 + \frac{1}{a} \tag{1.4.20}$$

Vide Example 1.6 page 23–24.

1.4.5 The Discrete Branching Process

The process evolved by Francis Galton and W.H. Watson (1874), now described, may have several applications in Population problems (vide art 1.5.4) as restricted population processes. We consider both continuous and discrete forms. Let us assume that every individual (female) in a populations gives rise to X number of births where X is a discrete random variable having its p.g.f.

$$\phi(s) = E(s^X), \quad \text{for } |s| < 1 \tag{1.4.21}$$

Further, we assume that individuals reproduce independently and finally die (or make themselves off from the reproductive span). For the sake of simplicity, we assume that all the females give births (female) at a particular point of time (which is the mean age of child bearing) and cease to give birth after that period which is considered to be the end of fertility span.

Let Z_n be the number of female children in the nth generation $n = 1, 2, 3\ldots$ and $\psi_n(s)$ be the p.g.f. of Z_n. Denoting $\phi(s)$ as the p.g.f. of X, it is easy to see

$$\psi_{n+1}(s) = \psi_n(\phi(s)) = \phi(\psi_n(s)) \tag{1.4.22}$$

To prove the same, we decompose

$$Z_{n+1} = X_1 + X_2 + \ldots + X_{Z_n} \tag{1.4.23}$$

which is obtained directly by assuming that Z_n be the number of females in the nth generation of which the first one gives X_1 number of female births, second one gives X_2 number of female births and Z_nth gives X_{Z_n} number of female births and hence (1.4.23) is an identity.

Therefore, $\psi_{n+1}(s) = E(s^{Z_{n+1}}) = E_{Z_n}(E_X(s^{X_1+X_2+\cdots+X_{Z_n}})|Z_n)$

$= E_{Z_n}(\phi(s)^{Z_n})$ $(\because X_1, X_2, \ldots X_{Z_n})$ are all i.i.d. r.v.s

$$= \psi_n(\phi(s)) \tag{1.4.24}$$

Also $$\psi_{n+1}(s) = E_X[E_{Z_n}(s^X)^{Z_n}|X] = E_X[\psi_n(s^X)] = \phi(\psi_n(s)) \tag{1.4.25}$$

(1.4.24) and (1.4.25) establish (1.4.22).

Next we are required to obtain $E(Z_n)$ and $\text{var}(Z_n)$ for population in the nth generation.

We know $E(X) = \phi'(1) = \alpha$, say and $\text{var}(X) = \phi''(1) + \alpha - \alpha^2 = \beta$, say (1.4.26)

If $$Z_0 = 1, E(Z_0) = 1 \quad \text{and} \quad \text{var}(Z_0) = 0 \tag{1.4.27}$$

$E(Z_{n+1}|Z_n = k) = E(X_1 + X_2 + \ldots + X_k)$ from (1.4.23)

$= kE(X) = k\alpha$ from (1.4.26) (1.4.28)

Therefore, $E(Z_{n+1}) = E_{Z_n}[E(Z_{n+1}|Z_n)] = E_{Z_n}(Z_n\alpha)$

$= \alpha E(Z_n)$ from (1.4.28) since $k = Z_n$

$$\Rightarrow \quad E(Z_{n+1}) = \alpha E(Z_n) = \alpha^2 E(Z_{n-1}) = \ldots = \alpha^{n+1}E(Z_0) = \alpha^{n+1} \tag{1.4.29}$$

Again, $\text{var}(Z_{n+1}|Z_n = k) = \text{var}(X_1 + X_2 + \ldots + X_k)$

$$= k\text{var}(X) = k\beta, \tag{1.4.30}$$

since X_i's are i.i.d. r.v.s and distributed as X.

Now, $\text{var}(Z_{n+1}) = E[\text{var}(Z_{n+1}|Z_n)] + \text{var}[E(Z_{n+1}|Z_n)]$

$= E[Z_n\beta] + \text{var}[\alpha Z_n]$, from (1.4.29) since $k = Z_n$

$\Rightarrow \quad \text{var}(Z_{n+1}) = \beta E(Z_n) + \alpha^2\text{var}(Z_n)$

This is a recurrence relation.

Now $E(Z_n) = \alpha^n$, from (1.4.28).

Therefore, $$\text{var}(Z_{n+1}) = \beta\alpha^n + \alpha^2\text{var}(Z_n) \tag{1.4.31}$$

Putting $n = 0$ in (1.4.31), we get

$\text{var}(Z_1) = \beta\alpha^0 + \alpha^2\text{var}(Z_0) = \beta \quad (\because \text{var}(Z_0) = 0)$

from (1.4.27) (1.4.32)

Putting $n = 1$ in (1.4.31), we get

$$\text{var}(Z_2) = \beta\alpha + \alpha^2 \text{var}(Z_1) = \beta\alpha + \alpha^2\beta = \beta\alpha(1+\alpha) \tag{1.4.33}$$

Proceeding in this way,

$$\begin{aligned}\text{var}(Z_3) &= \beta\alpha^2 + \alpha^2\text{var}(Z_2) = \beta\alpha^2 + \alpha^2(\beta\alpha + \alpha^2\beta)\\ &= \beta\alpha^2 + \beta\alpha^3 + \beta\alpha^4 = \frac{\beta\alpha^2(1-\alpha^3)}{(1-\alpha)}\end{aligned} \tag{1.4.34}$$

Proceeding similarly, for any finite n

$$\text{var}(Z_n) = \frac{\beta\alpha^{n-1}(1-\alpha^n)}{(1-\alpha)}, \text{ provided } \alpha \neq 1 \tag{1.4.35}$$

$$= n\beta \text{ if } \alpha = 1 \tag{1.4.36}$$

1.4.6 Some Results on Discrete Branching Process

We can easily see that the Population process $\{Z_n;\ n = 1, 2, 3, \ldots\}$ conforms to a Markov chain (M.C.).

Since $Z_{n+1} = X_1 + X_2 + \ldots + X_{Z_n}$ if $Z_n = 1, 2, 3, \ldots$

$= 0$ if $Z_n = 0$

and X_i's are i.i.d. r.v.'s each having a common p.g.f. $\phi(s)$; the p.g.f. of the conditional r.v. $Z_{n+1}|Z_n$ is given by

$$\psi_{n+1}(s) = [\phi(s)]^{Z_n}, \text{ from (1.4.24)} \tag{1.5.37}$$

which entirely depends on Z_n (so that the distribution of Z_{n+1} depends only on Z_n). Hence $\{Z_n\}$ conforms to a simple M.C.

Next we consider the probability of extinction of the Population process. Let $p_{ij} = P\{Z_{n+1} = j|Z_n = i\} = P\{Z_{n+1} = j|Z_n = i,\ Z_{n-1} = {}^*, Z_{n-2} = {}^* \ldots, Z_1 = {}^*\}$ where * indicates arbitrary values.

We have clearly $p_{0j} = \delta_{0j}$ (the Kronecker delta) since the state 0 is absorbing. Now, let us denote $P[X = j] = p_j$, say, $j = 0, 1, 2, \ldots$

For $p_{ij}, i = 2, 3, 4, \ldots$ we have p.g.f. of $[j|i]$ is given by

$$\widetilde{\phi}_i(s) = \sum_{j=0}^{\infty} p_{ij}s^j \quad \text{for } |s| < 1 \tag{1.4.38}$$

Given $Z_n = i$ and $Z_{n+1} = X_1 + X_2 + \ldots + X_{Z_n}$, if $Z_n = 1, 2, 3 \ldots$

$= 0$ if $Z_n = 0$,

it follows that, $\widetilde{\phi}_i(s) = E[s^{Z_{n+1}}|Z_n = i] = [\phi(s)]^i$ (1.4.39)

Since, $Z_{n+1} = X_1 + X_2 + \ldots + X_{Z_n}$

$$E(s^{Z_{n+1}}|Z_n = i) = [E(s^{X_i})]^{Z_n} = [\phi(s)]^{Z_n} = [\phi(s)]^i$$

Hence comparing (1.4.38) and (1.4.39), it follows that p_{ij}, the (i, j)th element of the transition matrix of the Markov chain is given by the coefficient of s^j in $[\phi(s)]^{Z_n} = i$. If $Z_0 = 1$, the probability that the population gets extinct on or before the nth generation is

$$p_{10}^{(n)} = P[Z_n = 0|Z_0 = 1] = \pi_n, \text{ say.}$$

Behaviour of π_n as n increases

To observe the behaviour of π_n as n increases, we assume that

$$0 < p_0 = p_{10} = \pi_1 \leq 1,$$

because if $p_0 = \pi_0$, no extinction is possible. The sequence $\{\pi_n\}$ is therefore bounded above and also increasing. Hence by Bolzano-Weirstrass's theorem (see A.1.4(1)), it must possess a limit.

$(\because \pi_1 = p_{10}^{(1)}, \pi_2 = p_{10}^{(2)} = p_{10}^{(1)} + P[Z_2 = 0|Z_0 = 1, Z_1 \neq 0] \Rightarrow \pi_2 \geq \pi_1.$

Similarly, $\pi_3 \geq \pi_2$. etc.)

$\therefore \quad \lim_{n\to\infty} \pi_n = \pi$, say.

$\Rightarrow \quad \pi$ satisfies the equation $\pi = \phi(\pi)$ as $n \to \infty$

$(\because \pi_n = \psi_n(0) = \phi(\psi_{n-1})) = \phi(\pi_{n-1})$ by (1.4.22).

Also, $\phi(s) = E(s^X) \Rightarrow \phi(1) = 1$ satisfies $|s| = 1$ is always a solution of $\phi(\pi) = \pi$

To search another solution say π^* of the same, let $\phi(\pi)$ be a power series in π with necessarily non-negative coefficients, (because the coefficients represent probabilities and as such they should be non-negative).

Therefore, $\phi(\pi)$ is an increasing function of π, $\forall \pi \in (0,1)$. Again, we have for any non-zero solution π^* while denoting $\pi_1 = \phi(\pi_0) = \phi(0), \pi_1 = \phi(0) < \phi(\pi^*) = \pi^*$ since $\pi^* > 0$. Also, $\pi_2 = \phi(\pi_1) < \phi(\pi^*) = \pi^*$ since $\pi_1 < \pi^* \Rightarrow \phi(\phi_1) < \phi(\pi^*)$, ϕ being an increasing function of π in $0 < \pi < 1$. Hence it follows that the chance of extinction given by π is the smallest positive root of the equation $\phi(\pi) = \pi$.

Next, let us consider the case when the process starts with i number of females instead of 1. Obviously, it can be seen that the chance of extinction is the chance that each of the i lines independently dies out; which is nothing but given by π^i where π represents the smallest root of the equation $\phi(\pi) = \pi$.

This leads to a simple iterative process of solution given by

$$\pi_{n+1} = \phi(\pi_n),\ n = 1, 2, 3, \ldots,$$

where π_n is the nth iterative solution and $\pi_1 = p_0$ satisfies the solution. This may be solved by successive iteration process using Newton-Raphson method or the method of false position (see Scarborough (1966)).

1.4.7 Pure Death Process

It is analogous to birth process except that here $X(t)$ decreases with increasing t.

$$P[X(t) = k|X(0) = k_0] = p_k(t)$$
$$= \binom{k_0}{k} \exp\left[-k\int_0^t \mu(\tau)d\tau\right]\left[1 - \exp\left(\int_0^t \mu(\tau)d\tau\right)\right]^{k_0-k} \tag{1.4.40}$$

where $\mu(t)$ is the intensity of death at time t.

1.4.8 Birth and Death Process

This is given by Kolmogorov's equation

$$p_k(t+\Delta) = p_k(t)(1 - (\lambda_k(t)\Delta + \mu_k(t)\Delta + 0(\Delta))) + p_{k-1}(t)\lambda_{k-1}(t)\Delta$$
$$+ p_{k+1}(t)\mu_{k+1}(t)\cdot\Delta + 0(\Delta)$$

Then Kolmogorov's differential equations are

$$\left.\begin{aligned}\frac{d}{dt}p_0(t) &= -[\lambda_0(t)+\mu_0(t)]p_0(t)+\mu_1(t)p_1(t)\\ \frac{d}{dt}p_k(t) &= -[\lambda_k(t)+\mu_k(t)]p_k(t)+\lambda_{k-1}(t)p_{k-1}(t)+\mu_{k+1}(t)p_{k+1}(t)\\ &\qquad +\mu_{k+1}(t)p_{k+1}(t),\ t\geq 0\end{aligned}\right\}\tag{1.4.41}$$

The initial conditions are $p_{k_0}(0)=1, p_k(0)=0\ \forall\ k\neq k_0$. The steady state solution is obtainable from Ex 1.2 where $t\to\infty$.

Vide example 1.6 page 23–24.

1.4.9 Chiang's Illness Death Process

This process has two illness states S_1 and S_2.

S_1 = State of remaining ill and S_2 = State of remaining well, and there are r death states (absorbing states) viz. $R_1, R_2, \ldots, R_\delta, \ldots, R_r$ $(\delta = 0, 1, 2, \ldots, r)$.

Two intensities have been defined viz., $v_{\alpha,\beta}$ and $v_{\alpha\delta}$, where

$$v_{\alpha\beta}\Delta + 0(\Delta) = P \text{ [an individual in state } S_\alpha \text{ at time } t \text{ will be in state } S_\beta \text{ at time } t+\Delta.] \tag{1.4.42}$$

$$\mu_{\alpha\delta}\Delta + 0(\Delta) = P \text{ [an individual in State } S_\alpha \text{ at time } t \text{ be in state } R_\delta \text{ at time } t+\Delta] \tag{1.4.43}$$

$$\text{and}\quad v_{\alpha\alpha} = -\left(v_{\alpha\beta}+\sum_{\delta=1}^{r}\mu_{\alpha\delta}\right),\ \alpha,\beta=1,2;\ \delta=1,2,\ldots,r. \tag{1.4.44}$$

Illness transition probability is defined as

$$P_{\alpha\beta}(\tau,t) = P \text{ [an individual in state } S_\alpha \text{ at time } \tau \text{ will be in state } S_\beta \text{ at time } t]; \tag{1.4.45}$$

and the death transition probability

$$Q_{\alpha\delta}(\tau,t) = P \text{ [an individual in state } S_\alpha \text{ at time } \tau \text{ will be in state } R_\delta \text{ at time } t];$$

$$\alpha,\beta = 1,2;\ \delta = 1,2,\ldots,r \tag{1.4.46}$$

and the initial conditions are

$$\left.\begin{aligned}P_{\alpha\alpha}(\tau,\tau) &= 1\\ P_{\alpha\beta}(\tau,\tau) &= 0\\ \text{and}\quad Q_{\alpha\delta}(\tau,\tau) &= 0;\ \alpha,\beta=1,2;\ \delta=1,2,\ldots,r.\end{aligned}\right\}\tag{1.4.47}$$

The Kolmogorov equations are

$$P_{\alpha\alpha}(\tau,t+\Delta) = P_{\alpha\alpha}(\tau,t)[1+v_{\alpha\alpha}\Delta+0(\Delta)]+P_{\alpha\beta}(\tau,t)[v_{\beta\alpha}\Delta+0(\Delta)] \tag{1.4.48}$$

$$P_{\alpha\beta}(\tau,t+\Delta) = P_{\alpha\beta}(\tau,t)[1+v_{\alpha\beta}\Delta+0(\Delta)]+P_{\alpha\alpha}(\tau,t)[v_{\alpha\beta}\Delta+0(\Delta)] \tag{1.4.49}$$

The process has been found to be homogeneous and the solution is of the form

$$P_{\alpha\alpha}(\tau,t) = P_{\alpha\alpha}(t-\tau) = P_{\alpha\alpha}(t) = \sum_{i\neq j}\left(\frac{\rho_i - v_{\beta\beta}}{\rho_i-\rho_j}e^{\rho_i t}\right) \tag{1.4.50}$$

and $$P_{\alpha\beta}(\tau,t) = P_{\alpha\beta}(t-\tau) = P_{\alpha\beta}(t) = \sum_{i\neq j}\left(\frac{v_{\alpha\beta}}{\rho_i-\rho_j}e^{\rho_i t}\right), \tag{1.4.51}$$

where ρ_1 and ρ_2 are two roots of the following quadratic equation

$$\rho^2 - (v_{\alpha\alpha}+v_{\beta\beta})\rho + (v_{\alpha\alpha}v_{\beta\beta} - v_{\alpha\beta}v_{\beta\alpha}) = 0 \tag{1.4.52}$$

$Q_{\alpha\delta}$ is given by

$$Q_{\alpha\delta} = \int_0^t P_{\alpha\alpha}(\tau)\mu_{\alpha\delta}d\tau + \int_0^t P_{\alpha\beta}(\tau)v_{\beta\delta}d\tau. \tag{1.4.53}$$

Since an individual in illness state S_α may reach the state $R_\delta(\delta = 1, 2, \ldots, r)$ directly from S_α or by way of $S_\beta(\beta \neq \alpha)$. An individual in R_δ at time t may have reached that state any time prior to t. Let us consider an infinitesimal time interval $(\tau, \tau + d\tau)$ for fixed $\tau, 0 \le \tau \le t$. The probability that an individual in state S_α at time '0' will reach the state R_δ in the interval $(\tau, \tau + d\tau)$ is

$$P_{\alpha\alpha}(\tau)\mu_{\alpha\delta}d\tau + P_{\alpha\beta}(\tau)\mu_{\beta\delta}d\tau \tag{1.4.54}$$

As τ varies over the interval $(0, t)$ the corresponding events, whose probabilities are given in (1.4.54) are mutually exclusive.

1.5 Renewal Theory

Renewal theory deals with a sequence of i.i.d. non-negative r.v.'s $\{X_i\}$ which constitute a Renewal Process. Renewal Process is a Stochastic Process, which is a generalization of the Poisson Process. In a Poisson process the intervals between successive occurrences are i.i.d. exponential variates whereas in a Renewal process they are not necessarily exponential, which makes it non-Markovian.

Let us assume that a working unit, after failure, is replaced with a new identical unit (i.e. renewed). Suppose the unit start working at time $t = 0$ and fails after a random period of time X_i, it is instantly renewed and operates for a period X_2 and so no.

Let $S_n = X_1 + X_2 + \ldots + X_n$.

It represents the waiting time for the nth renewal, X_i represents the waiting time from $(i-1)$th renewal to ith renewal, X_i's are independent random variables.

Now $P[S_n \le t] = P[N_t \ge n] = F_n(t)$, say, $\quad$ (1.5.1)

where N_t represents the number of renewals in $(0, t)$ and $F_n(t)$ is the cumulative distribution function of S_n.

The sequences $\{X_i\}$, $\{S_n\}$ and $\{N_t\}$ are all renewal processes.

Now $P[N_t = n] = P[N_t \ge n] - P[N_t \ge n+1] = F_n(t) - F_{n+1}(t)$

$$E(N_t) = \sum_{n=0}^{\infty} n \cdot P[N_t = n] = \sum_{n=0}^{\infty} n \cdot [F_n(t) - F_{n+1}(t)] = \sum_{n=0}^{\infty} F_n(t) \tag{1.5.2}$$

We define $E(Nt) = U(t) =$ Renewal function or the average number of renewals in $(0, t)$ $\quad$ (1.5.3)

$$\Rightarrow \quad \frac{dE(N(t))}{dt} = \frac{dU(t)}{dt} = \frac{d}{dt}\sum_{n=1}^{\infty} F_n(t) = \sum_{n=0}^{\infty}\frac{d}{dt}F_n(t) = \sum_{n=0}^{\infty} f_n(t) = u(t), \text{ say}, \tag{1.5.4}$$

where $u(t)$ is called the renewal density of the nth renewal.

The Laplace transform of $u(t)$ is given by

$$L(u(t)) = \int_0^\infty e^{-st}u(t)dt = \int_0^\infty e^{-st}\sum_{n=1}^\infty f_n(t)dt, \text{ from (1.5.4)}$$

$$= \sum_{n=1}^\infty \int_0^\infty e^{-st} f_n(t)dt = \sum_{n=1}^\infty L(f_n(t)). \tag{1.5.5}$$

Let the distribution of the waiting time X_1 be $f_1(t)$ and the common distribution of $X_i (i = 2, 3, 4, \ldots)$ be $f(t)$.

Then $f_n(t) = f_1(t) * [f(t)]^{(n-1)}$

where $*$ stands for convolution (vide Sec. A.2.1). (1.5.6)

Now, $L[f_n(t)] = L[f_1(t)][L(f(t))]^{n-1}$ (1.5.7)

$$L(u(t)) = \sum_{n=1}^\infty L(f_n(t)) = \sum_{n=1}^\infty L[f_1(t)][L(f(t)]^{n-1}, \text{ from (1.5.5) and (1.5.7)}$$

$$= L[f_1(t)]\sum_{n=1}^\infty [L(f(t))]^{n-1}$$

$$= \frac{L(f_1(t))}{1 - L(f(t))}. \tag{1.5.8}$$

Next $L(U(t)) = \int_0^\infty e^{-st}U(t)dt$, where $U(t)$ is as defined in (1.5.3)

and $L(u(t)) = \int_0^\infty e^{-st}u(t)dt$

$$= e^{-st}U(t)|_0^\infty + s\int_0^\infty e^{-st}U(t)dt, \text{ from (1.5.3)}$$

$$= sL(U(t))) \quad (\because U(0) = 0) \tag{1.5.9}$$

From (1.5.8) and (1.5.9) we have,

$$sL(U(t)) = \frac{L(f_1(t))}{1 - L(f(t))}$$

$$\Rightarrow \quad sL(U(t)) - sL(U(t))L(f(t)) = L((f_1(t))$$

$$\Rightarrow \quad L(U(t)) - L(U(t))L(f(t)) = \frac{1}{s}L((f_1(t)) \tag{1.5.10}$$

$$\text{Again, } L(f_1(t)) = \int_0^\infty e^{-st}f_1(t)dt = e^{-st}F_1(t)|_0^\infty + s\int_0^\infty e^{-st}F_1(t)dt,$$

where $F_1(t)$ is the c.d.f. corresponding to the density function $f_1(t)$.

$$\Rightarrow \quad L(f_1(t)) = sL(F_1(t)) \quad (\because F_1(0) = 0)$$

$$\Rightarrow \quad \frac{1}{s}L(f_1(t)) = L(F_1(t)) \tag{1.5.11}$$

Putting (1.5.11) in (1.5.10), we get

$$L(U(t)) = L(U(t))L(f(t)) + L(F_1(t)) \tag{1.5.12}$$

Taking inverse Laplace transform on both sides of (1.5.12) we get

$$U(t) = F_1(t) + \int_0^t U(t-\tau)f(\tau)d\tau = F_1(t) + U(t) * f(t). \tag{1.5.13}$$

Differentiating both sides with respect to t, we get

$$u(t) = f_1(t) + \int_0^\infty u(t-\tau)f(\tau)d\tau \tag{1.5.14}$$

(1.5.13) and (1.5.14) both are forward renewal equations.

Remark 1.1: From (1.5.13) and (1.5.14) it follows that $u(t)$ (or $U(t)$) depends not only on $u(t-\tau)$ (or $U(t-\tau)$) but also on f and f_1 (or F and F_1) too. This shows that a renewal process is essentially Non-Markovian in character.

1.5.1 Alternative Technique (Method of Probability Generating Function) for Forward and Backward Renewal equations

Let τ be the time for the first renewal and N_t represent the number of renewals in $(0,t)$. If

$$\left.\begin{array}{lll} \text{(i) } t < \tau & \text{then} & N_t = 0 \text{ clearly} \\ \text{(ii) } t \geq \tau & \text{then} & N_t = 1 + N_{t-\tau} \end{array}\right\} \tag{1.5.15}$$

Denote the probability generating function i.e. $E(s^{N_t})$ for the real variable N_t with $|s| < 1$ as $\psi_t(s)$. Now from (1.5.15)

$$\left.\begin{array}{ll} s^{N_t} = 1 & \text{if } t < \tau \\ s^{N_t} = s^{N_{t-\tau}+1} & \text{if } t \geq \tau \end{array}\right\} \tag{1.5.15}$$

$$\text{Also,}\quad \psi_t(s) = E(s^{N_t}) = E_T(E(s^{N_t}|\tau)) = \int_0^\infty E(s^{N_t}|\tau)f(\tau)d\tau$$

$$= \int_0^t E(s^{N_t}|\tau)f(\tau)d\tau + \int_t^\infty E(s^{N_t}|\tau)f(\tau)d\tau$$

$$= \int_0^t E(s^{N_{t-\tau}+1})f(\tau)d\tau + \int_t^\infty f(\tau)d\tau, \text{ from (1.5.16)}$$

$$\Rightarrow \quad \psi_t(s) = s\int_0^t \psi_{t-\tau}(s)f(\tau)d\tau + \int_t^\infty f(\tau)d\tau, \tag{1.5.17}$$

which is recurrence relation between $\psi_t(s)$ and $\psi_{t-\tau}(s)$.

Differentiating both sides w.r.t. s and putting $s = 1$

$$\Rightarrow \quad \frac{d\psi_t(s)}{ds}|_{s=1} = \int_0^t \psi_{t-\tau}(s)|_{s=1}f(\tau)d\tau + s\int_0^t \frac{d\psi_{t-\tau}(s)}{ds}|_{s=1}f(\tau)d\tau$$

$$\Rightarrow \quad U(t) = F(t) + \int_0^t U(t-\tau)f(\tau)d\tau, \text{ from (1.4.26)} \quad (\because \psi(1) = 1). \tag{1.5.18}$$

This equation is a special case of (1.5.13) when $f_1 = f$ and $F_1 = F$ and the process is a simple renewal process whereas (1.5.13) corresponds to that of a general recurrent process.

Differentiating (1.5.18) with respect to t, we get

$$u(t) = f(t) + \int_0^t u(t-\tau)f(\tau)d\tau, \tag{1.5.19}$$

which is again a special case of (1.5.14).

Example 1.4: Obtain the waiting time distribution of a simple Poisson process with parameter λt.

A Poisson process is given by

$$P[X = x|(0,t)] = \frac{e^{-\lambda t}(\lambda t)^x}{x!}, \lambda > 0,\ x = 0, 1, 2, \ldots$$

$P[X = 0|t] = e^{-\lambda t}$

$\Rightarrow\ P[T_1 > t] = e^{-\lambda t}$, where T_1 is the waiting time for the first renewal.

$\Rightarrow\ F(t) = P[T_1 \le t] = 1 - e^{-\lambda t}$.

$\therefore\ \ f(t) = \frac{d}{dt}F(t) = \lambda e^{-\lambda t},\ \lambda > 0;\ 0 \le t < \infty.$

Thus the interval density function of the waiting time in a Poisson process is exponential with parameter λ.

Example 1.5: Obtain the probability generating function (p.g.f.) of the simple Poisson process

$$P[X = x|t] = \frac{e^{-\lambda t}(\lambda t)^x}{x!} \tag{I}$$

Solution: The p.g.f. of (I) is

$$\psi_t(s) = E(s^X) = \sum_{x=0}^{\infty} s^x \frac{e^{-\lambda t}(\lambda t)^x}{x!} = e^{-\lambda t}\sum_{x=0}^{\infty}\frac{(s\lambda t)^x}{x!} = e^{-\lambda t}e^{s\lambda t}$$

$$= e^{\lambda t(s-1)}$$

Further, $E(N(t)) = U(t) = \frac{d\psi_t(s)}{ds} = \lambda t e^{\lambda t(s-1)}|_{s=1} = \lambda t$

Also $\frac{d^2\psi_t(s)}{ds^2}|_{s=1} = E(N_t(N_t - 1)) = E(N_t^{(2)})$ the second factorial moment

$$= \lambda^2 t^2 e^{\lambda t(s-1)}|_{s=1} = \lambda^2 t^2$$

$$\text{var}(N_t) = E(N_t(N_t-1)) + E(N_t) - [E(N_t)]^2 = \lambda^2t^2 + \lambda t - (\lambda t)^2 = \lambda t$$

Thus $E(N(t))) = \text{Var}(N_t) = \lambda t$ for a Poisson Process.

Example 1.6: A linear growth model with immigration Such process are important in the analysis of Population growth. Here we have, $\lambda_n = n\cdot\lambda + \alpha$ and $\mu_n = n\cdot\mu$; λ_n represents the intensity of change in the positive direction as a result of the instantaneous birth rate 'λ' applied over population of size n and 'α' stands for immigration; μ_n stands for intensity of death in infinitesimal small period of time Δt. μ_n is given by

$$\mu_n\Delta t = n\mu\Delta t + 0(\Delta t)$$

The Kolmogorov equations become

$$p'_{i0}(t) = -\alpha p_{i0}(t) + \mu p_{i1}(t)$$

$$p'_{ij}(t) = -[(\lambda(j-1)+\alpha)p_{i,j-1}(t) - ((\lambda+\mu)j+\alpha)p_{ij}(t)$$
$$+\mu(j+1)p_{i,j+1}(t)] \quad \text{for } j \geq 1.$$

If we denote $E(X(t)) = \sum_{j=1}^{\infty} j p_{ij}(t) = g(t)$, (say) then one may check that

$$g'(t) = \alpha + (\lambda - \mu)g(t)$$

The solution of the differential equation with initial condition $g(0) = k$, say $\Rightarrow X(0) = k$) is given by

$$g(t) = \alpha t + k, \quad \text{when} \quad \lambda = \mu$$

and $$g(t) = \frac{\alpha}{\lambda - \mu}[e^{(\lambda-\mu)t} - 1] + ke^{(\lambda-\mu)t}, \quad \text{when} \quad \lambda \neq \mu.$$

Remarks. This is a more general process. Special cases of this process are;

(i) $\lambda = \mu = 0 \Rightarrow$ Poisson process.

(ii) $\lambda = 0 \Rightarrow$ Immigration-Emigration process.

(iii) $\alpha = \mu = 0 \Rightarrow$ Linear growth process or Yule's Process.

(iv) $\alpha = 0 \Rightarrow$ Linear birth and death process.

1.5.2 Basic Renewal Theorems

We present here (without proof) the results of two important renewal theorems.

(i) Asymptotic Renewal Theorem

If N_t represents the number of renewals up to time t, S_{N_t} represents the waiting time for N_t, then

$$P\left[\lim \frac{S_{N_t}}{N_t}\right] = \mu \text{ as } t \to \infty \tag{1.5.20}$$

where μ represents the average time between renewals. This

$$\Rightarrow \frac{N_t}{t} \xrightarrow{P} \frac{1}{\mu} \quad \text{as} \quad t \to \infty.$$

(ii) Elementary Renewal Theorem

Given N_t/t converges almost surely (a.s.) to $1/\mu$, then

$$\lim \frac{E(N_t)}{t} = \frac{1}{\mu} \quad \text{as} \quad t \to \infty \tag{1.5.21}$$

Note that the asymptotic renewal density $u(t)$ converges to $1/\mu$ as t tends to infinity.

1.5.3 Distribution of Forward and Backward Recurrence Time

A. Forward Recurrence time

Let S_n be the waiting time for the nth renewal and define

$$T_t^+ = S_n - t \quad \text{and} \quad T_t^- = t - S_{n-1} \tag{1.5.23}$$

then T_t^+ and T_t^- are called the forward and back ward recurrence time, respectively.

The problem is to consider their joint distribution. Let $f(.)$ and $u(.)$ denote the interval and the renewal densities, respectively, of the renewal process.

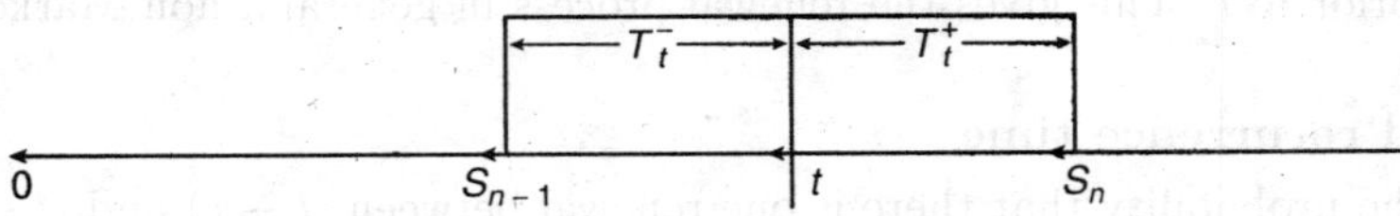

Fig. 1.1

Let $g^+(x)$ and $g^-(x)$ denote density of the forward and backward recurrence times respectively. Then

$$P[x \le T_t^+ \le x + dx] = g^+(x)dx$$

Further, let a renewal take place between v to $v + dv$ prior to t, (without any consideration of the backward recurrence time) with probability

$$u(v)dv, \; 0 < v < t.$$

$$\begin{aligned} P[T_t^+ \in (x, x+\Delta)] &= g^+(x)\Delta + 0(\Delta) \\ &= \{f(t+x)\Delta + 0(\Delta)\} + \int_0^t u(v)[f(t+x-v)\Delta + 0(\Delta)]dv \end{aligned}$$

$$\Rightarrow \quad g^+(x) = f(t+x) + \int_0^t u(v) f(t+x-v)du \tag{1.5.24}$$

Putting $u' = t - v$

$$\Rightarrow \quad g^+(x) = f(t+x) + \int_0^t u(t-u')f(u'+x)du' \text{ holds} \tag{1.5.25}$$

The first term on the right hand side gives the probability of first renewal at $(t+x)$ and the second term gives the probability of a further renewal at $(t+x)$ given that the immediate past renewal took place at v, $0 < v < t$. Note that (1.5.24) is true for fixed t.

As $\quad t \to \infty, \; u(t) \to \dfrac{1}{\mu} \; \forall \, t$

$$\Rightarrow \quad g^+(x) = \int_0^\infty \frac{1}{\mu} f(u'+x)du' = \frac{1}{\mu}\int_x^\infty f(\omega)d\omega = \frac{1-F(x)}{\mu} \tag{1.5.26}$$

Example 1.7: Let $f(t) = \lambda e^{-\lambda t}$; $u(t) = \lambda$

$$\begin{aligned} g^+(x) &= \lambda e^{-\lambda(t+x)} + \lambda \int_0^t \lambda e^{-\lambda(t+x-v)} dv, \text{ from (1.5.24)} \\ &= \lambda e^{-\lambda(t+x)} + \lambda^2 \int_0^t e^{-\lambda(t+x-v)} dv \\ &= \lambda e^{-\lambda(t+x)} + \frac{\lambda^2}{\lambda}\int_{t+x}^{x} e^{-\eta} d\eta, \quad \text{when} \quad \eta = \lambda(t+x-v) \\ &= \lambda e^{-\lambda(t+x)} - \lambda e^{-\lambda(t+x)} + \lambda e^{-\lambda x} = \lambda e^{-\lambda x} \end{aligned}$$

$$\Rightarrow \quad E(T_t^+) = \int_0^\infty e^{-\lambda x} dx = \frac{1}{\lambda}$$

Note that $g^+(x) = \lambda e^{-\lambda x}$ does not depend on t; because of the memoryless property of the distribution. But in general $g^+(x)$ will depend not only on n but also on the value prior to t. This gives the renewal process in general a non Markovian character.

B. Backward recurrence time

We have the probability that there is one renewal between $(t-x)$ and $(t-x+dx) = u(t-x)dx$

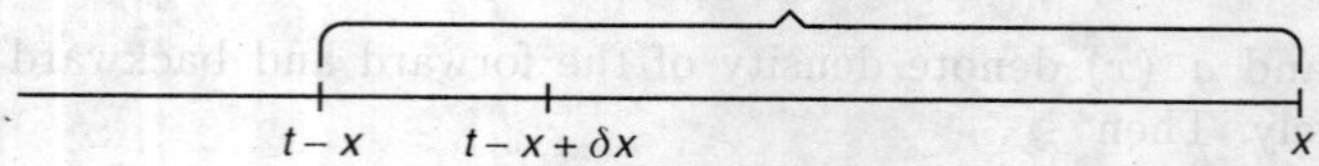

Fig. 1.2

Further, the probability of not having a renewal for a further length of time x

$$= (1 - F(x)), \quad \text{where } F(x) = \int_0^x f(t)dt.$$

$F(t)$ is the c.d.f. and $f(t)$ the density of the interval.

$$\Rightarrow \quad g^-(x)dx = u(t-x)(1-F(x))dx,$$

where $g^-(x)$ is the distribution of backward recurrence time.

$$\Rightarrow \quad g^-(x) = u(t-x)(1-F(x)) \tag{1.5.27}$$

$$\text{As} \quad t \to \infty,\ u(t-x) \quad \to \quad \frac{1}{\mu} \Rightarrow \lim g^-(x) = \frac{1}{\mu}(1-F(x)), \tag{1.5.28}$$

which gives the distribution of the backward recurrence time as $t \to \infty$. One may compare the result (1.5.27) with (1.5.25) as well as (1.5.28).

1.5.4 Joint Distribution of T_t^+ and T_t^-, the Forward and Backward Recurrence Time

Let T_t^+ and T_t^- be the forward and backward recurrence time, respectively, measured from an arbitrary point t.

The joint distribution of T_t^+ and T_t^- is obtained in the same way as the distribution of marginals derived earlier.

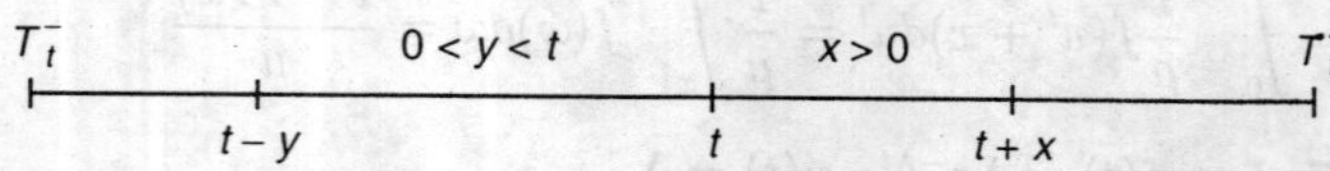

Fig. 1.3

In fact, for any $x > 0$ and $0 < y < t$ the event $E^* = [T_t^+ > x \cap T_t^- > y]$ occurs if there are no renewals in the intervals $(t-y, t+x)$ and this interval has probability that the time of a particular renewal, say the nth renewal, $Z_n > t + x$ given that the nth renewal has not taken place upto $(t-y)$

$$P[T_t^+ > x \cap T_t^- > y] = \left\{ \frac{\overline{F}_n(t+x)}{\overline{F}_n(t-y)} \right\},\ n = 1, 2, \ldots \tag{1.5.29}$$

It is the probability that a renewal takes place at time $> t + x$ given that it has not taken upto $(t-y)$.

For a Poisson process the same is given by

$$P[T_t^+ > x \cap T_t^- > y] = \frac{e^{-\lambda(t+x)}}{e^{-\lambda(t-y)}} = e^{-\lambda(x+y)}, \tag{1.5.30}$$

where $P[T_t^+ > x] = e^{-\lambda x}$ (1.5.31)

$$P[T_t^- > y] = e^{-\lambda y} \tag{1.5.32}$$

$$\therefore \quad P[T_t^+ > x]P[T_t^- > y] = e^{-\lambda(x+y)}. \tag{1.5.33}$$

Thus for a Poisson process

$$P[T_t^+ > x \cap T_t^- > y] = P[T_t^+ > x]P[T_t^- > y]. \tag{1.5.34}$$

Thus for a Poisson process, distribution of forward and backward recurrence times are independent. In fact, it can be shown that this property is a characterisation property of the Poisson process.

For further study on Renewal Theory, one may see Cox (1962).

1.5.5 Continuous Time Branching Process

Suppose that at time $t = 0$, the process starts with a single female birth, who may be assumed to be the founder of the generation and future female births are not considered over generations but over continuous time. Let Z_t be the population size at any time t. Let $\psi_t(s)$ be the probability generating function (p.g.f) of Z_t. Let U be the time of founder's death (or the end of her fertility span). Let N be the number of female offsprings or children given by the founder. Both U and N are r.v.'s. We denote

$$E_N(s^N) = \phi(s)$$

Given that $U = u$, we have $\forall\, t < u$

$$Z_t = Z_0 = 1, \tag{1.5.35}$$

while for $t > u$, we have,

$$Z_t = Z_{t-u}^{(1)} + Z_{t-u}^{(2)} + \ldots + Z_{t-u}^{(N)} \quad \text{for} \quad N = 1,\ 2,\ 3\ldots \tag{1.5.36}$$

$$= 0, \quad \text{for} \quad N = 0,$$

where $Z_{t-u}^{(i)}$ represents the number of female births to the ith female during $(t-u)$ to t, $(\forall\, t > u)$, $(i = 1, 2, 3, \ldots, N)$.

Each $Z_{t-u}^{(i)}$ are i.i.d. r.v.'s and distributed as Z_{t-u}. The p.g.f. of Z_{t-u} is denoted as $\psi_{t-u}(s)$. Then

$$\begin{aligned} E[s^{Z_t}|U = u < t] &= E_N[E(s)^{NZ_{t-u}}] = E_N[(\psi_{t-u}(s)^N)] \\ &= \phi(\psi_{t-u}(s)) \end{aligned} \tag{1.5.37}$$

Hence, $\psi_t(s) = E_{Z_t}(s^{Z_t}) = E_U[E(s^{Z_t}|U = u)]$

$$= \int_0^\infty f(u)\{E(s^{Z_t}|U = u)\}du, \quad \text{where } f(u) \text{ is p.d.f. of } U.$$

$$= \int_0^t f(u)E(s^{Z_t}|U = u)du + \int_t^\infty f(u)E(s^{Z_t}|U = u)du$$

$$= \int_0^t f(u)\phi(\psi_{t-u}(s))du + \int_t^\infty f(u)E(s)du,$$

$$\because Z_t = 1 \quad \text{if} \quad t < u \text{ from } (1.5.35)$$

$$\Rightarrow \quad \psi_t(s) = \int_0^t f(u)\phi(\psi_{t-u}(s))du + s(1 - F(t)), \text{ from } (1.5.37)$$

and $\quad \because E(s) = s \quad (1.5.38)$

Example 1.8: Obtain $\psi_t(s)$ when $N = 2$ (the number of offsprings) given $f(u) = \lambda c^{-\lambda u}$; $0 < \lambda < \infty$, $u > 0$ and show that $(Z_t - 1)$ is a geometrically distributed r.v.

Solution: $\psi_t(s) = \dfrac{sc^{-\lambda t}}{1 - (1 - c^{-\lambda t})s}$

1.5.6 Diffusion Process

Diffusion processes belong to the class of Markov Process $\{X_t : t \in T\}$ with both parameter space as well as state space continuous while a small change in t leading to a small change in X_t (or $X(t)$). A realization of the process can be thought of a system, moving quite erratically in a continuous medium while its progress depends only on its current position.

An important class of diffusion processes is known as '*Gaussian process*'. If $\{X(t) : t \in T\}$ is a process whose state space is real line $R^{(1)}$ (one dimensional and the parameter space is a subset of $R^{(1)}$ then it is known as a Gaussian process if $\forall$ finite n the joint distribution of $(X(t_1), \ldots, X(t_n))$ is multivariate normal. Again a Gaussian process is *stationary* if its covariance function viz. Cov $(T_i, T_j) = g(T_i, T_j) = g(T_i - T_j)$. Finally, the Brownian Motion (Wienner process) with parameter $\lambda(\lambda > 0)\{X(t);\ t \in R\}$ is the Gaussian process with independent increments with

(i) $E(X(t)) = 0$
(ii) $g(T_i, T_j) = \lambda \min(T_i, T_j)$

The Brownian motion is useful in several biological situations concerning stochastic models especially in Diffusion theory of Populations models.

1.6 Counter Theory (Geiger Müller (G.M.) Counter Models)

A G.M. counter is a registering mechanism that detects the presence of radioactive material. Impulses emitted by the radioactive material arrive at the counter, but because of inertia, the counter does not register some of the impulses. Suppose that an impulse arrives at a fixed time t the counter registers the impulse. The registration results in a dead time of duration say π_1, during which impulses will not be registered by the counter. In general, the first impulse to arrive after the termination of the dead time π_1, will be registered by the counter resulting in a dead time again of length (say) π_2 and so on. However, in simpler cases, impulses arriving during dead time do not prolong it so that each period of dead time is caused by a registered impulse only. This is called a G.M. counter of Type I (Pyke 1961).

In a counter of Type II (Smith (1958)) each arriving impulse causes a dead time so that arrivals during a period of dead time prolong it further. Theory of counter model arose from certain physical processes. It can be applied to many biological, social, demographic and industrial processes.

1.6.1 Demographic Problem of Counter Model with Fixed Dead Time π (Counter Model of Type I with Fixed Dead Time)

We describe the G.M. Counter Model Type I with the help of the following example of a demographic problem:

If a conception takes place with intensity λ subject to the condition that every conception is followed by fixed infecundable exposure π (dead time) during which no further conception takes place, then obtain the probability distribution of the number of conceptions in $(0, t)$.

Let $\tau_1 < \tau_2 < \ldots < \tau_n < \ldots$ be the renewal times (waiting time for conceptions) and with conception rate λ (Poisson intensity) with negative exponential density function,

$$f(t) = \lambda e^{-\lambda t}; 0 \leq t < \infty,\ \lambda > 0$$

$$\Rightarrow \quad P[\tau_1 < x] = 1 - e^{-\lambda x},\ \lambda \geq 0,\ 0 \leq x < \infty \tag{1.6.1}$$

$$\Rightarrow \quad P[\tau_n - \tau_{n-1} \leq x] = 1 - e^{-\lambda(x-\pi)} \quad \text{for} \quad x \geq \pi \tag{1.6.2}$$

Let $W(t, n)$ = probability of not more than n events or conceptions upto time t.

Then, $W(t, n) = P[\tau_{n+1} > t] = 1 - P[\tau_{n+1} \leq t]$

$$= 1 - F_{n+1}(t) = R_{n+1}(t) \tag{1.6.3}$$

where $F_{n+1}(t)$ is the cumulative distribution function of the random variable τ_{n+1} and $R_{n+1}(t) = 1 - F_{n+1}(t)$ is the corresponding survival function. Taking the Laplace transform of (1.6.3) we get

$$L[W(t, n)] = \int_0^\infty e^{-st}(1 - F_{n+1}(t))dt$$

$$= \frac{1}{s} - \frac{1}{s} \cdot L[f_{n+1}(t)], \tag{1.6.4}$$

$$\because \quad L[F_{n+1}(t)] = \frac{1}{s} L[f_{n+1}(t)], \text{ from (A.1.24)},$$

where $f_{n+1}(t)$ is the interval density function of $(n+1)$th renewal (order of conceptions).

We have $\quad f_{n+1}(t) = f_1 * \{f^{(n)}\}$

where f_1 is the density function of τ_1 and f the density function of $(\tau_r - \tau_{r-1})$; $r = 2, 3, \ldots$ and $*$ stands for convolution; $\{f^{(n)}\}$ stands for n-fold convolution of f with itself.

Hence, $\quad L[f_{n+1}(t)] = L[f_1(t)]\{L[f(t)]\}^n \quad (1.6.5)$

Now $\quad L[f_1(t)] = \dfrac{\lambda}{\lambda + s} \quad (1.6.6)$

and $\quad L[f(t)] = \displaystyle\int_\pi^\infty e^{-st} \lambda e^{-\lambda(t-\pi)} dt = \frac{\lambda e^{-\pi s}}{(\lambda + s)} \quad (1.6.7)$

$\therefore \quad L[f_{n+1}(t)] = \dfrac{\lambda^{n+1}}{(\lambda + s)^{n+1}} \cdot e^{-n\pi s}$, from (1.6.5), (16.6) and (1.6.7) $\quad (1.6.8)$

Putting (1.6.8) in (1.6.4), we get

$$L[W(t,n)] = \frac{1}{s} - \frac{1}{s}\cdot\frac{\lambda^{n+1}}{(\lambda+s)^{n+1}}e^{-n\pi s} \tag{1.6.9}$$

By taking inverse Laplace transform (see Table A.1 in Appendix)

$$W(t,n) = 1 - \lambda\int_{n\pi}^{t}\frac{e^{-\lambda(u-n\pi)}\lambda^n(u-n\pi)^n du}{\Gamma(n+1)} \tag{1.6.10}$$

Also using result of incomplete Poisson distribution

$$\sum_{j=0}^{n} e^{-\mu}\frac{\mu^j}{j!} = 1 - \int_0^{\mu}\frac{e^{-z}z^n dz}{\Gamma(n+1)} \tag{1.6.11}$$

and putting $\mu = \lambda(t-n\pi)$ on both sides of (1.6.11), we get

$$\sum_{j=0}^{n} e^{-\lambda(t-n\pi)}\frac{[\lambda(t-n\pi)]^j}{j!} = 1 - \frac{1}{\Gamma(n+1)}\int_0^{\lambda(t-n\pi)} e^{-z}z^n dz \tag{1.6.12}$$

Further, on substitution of $z = \lambda(u-n\pi)$

$z = 0 \Rightarrow u = n\pi, \quad z = \lambda(t-n\pi) \Rightarrow u = t$

$$\Rightarrow \sum_{j=0}^{n}\frac{e^{-\lambda(t-n\pi)}}{j!}\frac{[\lambda(t-n\pi)]^j}{j!}$$

$$= 1 - \frac{\lambda}{\Gamma(n+1)}\int_{n\pi}^{t} e^{-\lambda(u-n\pi)}[\lambda(u-n\pi)]^n du \tag{1.6.13}$$

Comparing (1.6.10) with (1.6.13), we have

$$W(t,n) = \sum_{j=0}^{n} e^{-\lambda(t-n\pi)}\frac{[\lambda(t-n\pi)]^j}{j!} \tag{1.6.14}$$

Now $P[x=n] = W(t,n) - W(t,n-1)$

$$= \sum_{j=0}^{n} e^{-\lambda(t-n\pi)}\frac{[\lambda(t-n\pi)]^j}{j!} - \sum_{j=0}^{n-1} e^{-\lambda(t-(n-1)\pi)}\frac{[\lambda(t-(n-1)\pi)]^j}{j!},$$

from (1.6.12) (1.6.15)

and $n \le \left[\frac{t}{\pi}\right]$ where $\left[\frac{t}{\pi}\right]$ refers the greatest integer contained in $\frac{t}{\pi}$.

(1.6.15) describes the G.M. Counter I model.

1.6.2 A Problem on Counter Model (Biswas, Nair and Nautiyal, (1983))

Theorem 1.6.1: Show that the conditional probability of having n conceptions at $x_1 < x_2 < \ldots < x_n$ subject to the condition that no conception can occur between $(x_i, x_i+\pi)$, $i = 1, 2, \ldots, n$, is given by

$$\prod_{i=1}^{n-1}\frac{\lambda(x_i)f(x_n)}{\prod_{i=1}^{n}\{1-(F(x_i+\pi)-F(x_i))\}} \tag{1.6.16}$$

where $\lambda(x_i)$ is the hazard rate at $x = x_i$,

$f(x_i)$ is the waiting time density function and

$F(.)$ is the cumulative distribution function.

Proof: We have $P[X_1 < x_1, X_2 < X_2] = F(x_1, x_2)$

$$= \int_0^{x_1} \frac{F(x_2) - F(x)}{1 - F(x)} dF(x)$$

$$= \int_0^{x_1} \frac{F(x_2)}{\overline{F}(x)} dF(x) - \int_0^{x_1} \frac{F(x)}{\overline{F}(x)} dF(x), \quad \text{where } \overline{F}(x) = 1 - F(x).$$

$$= F(x_2) \int_0^{x_1} \frac{dF(x)}{\overline{F}(x)} + [F(x_1) + \log(1 - F(x_1)], \quad \because F(0) = 0.$$

$$= F(x_2) \int_0^{x_2} \lambda(x) dx + [F(x_1) + \log(1 - F(x_1))]$$

where $\lambda(x) = \dfrac{f(x)}{\overline{F}(x)}$

$$\Rightarrow \quad F(x_1, x_2) = F(x_2)\Lambda(x_1) + F(x_1) - \int_0^{x_1} \lambda(x) dx, \text{ from (A.5.4)},$$

where $\displaystyle\int_0^{x_1} \lambda(x) dx = \Lambda(x_1)$

$$\Rightarrow \quad F(x_1, x_2) = F(x_2)\Lambda(x_1) + F(x_1) - \Lambda(x_1)$$

$$\Rightarrow \quad f(x_2, x_1) = \frac{\partial^2 F(x_1, x_2)}{\partial x_1 \partial x_2} = f(x_2)\lambda(x_1)$$

Similarly, one can show

$$f(x_1, x_2, x_3) = \lambda(x_1)\lambda(x_2)f(x_3)$$

and in general $\quad f(x_1, x_2, \ldots, x_n) = \lambda(x_1)\lambda(x_2)\ldots\lambda(x_{n-1})f(x_n).$

Now the conditional probability of n conceptions at $x_1 < x_2 < \ldots < x_n$ subject to the condition that no conception (or event) can occur between $(x_1, x_1 + \pi)$, $i = 1, 2, 3, \ldots, n$, is given by

$$= \prod_{i=1}^{n-1} \frac{\lambda(x_i) f(x_n)}{\prod_{i=1}^{n} [1 - (F(x_i + \pi) - F(x_i))]} \quad // \qquad (1.6.16)$$

1.6.3 Counter Model Type I with Random Dead Time

A particle arrives at $t = 0$, is registered and the counter is locked for a dead time of duration, say Y_1. The next particle to be registered is that of the first arrival once the counter is freed. With the registration of the 2nd particle the counter is blocked for a time of length, say Y_2. The process is repeated where the successive locking times, denoted by $Y_1, Y_2, Y_3 \ldots$, are assumed to be independent with a common distribution $P\{Y_k < y\} = G(y)$ and are also independent of the arrival process.

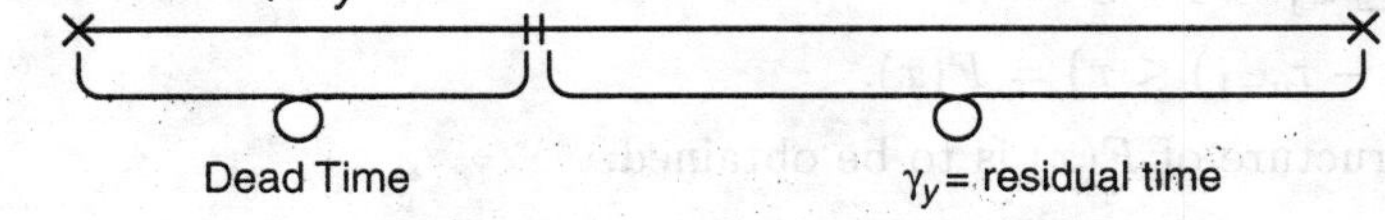

Fig. 1.4

We want to obtain the waiting time distribution of the inter-arrival of the process.

Let $Z = Y + \gamma_y$ where Y is the dead time following the registration and γ_y is the residual life time before the counter is locked.

Let $g(\cdot)$ and $f(\cdot)$ denoting the density functions of the dead time and the residual free time before the counter is locked again.

We have,

$$P\{z \leq Z = Y + \gamma_y \leq z + dz\} = h(z)dz = \left\{ \int_0^z g(y)f(z-y)dy \right\} dz \tag{1.6.17}$$

A special case

If $g(y) = \mu e^{-\mu y}$ and $f(z-y) = \lambda \exp[-\lambda(z-y)]$, then

$$h(z) = \frac{\lambda\mu \exp(-\mu z)}{(\mu-\lambda)} \{1 - \exp[-(\mu-\lambda)z]\}, \quad \lambda \neq \mu, \tag{1.6.18}$$

represents the waiting time distributions between two consecutive registrations.

Biswas, Nautiyal and Tyagi (1983) obtained the p.g.f. of the process as

$$\psi_t(s) = \frac{1}{2}\left[1 + \frac{(\lambda+\mu)}{(\lambda^2+\mu^2-2\lambda\mu+4\lambda\mu s)}\right]$$

$$\exp\left\{-\left(\frac{\lambda+\mu}{2}\right)t - \left(\frac{\mu-\lambda}{2}\right)t\left[1+\frac{4\lambda\mu s}{(\mu-\lambda)^2}\right]^{\frac{1}{2}}\right\}$$

$$+\frac{1}{2}\left[1 - \frac{(\lambda+\mu)}{(\lambda^2+\mu^2-2\lambda\mu+4\lambda\mu s)^{1/2}}\right]$$

$$\exp\left\{-\left(\frac{\lambda+\mu}{2}\right)t - t\left(\frac{\mu-\lambda}{2}\right)\left[1+\frac{4\lambda\mu s}{(\mu-\lambda)^2}\right]^{\frac{1}{2}}\right\} \tag{1.6.19}$$

1.6.4 Counter Model Type II with Fixed Dead Time

Here the locking mechanism is more complicated. As before an incoming signal is registered if it arrives when the counter is free. In type II counter every arriving signal can prolong the dead time in the counter, increasing the associated locking times.

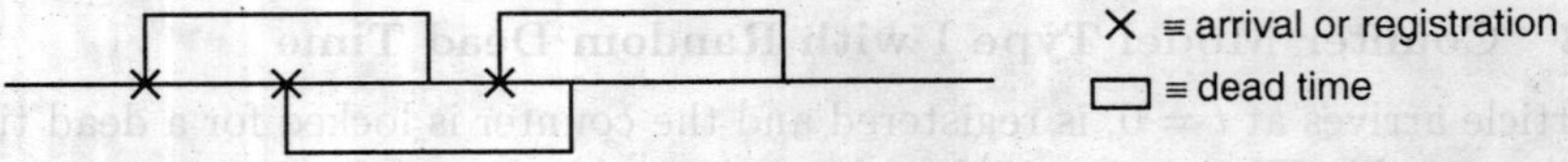

Fig. 1.5

Let $\tau_1 < \tau_2 < \ldots < \tau_n$ are the registration times of the first, second and nth renewal and arrivals take place with Poisson rate λ. The length of the dead time following every registration is π. We assume $\{\tau_n - \tau_{n-1}\}$ are the i.i.d.r.v.'s. Now

$$P\{\tau_1 \leq x\} = 1 - e^{-\lambda x}$$

$$P\{(\tau_n - \tau_{n-1}) \leq x\} = F(x), \tag{1.6.20}$$

where the structure of $F(x)$ is to be obtained.

Let $P\{x \leq \tau_n - \tau_{n-1} \leq x + dx\} = f(x)dx$

Let $\phi(s) = \int_0^\infty e^{-st} f(t)dt = L(f(t))$ (1.6.21)

Let $u(t)$ be the renewal density of the process and $U(t)$ be the renewal function.

Then $L[u(t)] = \dfrac{L(f_1(t))}{1 - L(f(t))}$, from (1.5.8)

Now, $$L[f_1(t)] = \lambda \int_0^\infty e^{-st} e^{-\lambda t} dt = \frac{\lambda}{\lambda + s} \tag{1.6.22}$$

Hence, $L[u(t)] = L\left[\dfrac{dU(t)}{dt}\right] = \dfrac{\left(\dfrac{\lambda}{\lambda+s}\right)}{1 - \phi(s)}$, from (1.6.22) and (1.6.21)

$$\Rightarrow \quad 1 - \phi(s) = \frac{\lambda}{\lambda + s}\left[\int_0^\infty e^{-st}\frac{dU(t)}{dt} \cdot dt\right]^{-1}$$

$$\Rightarrow \quad \phi(s) = 1 - \frac{\lambda}{\lambda + s}\left[\int_0^\infty e^{-st}\frac{dU(t)}{dt} \cdot dt\right]^{-1} \tag{1.6.23}$$

Next, to obtain $\dfrac{dU(t)}{dt}$ we proceed as follows:

We have

$$U(t + \delta t) - U(t) = e^{-\lambda t}\lambda \cdot \delta t + 0(\delta t), \quad \text{if} \quad t < \pi \tag{1.6.24}$$

given that initially the counter is free and

$$U(t + \delta t) - U(t) = e^{-\lambda \pi}\lambda \cdot \delta t + 0(\delta t), \quad \text{if } t \geq \pi \tag{1.6.25}$$

$$\Rightarrow \quad \left.\begin{aligned} \frac{dU(t)}{dt} &= \lambda e^{-\lambda t}, \quad \text{if} \quad l < \pi \\ &= \lambda e^{-\lambda \pi}, \quad \text{if} \quad t \geq \pi \end{aligned}\right\} \tag{1.6.26}$$

$$\therefore \quad L\left[\frac{dU(t)}{dt}\right] = \int_0^\infty e^{-st}\frac{dU(t)}{dt} \cdot dt = \int_0^\pi e^{-st}\frac{dU(t)}{dt}dt + \int_\pi^\infty e^{-st}\frac{dU(t)}{dt} \cdot dt$$

$$= \lambda \int_0^\pi e^{-st}e^{-\lambda t}dt + \int_\pi^\infty e^{-st}\lambda e^{-\lambda\pi}dt$$

$$L[u(t)] = \frac{\lambda}{\lambda + s} - \frac{\lambda \exp(-(\lambda + s)\pi)}{(\lambda + s)} + \frac{\lambda \exp(-(\lambda + s)\pi)}{s} \tag{1.6.27}$$

$$\Rightarrow \quad u(t) = \frac{dU(t)}{dt} = L^{-1}\left[\frac{\lambda}{\lambda + s} + \lambda e^{-(\lambda+s)\pi}\left(\frac{1}{s} - \frac{1}{\lambda + s}\right)\right] \tag{1.6.28}$$

Now, $$L^{-1}\left(\frac{\lambda}{\lambda + s}\right) = \lambda e^{-\lambda t},$$

$$L^{-1}\left(\lambda e^{-\lambda\pi}\left(\frac{1}{s} - \frac{1}{\lambda + s}\right)\right) = \lambda e^{-\lambda\pi}(1 - e^{-\lambda t})$$

and $$L^{-1}\left[\lambda e^{-\lambda\pi}\left(\frac{1}{s} - \frac{1}{\lambda + s}\right)e^{-s\pi}\right] = \begin{cases} \lambda e^{-\lambda\pi}\{1 - \exp[-\lambda(t - \pi)]\}, & \text{if} \quad t > \pi \\ \lambda e^{-\lambda\pi}, & \text{if} \quad t \leq \pi \end{cases}$$

by the second shifting property of Laplace transform (A.1.20)

$$\therefore \quad \left.\begin{aligned} u(t) &= \lambda e^{-\lambda t} + \lambda e^{-\lambda\pi}(1-\exp[-\lambda(t-\pi)]) \quad \text{if} \quad t \geq \pi \\ &= \lambda e^{-\lambda t} + \lambda e^{-\lambda\pi} \quad \text{if} \quad t < \pi \end{aligned}\right\} \tag{1.6.29}$$

Now,

$$\because \quad L(U(t)) = \frac{1}{s} L\left(\frac{dU(t)}{dt}\right) = \frac{1}{s} L(u(t))$$

$$L(U(t)) = \frac{1}{s}\left[\frac{\lambda}{\lambda+s} + \lambda\left(\frac{1}{s} - \frac{1}{\lambda+s}\right)\exp(-(\lambda+s)\pi)\right]$$

$$\Rightarrow \quad \left.\begin{aligned} U(t) &= 1 - e^{-\lambda t} + \lambda e^{-\lambda\pi}\left[(t-\pi) - \frac{1}{\lambda}(1-\exp\{-\lambda(t-\pi)\})\right], \text{ if } t \geq \pi \\ &= 1 - e^{-\lambda t}, \quad \text{if} \quad t < \pi. \end{aligned}\right\} \tag{1.6.30}$$

For the waiting time distribution $f(.)$ the reader is referred to Takacs (1960).

Biswas (1980) has shown an application of the model to describe the morbid condition in the human system, in which one morbid condition in presence of another only prolongs the morbidity period (dead time). The average number of fresh morbid spells in $(0, t)$ under a given morbidity rate λ and sickness period π (which may get extended when further complication occurs with the same rate λ) is given by (1.6.30).

1.6.5 Counter Model Type II with Random Dead Time

This counter process is quite complicated and difficult to analyze in general form. However, the results are known only for process with Poisson inputs with arrival rate λ.

Theorem 1.6.2: Let $p(t)$ the probability that the counter is open at a time t (i.e. registration is possible).

We require to show that

$$p(t) = exp\left(-\lambda \int_0^t [1-G(y)]dy\right), \tag{1.6.31}$$

where $G(y)$ represents the c.d.f. of the dead time distribution.

Proof: To Prove (1.6.31), we assume the following result (without proof) viz.

[Given n occurrences of a Poisson process in the interval $(0, t)$ the distribution of occurrence times is the same as that of n independent random variables taken from a uniform distribution in $(0, t)$]

The counter is free at time $T = t$ if all dead periods engendered by the signals in this period have been terminated before t.

Let $G(t-y) = P$ [dead time commencing in y will end before time t]. Then,

$$P\,[\text{An induced dead period terminated prior to } t] = \frac{\int_0^t G(t-y)dy}{\int_0^t dy} \tag{1.6.32}$$

Since the locking times are assumed to be independent of the arrival process, we have

$$\begin{aligned} p(t) &= P\{\text{Counter is free at time } t|n \text{ signals in } (0,t)\} \\ &= P\{\text{All the } n \text{ dead periods have terminated before time } t\} \\ &= \left[\frac{\int_0^t G(t-y)dy}{\int_0^t dy}\right]^n. \end{aligned} \tag{1.6.33}$$

But the number of signals arriving during the interval $(0,t)$ has a Poisson distribution with mean λt.

From the law of total probability

$$\begin{aligned} p(t) &= \sum_{j=0}^{\infty}\left\{\frac{1}{t}\int_0^t G(t-y)dy\right\}^j \frac{(\lambda t)^j e^{-\lambda t}}{j!} = \sum_{j=0}^{\infty}\left\{\frac{\lambda t}{t}\int_0^t G(t-y)dy\right\}^j \frac{e^{-\lambda t}}{j!} \\ &= e^{-\lambda t}\sum_{j=0}^{\infty}\frac{\left\{\lambda\int_0^t G(t-y)dy\right\}^f}{j!} = e^{-\lambda t}\exp\left(+\lambda\int_0^t G(t-y)dy\right) \\ &= \exp\left(-\lambda\int_0^t dy\right)\exp\left(\lambda\int_0^t G(t-y)dy\right) \\ &= \exp\left(-\lambda\int_0^t dy\right)\exp\left(\lambda\int_0^t G(\tau)d\tau\right) \end{aligned}$$

$$\Rightarrow \quad p(t) = \exp\left(-\lambda\int_0^t (1-G(y))dy\right) \quad // \tag{1.6.34}$$

Further, continuing with our assumption which proves (1.6.31), $p(t)$ can be related to $U(t)$ and we have

$$P\ [\text{a signal appear in } (t, t+\delta t)] = p(t)\lambda\delta t + 0(\delta t)$$

$$\therefore \quad U(t+\delta t) = U(t) + \lambda\delta t p(t) + 0(1-p\lambda\delta t) + 0(\delta t)$$

$$\lim_{\delta t\to 0}\frac{U(t+\delta t)-U(t)}{\delta t} = \lambda p(t)$$

$$\frac{dU(t)}{dt} = \lambda p(t) \tag{1.6.35}$$

$$\Rightarrow \quad U(t) = \lambda\int_0^t p(\tau)d\tau, \quad (\because U(0)=0) \tag{1.6.36}$$

Putting (1.6.34) in (1.6.36), we get

$$U(t) = \lambda\int_0^t \exp\left(-\lambda\int_0^\tau [1-G(y)]dy\right)d\tau, \tag{1.6.37}$$

which gives the renewal function for the model.

1.6.6 A Problem of Counter Model Type II with Random Dead Time

Theorem 1.6.3: Obtain the expectation and variance of inter-registration time in a counter model Type II when dead time is a random variable. (Takacs (1960); Biswas, Nautiyal and Tyagi (1983)).

Proof: From (1.6.34) we have

$$P\text{ [the counter is open at time } \tau] = p(\tau) = \exp\left(-\lambda\int_0^{\tau}(1-G(y))dy\right) \tag{1.6.38}$$

Renewal density $u(t)$ at time t, is obtained by differentiating (1.6.37) w.r.t. 't' using (A.1.15) as

$$u(t) = \lambda \exp\left(-\lambda\int_0^{t}(1-G(y))dy\right) \tag{1.6.39}$$

$$u(\infty) = \lambda \exp\left(-\lambda\int_0^{\infty}(1-G(y)dy)\right) = \lambda e^{-\lambda\cdot\alpha}, \tag{1.6.40}$$

$$\text{where } \int_0^{\infty}(1-G(y))dy = \alpha \text{ (say)} = \text{mean of the dead time.} \tag{1.6.41}$$

By asymptotic renewal theorem, denoting μ as the mean of the inter arrival time, we have

$$\mu \to \frac{1}{u(\tau)} \quad \text{as } \tau\to\infty, \text{ from (1.5.22)}$$

$$\Rightarrow \quad \mu \to \frac{1}{u(\infty)} = \frac{1}{\lambda e^{-\lambda\cdot\alpha}} = \frac{e^{\lambda\cdot\alpha}}{\lambda} \tag{1.6.42}$$

$$\text{Thus} \quad \mu = \frac{e^{\lambda\cdot\alpha}}{\lambda}, \tag{1.6.43}$$

where $\alpha = \int_0^{\infty}(1-G(y))dy$, from (1.6.41).

For the variance σ^2 of the inter-arrival time, we have

$$U(t) = \left[\frac{t}{\mu} + \frac{\sigma^2-\mu^2}{2\mu^2} + 0(1)\right] \tag{1.6.44}$$

holds for large values of t, [see Cox (1962)].

$$\Rightarrow \quad 2\mu^2 U(t) = 2\mu t + (\sigma^2-\mu^2) + 0(2\mu^2)$$

$$\Rightarrow \quad 2\mu^2\lambda\int_0^t \exp\left[-\lambda\int_0^{\tau}(1-G(y))dy\right]d\tau, \text{ from (1.6.37)}$$

$$= 2\mu\int_0^t d\tau + (\sigma^2-\mu^2) + 0(2\mu^2)$$

$$\Rightarrow \quad 2\mu^2\lambda\left[\int_0^t\left(\exp\left(-\lambda\int_0^{\tau}(1-G(y))dy\right) - \frac{1}{\lambda\mu}\right)d\tau\right] + \mu^2 + 0(2\mu^2) = \sigma^2$$

Denoting $\mu^2 + 0(2\mu^2) = \sigma_0^2$, we have

$$\sigma^2 = \sigma_0^2 + 2\mu^2\lambda\left[\int_0^t\left(\exp\left(-\lambda\int_0^{\tau}(1-G(y))dy\right) - \frac{1}{\lambda\mu}\right)d\tau\right]$$

$$\sigma^2 = \sigma_0^2 + 2\mu^2\lambda\left[\int_0^t\left(p(\tau) - \frac{1}{\lambda\mu}\right)d\tau\right], \text{ from (1.6.38)} \tag{1.6.45}$$

Again as $\tau \to \infty$, from (1.6.38)

$$p(\tau) \to \exp\left[-\lambda \int_0^\infty (1 - G(y))dy\right] = e^{-\lambda\alpha} = \frac{1}{e^{\lambda\alpha}} = \frac{1}{\lambda\mu} \text{ from (1.6.43)}$$

Hence, as $\tau \to \infty$, $\sigma^2 = \sigma_0^2$

To evaluate σ_0^2 we proceed as follows:

$$L(u(t)) = \frac{L(f_1(t))}{1 - L(f(t))}, \text{ from (1.5.58),}$$

where $L(f(t)) = \phi(s)$ and $L(f_1(t)) = \dfrac{\lambda}{\lambda + s}$

$$\Rightarrow \quad L(u(t))(1 - \phi(s)) = \frac{\lambda}{\lambda + s}$$

$$\Rightarrow \quad (1 - \phi(s)) = \frac{\lambda}{\lambda + s}(L(u(t)))^{-1}$$

$$\Rightarrow \quad 1 - \frac{\lambda}{\lambda + s}(L(u(t)))^{-1} = \phi(s)$$

$$\Rightarrow \quad 1 - \frac{\lambda}{\lambda + s}\left[\int_0^\infty e^{-st}\lambda\left\{\exp\left(-\lambda\int_0^\tau (1 - G(y))dy\right)\right\}dt\right]^{-1} = \phi(s) \text{ from (1.6.39)}$$

$$\Rightarrow \quad \text{As } \tau \to \infty,\ \phi(s) = 1 - \frac{\lambda}{\lambda + s}\left[\int_0^\infty \exp\left(-\lambda\int_0^\infty (1 - G(y))dy\right)\lambda e^{-st}dt\right]^{-1}$$

$$= 1 - \frac{\lambda}{\lambda + s}\left[\int_0^\infty \lambda e^{-st}e^{-\lambda\alpha}dt\right]^{-1}, \text{ from (1.6.41)}$$

$$= 1 - \frac{\lambda}{\lambda + s}\left(\lambda e^{-\lambda\alpha}\frac{1}{s}\right)^{-1} = 1 - \frac{\lambda}{\lambda + s}\cdot\frac{1}{\lambda}e^{\lambda\alpha}\cdot s$$

$$\Rightarrow \quad \phi(s) = 1 - \frac{s}{\lambda + s}e^{\lambda\alpha} \tag{1.6.46}$$

$$\therefore \quad \phi'(s) = -\left(\frac{(\lambda + s) - s}{(\lambda + s)^2}\right)e^{\lambda\alpha} = -\frac{\lambda}{(\lambda + s)^2}e^{\lambda\alpha}$$

$$\Rightarrow \quad \phi'(0) = -\frac{e^{\lambda\alpha}}{\lambda} = -\mu, \text{ from (1.6.43)} \tag{1.6.47}$$

$$\text{Further, } \phi''(0) = -\left.\frac{(-2)\lambda e^{\lambda\cdot\alpha}}{(\lambda + s)^3}e^{\lambda\alpha}\right|_{s=0} = \frac{2}{\lambda^2}e^{\lambda\alpha} \tag{1.6.48}$$

$$\therefore \quad \sigma_0^2 = \phi''(0) - [\phi'(0)]^2, \text{ from (A.1.16)}$$

$$= \frac{2}{\lambda^2}e^{\lambda\alpha} - \mu^2 = \frac{2}{\lambda}\mu - \mu^2 = \frac{2\mu - \lambda\mu^2}{\lambda} \tag{1.6.49}$$

Putting (1.6.49) in (1.6.45), we have

$$\sigma^2 = \left(\frac{2\mu - \lambda\mu^2}{\lambda}\right) + 2\mu^2\lambda\left[\int_0^t \left(p_0(\tau) - \frac{1}{\lambda\mu}\right)d\tau\right] \tag{1.6.50}$$

The present technique of derivation of (1.6.50) is due to Biswas *et. al.* (1983).

1.7 Palm Probability

The stochastic model can be used,

(i) In predicting the number of events (like births, accidental shocks etc.) during a fixed period of time say $(0, t)$,

(ii) In predicting the waiting time distribution between two consecutive events or between any two events say ith and jth event.

The same type of problems can be solved by an entirely new technique known as "*palm probability*" which was hitherto used in "queueing processes" (Khintchine (1960)). Palm probability is defined as the conditional probability of a specified number of events in a time interval given that an event has happened at the beginning of the interval. Cox and Isham's (1980) treatment of palm-distributions is described below.

For $u < v$, let $N(u, v)$ be a random variable giving the number of events occurring in (u, v) and $\{X_i\}$ be the sequence of the intervals between successive events in a process starting from an arbitrary point.

Consider the survival function

$$K_X(x) = P[X > x] = \lim_{\delta \to 0^+} P[N(0, x) = 0 | N(-\delta, 0) > 0] \tag{1.7.1}$$

which is the limiting probability that, given that an event occurs immediately before the origin; the next event of the process occurs after the instant x (i.e. the system will survive more than a period x).

Now, by stationarity condition

$$\begin{aligned} P[N(0, x) &= 0 \cap N(-\delta, 0) > 0] \\ &= P[N(0, x) = 0] - P[N(-\delta, x) = 0] \\ &= P(N(x) = 0) - P[N(x + \delta) = 0]. \end{aligned} \tag{1.7.2}$$

$$\begin{aligned} \Rightarrow \quad P[N(0, x) &= 0 | N(-\delta, 0) > 0] \\ &= \frac{P[N(0, x) = 0 \cap N(-\delta, 0) > 0]}{P[N(-\delta, 0) > 0]} \\ &= \frac{P[N(x) = 0] - P[N(x + \delta) = 0]}{P[N(\delta) > 0]}, \text{ from (1.7.2)} \end{aligned} \tag{1.7.3}$$

$$\begin{aligned} \Rightarrow \quad P[N(0, x) &= 0 | N(-\delta, 0) > 0] \delta^{-1} P[N(\delta) > 0] \\ &= \delta^{-1} \{P(N(x) = 0) - P(N(x + \delta) = 0)\} \\ &= -\delta^{-1} \{P[N(x + \delta) = 0] - P[N(x) = 0]\} \end{aligned} \tag{1.7.4}$$

We define $\lim\limits_{\delta \to 0^+} \delta^{-1} P[N(\delta) > 0] = \lambda$ (1.7.5)

as the occurrence parameter λ of the process, which we assume to be finite.

Further, we denote,

$$P[N(x) = k] = p_k(x); \; k = 0, 1, 2, \ldots \tag{1.7.6}$$

as the distribution of $N(x)$, then in the limit as $\delta \to 0^+$

$$K_X(x) = P[X > x] = \lim_{\delta \to 0^+} P[N(0, x) = 0 | N(-\delta, 0) > 0]$$

$$\Rightarrow \quad P[N(0, x) = 0 | N(-\delta, 0) > 0] \delta^{-1} P[(\delta) > 0]$$

$$= K_X(x)\delta^{-1}P[N(\delta) > 0]$$
$$= -\delta^{-1}\{P(N(x+\delta) = 0) - P(N(x) = 0)\}, \text{ from (1.7.4)}$$
$$\Rightarrow \quad \lambda K_X(x) = -D_x\{p_0(x)\} \quad \text{as} \quad \delta \to 0, \tag{1.7.7}$$

where D_x denotes the derivative w.r.t. x.

The equation (1.7.7) links the distribution of the interval between successive events with survival function $K_X(x)$, to that of forward recurrence time survival function $p_0(x)$ (where $p_0(x)$ is the probability that, starting from an arbitrary time instant, there are no events in the following interval of length x).

1.8 A Technique of Converting Deterministic Population Model to Stochastic Model

Considerable effort has been directed to construct Stochastic Population models by Kendal (1949), Bartlett (1955). However, whenever an otherwise deterministic model is made Stochastic, Mathematical complications becomes often too great to run into the analytic solution of the models. But if both fertility and mortality components are assumed to be independent of age as well as the size of the population then, of course, reasonably adequate solution exists but again the same is far from realistic. Even when the logistic model is made Stochastic considerable complications arise. However, a simple exercise relating to the conversion of deterministic into stochastic model due to Bartlett is useful in this respect to throw light in this respect. Let us consider the following transitions with the probability rates in infinitesimal time dt given below:

Transition	Probability rate
$N \to N+1$	$a_1N - b_1N^2$
$N \to N-1$	$a_2N + b_2N^2$

The above transitions are based on the differential equation of the logistic curve (see Sec. 4.2.1)

$$\frac{dN(t)}{dt} = N(B(N) - D(N)) = N(\alpha - \beta N) \tag{1.8.1}$$

where $B(N) - D(N) = \alpha - \beta N$

$$\Rightarrow \quad dN(t) = N(\alpha - \beta N)dt \tag{1.8.2}$$

We make this deterministic differential equation by introducing an error term, converting the above equation in the form

$$dN = [(\alpha N - \beta N^2)dt] + dz_1 - dz_2, \tag{1.8.3}$$

where the 1st term under [ ] represents the systematic part of the random or stochastic changes in dN(dN can take really the value 0 or 1) and dz_1 and dz_2 have consequently zero means. But since dz_1 and dz_2 may reasonably be regarded as Poisson variables with means $(a_1N - b_1N^2)dt$ and $(a_2N + b_2N^2)dt$, respectively, we may regard the variance of dz_1 and dz_2 to be the same as their means, as given above.

We have
$$\text{Var}(dz) = \text{Var}(dz_1 - dz_2) = \text{Var}(dz_1) + \text{Var}(dz_2)$$
$$= (a_1N - b_1N^2)dt + (a_2N + b_2N^2)dt$$
$$= [(a_1 + a_2)N - (b_1 - b_2)N^2]dt. \tag{1.8.4}$$

In the next place, we introduce a change of variable from N to u given by

$$N = \frac{\alpha(1+u)}{\beta}, \alpha \text{ and } \beta \text{ constants} \tag{1.8.5}$$

$$\Rightarrow \quad \alpha - \beta N = \alpha - \alpha(1+u) = -\alpha u \tag{1.8.6}$$

From (1.8.3) and (1.8.5)

$$\frac{dN}{dt} = \frac{\alpha}{\beta}\frac{du}{dt} = N(\alpha - \beta N) + \frac{dz}{dt} \quad \text{where} \quad dz = dz_1 - dz_2 \tag{1.8.7}$$

$$\Rightarrow \quad \frac{dN}{dt} = -N\alpha u + \frac{dz}{dt}, \text{ from (1.8.6)}$$

$$\Rightarrow \quad \frac{\alpha}{\beta}\frac{du}{dt} = -N\alpha u + \frac{dz}{dt}, \text{ from (1.8.7)}$$

$$\Rightarrow \quad du = -N\alpha u \frac{\beta}{\alpha} dt + \left(\frac{\beta}{\alpha}\frac{dz}{dt}\right) dt = \frac{\beta}{\alpha} dt \left[-N\alpha u + \frac{dz}{dt}\right]$$

$$= \frac{\beta}{\alpha} dt \left[-\frac{\alpha(1+u)\alpha u}{\beta} + \frac{dz}{dt}\right] = -\alpha(1+u)u\,dt + \left(\frac{\beta}{\alpha}\cdot dz\right)$$

$$du = -\alpha(1+u)u\,dt + \frac{\beta}{\alpha}dz \tag{1.8.8}$$

The representation (1.8.8) is quite useful. This may show that how the solution of stochastic equation differs from that of the deterministic equation (in which case we put $dz = 0$). Suppose the process has started down its ultimate value $\frac{\alpha}{\beta}$ when $dz = 0$

$$N \to \frac{\alpha}{\beta} \Rightarrow (1+u) \to 1 \quad \left(\because N = \frac{\alpha}{\beta}(1+u)\right). \tag{1.8.9}$$

We can write (1.8.8) as

$$du = -\alpha u\,dt + \frac{\beta}{\alpha}dz \tag{1.8.10}$$

$$\text{Further, Var (error component)} = \text{Var}\left(\frac{\beta}{\alpha}dz\right) = \frac{\beta^2}{\alpha^2}\text{Var}(dz)$$

$$= \frac{\beta^2}{\alpha^2}\left[(a_1+a_2)\frac{\alpha}{\beta} - (b_1-b_2)\frac{\alpha^2}{\beta^2}\right] dt$$

$$\text{from (1.8.4) for } N = \frac{\alpha}{\beta} \tag{1.8.11}$$

$$= \gamma\,dt, \text{ say,}$$

$$\text{where} \quad \gamma = \left\{\frac{\beta}{\alpha}(a_1+a_2) + (b_2-b_1)\right\}. \tag{1.8.12}$$

Strictly speaking for $a_2 > 0$ and $b_2 > 0$ as the death rate $a_2N + b_2N^2 > 0$, even for large N the population under these conditions become extinct. But at the same time it may be observed that the time of extinction becomes enormously large under this condition and the complication of Population getting extinct can be avoided in all practical sense.

Also $u + du = u - \alpha u\,dt + \frac{\beta}{\alpha}dz$ from (1.8.10).

Now under statistical equilibrium

$u(t + dt) = u + du$ is same as $u(t) = u$ which gives (denoting r.v. u as U)

$$\text{Var}\,(U) = \text{Var}\,(U + dU) = E(U + dU)^2 = E(U - \alpha U\,dt)^2 + E\left(\frac{\beta}{\alpha}dz\right)^2,$$

from (1.8.8)

$$= E(U^2) + \alpha^2 E(U^2)(dt)^2 - 2\alpha E(U^2)dt + \text{Var}\left(\frac{\beta}{\alpha}dz\right)$$

$$\because \quad \text{Var}\left(\frac{\beta}{\alpha}dz\right) = E\left(\frac{\beta}{\alpha}dz\right)^2 - \left[D\left(\underbrace{\frac{\beta}{\alpha}dz}_{=0}\right)\right]^2$$

$\because$ As assumed dz_1 and dz_2 have zero means so $dz = dz_1 - dz_2$ (see 1.8.7) has zero mean.

Neglecting the term $\alpha^2 E(U^2)(dt)^2$ because of the orderliness of the Poisson Process, we have

$\Rightarrow \quad \sigma_U^2 = \sigma_U^2 - 2\sigma_U^2\alpha dt + \gamma dt$ from (1.8.11)

$\Rightarrow \quad \sigma_U^2 = \sigma_U^2(1 - 2\alpha dt) + \gamma dt$

$\Rightarrow \quad 1 = 1 - 2\alpha dt + \dfrac{\gamma}{\sigma_u^2}dt$

$\Rightarrow \quad \dfrac{\gamma}{\sigma_U^2} = 2\alpha$

$$\Rightarrow \quad \sigma_U^2 = \frac{\gamma}{2\alpha} \tag{1.8.13}$$

Let us consider the asymptotic stochastic mean of the population given by $\overline{\eta}_\infty$. The first approximation of the mean is $\dfrac{\alpha}{\beta}$.

We next proceed to calculate the approximate mean. If μ is the true mean and

$$X(t) = N(t) - \mu. \tag{1.8.14}$$

We have approximately

$\Rightarrow \quad X(t + \Delta t) - X(t) = (\alpha N_t - \beta N_t^2)dt + dz_1 - dz_2$ from (1.8.14)

$E[X(t) + \Delta t) - X(t)] = [\alpha E(N_t) - \beta E(N_t^2)]dt + E(dz)$

$0 = \alpha\mu - \beta(\sigma_N^2 + \mu^2) + 0,$

$\Rightarrow \quad \alpha\mu = \beta(\sigma_N^2 + \mu^2)$

$$\Rightarrow \quad \alpha = \beta \cdot \frac{\sigma_N^2}{\mu} + \beta\mu \tag{1.8.15}$$

$$\Rightarrow \quad \left[\frac{\alpha}{\beta} - \frac{\sigma_N^2}{\mu}\right] = \mu \tag{1.8.16}$$

$$\overline{\eta}_\infty \sim \frac{\alpha}{\beta} - \frac{\sigma_N^2}{\mu}, \tag{1.8.17}$$

is the second approximation of the asymptotic mean of the population.

Hence, the population lies between

$$\left[\frac{\alpha}{\beta} - \frac{\sigma_N^2}{\mu} \pm 3\sqrt{\frac{\gamma}{2\alpha}}\right], \text{ from (1.8.16) and (1.8.13)}, \qquad (1.8.17)$$

with probability more than 99%.

Now if the stipulated population has to lie in the above range, then what should be the choice of α and β i.e. birth and death rates? (from (1.8.1))

In this case, we have

$$a_1 - a_2 = \alpha \quad \text{and} \quad -b_1 + b_2 = \beta$$

$$\therefore \quad \frac{\alpha}{\beta} = \frac{a_1 - a_2}{-b_1 + b_2}$$

$$\therefore \quad \sigma_U^2 = \frac{\gamma}{2}\left(\frac{\alpha}{\beta}\right)^2,$$

where $\gamma = (a_1 + a_2)\dfrac{\beta}{\alpha} + (b_2 - b_1)$, from (1.8.12)

$$\alpha = \beta\mu + \frac{\beta}{\mu}\sigma_\mu^2 \text{ from (1.8.14)}$$

and $\quad \overline{n}_\infty - \dfrac{\alpha}{\beta} - \dfrac{\sigma_u^2}{\mu}$, from (1.8.16) where $\overline{\eta}_\infty$ is the steady state population size.

References

1. Bellman, R; Kataba, R.E. and Lockett, J.A. (1965): Numerical Inversion of Laplace transforms, American Elesvier, N.Y.

2. Bailey, N.T.J. (1963): Elements of Stochastic Process with application to Natural Science; John Wiley and Sons. Inc. London, New York, Sydney.

3. Bhat, U.N. (1972): Elements of Applied Stochastic Processes; John Wiley and Sons, New York.

4. Bartlett, M.S. (1960): Stochastic Population models in Ecology and Epidemiology, Mathuen Monograph on Applied probability and Statistics, London, New York.

5. Biswas, S. (1973): A note on the generalisation of William Brass Model; Demography, August 1973, Vol. 10. No. 3, pp. 459–467.

6. Biswas, S. (1980): On the extension of some results of Counter Models with Poisson inputs and their applications: Journal of Indian Statistical Association, Vol. 18 page 45–53.

7. Biswas, S and Nair, G. (1984): A generalization of Freund's model for a repairable paired component based on a Bivariate Geiger Huller Counter; Micro-Electronics and Reliability, Vol. 24, No. 4, pp. 671–675.

8. Biswas, S. and Nair, G. (1985): On the development of successive damage model based on Palm Probability; Micro-Electronics and Reliability, Vol. 25 Nos. 5 page 271–274.

9. Biswas, S. and Nair, G. (1986): A palm probabilitistic technique on the prediction of the arrival time of the last fatal shock based on the data of earlier shocks. Journal of Agricultural Research Statistics Vol. XXXVIII. Nos. 2 August 1986 pp 240–248.

10. Biswas, S; Nair, G. and Nautiyal B.L. (1983): On a probability model of the number of conceptions classified by the nature of terminations based on Bivariate Geiger Muller Counter. Demography India, Vol 12 No. July–Dec. 1983 page 189–293.

11. Biswas, S. and Nauharia, Indu (1980): A note on the development of some interrupted waiting time distribution-Pure and applied Mathematica Sciences. Vol XI Nos. 1–2, March 1980, pp. 83–90.

12. Biswas, S; Nautiyal B.L. and Tyagi, R.N.S. (1981): A Stochastic epidemic model based on G.M. Counter Type II for the evaluation of Epidemic indices, Gujarat Statistical Review, Vol IX No, page 17–28.

13. Biswas, S; Nautiyal B.L. and Tyagi, R.N.S. (1983): Some results of a Stochastic process associated with Geiger Müller Counter model; Sankhya, Series B, Vol. 45, Part 2 page 271–283.

14. Biswas, S. and Pachal, T.K. (1982): On the applications of Palm probability for obtaining Inter-arrival time distribution in weighted Poisson Process; Calculate Statistical Association Bulletin, Vol. 32, March–June Nos. 125–126 page 111–115.

15. Biswas, S. and Pachal, T.K. (1987): A Multistage Markov Chain Model for Evaluating a Sterilization Policy, Biometrical Journal, Vol. 29, No. 1, pp. 57–64.

16. Chiang, C.L. (1968): Introduction to Stochastic Processes in Bio-statistics, John Wiley and Sons, New York.

17. Chiang, C.L. (1971): A Stochastic model on Human fertility, multiple transition probabilities; Biometrics, Vol. 27, pp. 345–354.

18. Cox, D.R. (1962): Renewal Theory, Methuen Monograph, London.

19. Cox, D.R. and Isham, V. (1980): Point Processes, Champam and Hall, London.

20. Cox, D.R. and Oakes, D. (1983): Analysis of survival data, Monograph on Statistics and Applied Probability, Chapman and Hall, London.

21. Dandekar, V.M. (1955): Certain modified forms of Binomial and Poisson distributions, Sankhyā, Series B, Vol. 15, pp. 237–50.

22. Dharmadhikari, S.W. (1964): A generalisation of a Stochastic model considered by V.M. Dandekar. Sankhyā, Series A, Vol. 26 pp 31–36.

23. Feller, Willian (1951): An introduction to Probability theory and its applications, Vol. I 3rd Edition, John Wiley and Sons, New York.

24. Feller, William (1968): An introduction to probability theory and its applications, Vol II John Wiley and Sons, New York.

25. Gill, R.D. (1984): Understanding of Cox's Regression Model—A Martinagale approach—Journal of American Statistical Association, Vol. 79, page 441–447.

26. Kendall, D. G. (1949)—Stochastic Processes and population growth—J. Royal Statistical Society, B, Vol. II, page 230–264.

27. Karlin, S. and Taylor, H. M. (1975): A first Course in Stochastic Processes, 2nd Edition Academic Press, New York.

28. Khintchine, A.Y. (1960): Mathematical models in theory of Queueing (Translated by D. M. Andrews and H.M. Quenoulle)–Charles Griffin and Company Ltd. London.

29. Pyke, Ronald (1958): A Renewal Process related to Type I and Type II Counter Model; Annals of Mathematical Statistics, Vol. 29, pp. 737–754.

30. Smith, Walter L. (1958): Renewal Theory and Its Ramifications; Jn. Royal Statistical Society, Ser. B, Vol. 20, pp. 243–302.

31. Takacs, Lajos (1951): On the occurence and coincidence phenomena in case of happenings with arbitrary distribution law of duration—Acta Mathematica Hungary Vol. 2 pp. 176–297.

32. Takacs, Lajos (1957): On some probability problems concerning the theory of counters, Acta Mathematica, Hungary, Vol. 8, pp. 127–138.

33. Takacs, Lajos (1960): Stochastic Processes (Problems and Solution) Translated by P. Zador, John Wiley and Sons. Inc. New York, Mathuen and Co. London, Butter and Tanner Ltd. Rome and London.

2

Techniques of Demographic Analysis: Mortality

2.0 Introduction

Mortality is a measure of the status of health condition of the Population. It is an important item of consideration in national planning. Not only it is a major component in projecting the population size at any time; it also provides a profile of the living condition of the people deciding the strategies in health management and reduction in Infant mortality. Therefore, a study on its impact on population by areas by age and sex composition etc. will be of great significance in the planning the health condition of the Population at large.

Remark 2.1: Distinction between a Rate and a Ratio

The basic characteristics of a Population are indicated by Mortality, Fertility and Migration (both Emigration and Immigration) and their joint effect which affects the growth of a Population. Now the changes in the structure and the composition of a Population may be ascribed as the changes in the Mortality, Fertility or Migration and should be measured quantitatively; as the probability of dying, probability of a new born individual or the probability of a person migrating from or moving to the population during a specified period of time.

Probability is a ratio in the classical sense. But the Probability of happening a vital event (viz. birth, death or migration) during a specific period of time is called "Rate" or "Probability rate". Thus the ratio or probability represents merely the chance of happening of an event; the rate differs from the ratio while representing the chance of the same during a fixed segment of time say a year or so.

2.1 Indices of Mortality Measure

Following are the common indices used for measurement of mortality:

(i) Crude death rate

(ii) Age specific death rates

(iii) Infant mortality, Neonatal and Perinatal mortality rates.

2.1.1 Crude Death Rate (CDR)

This is the simplest of the mortality rates. It gives an overall measure of the mortality situations. It is defined as

$$\text{CDR} = \frac{D^Z}{P^Z} \cdot K, \tag{2.1.1}$$

where D^Z = Number of deaths which occurred in the calender year Z.

P^Z = Mean (since population changes over different time of the calendar year Z) population in the calendar year Z.

CDR is very simple to calculate and easy to understand. It has the simple interpretation, for it gives the number of deaths that occurs on the average per K persons in the given calendar year Z. Of course, it gives a very rough overall measure of the mortality situation. It does not point out the mortality situation is different segment of the population viz. Age, Sex, Socio-Economic and Occupational groups.

CDR is a probability rate because for $K = 1$, it represents the chance of dying fór a person belonging to the given population in unit time, (say in a year), while the entire population is exposed to the risk of mortality.

Note 2.1: A vital rate is (generally) probability (or chance) of a vital phenomenon (birth and death) in unit time and as such lie between 0 and 1 for $K = 1$. But in practice they are multiplied by a constant $K(= 1000$ usually) and then it gives number of persons per thousand of the population under that risk during the given period of time. Thus

$$\text{CDR} = \frac{D^Z}{P^Z}.1000 \tag{2.1.2}$$

and then it gives number of deaths per thousand population in the calendar year Z.

Thus when we are talking about a rate as probability, it is understood that we are considering it with the factor. K taken as unity. There will be no ambiguity and the meaning will be clear from the context.

Note 2.2: As all the rates are generally considered per 1000 population, hence, we will understand the rates are referred to per 1,000 population in unit period of time.

Note 2.3: The super script Z always represents the specific quantity for or during the calendar year (i.e. year under consideration) Z e.g. D^Z and P^Z in (2.1.1), so we shall stick to this notation throughout the text to denote the same.

2.1.2 Age-Sex Specific Death Rates(ASSDR)

To differentiate the mortality by Age, Sex, Socio-Economic and Occupational groups, we should have different indices of mortality. One such mortality situation by Age and Sex is Age-Sex specific death rate (ASSDR).

The age-specific death rate(ASDR)for the age group $[x \text{ to } (x+n)]$ is denoted as ${}_nM_x$ and is defined as

$${}_nM_x^Z = \frac{{}_nD_x^Z}{{}_nP_x^Z} \cdot K \tag{2.1.3}$$

where ${}_nD_x^Z$ and ${}_nP_x^Z$ is the number of deaths and number of persons respectively, in the age group $[x \text{ to } (x+n)]$.

For annual ASDR, M_x^Z, $n = 1$, is then

$$M_x^Z = \frac{D_x^Z}{P_x^Z} \cdot K. \tag{2.1.4}$$

ASDR for Males, aged between x and $(x+n)$ is defined as

$${}_n^mM_x^Z = [{}_n^mD_x^Z / {}_n^mP_x^Z] \cdot K \tag{2.1.5}$$

Similarly, ASDR for Females for the age group $[x \text{ to } (x+n)]$ is defined as

$$ {}_n^f M_x^Z = \frac{{}_n^f D_x^Z}{{}_n^f P_x^Z} \cdot K \tag{2.1.6} $$

Here the super script 'm' in (2.1.5) and 'f' in (2.1.6), respectively, stands for 'male' and 'female'.

2.1.3 Infant Mortality Rate (IMR)

The IMR is defined as

$$ \text{IMR} = \frac{D_o^Z}{B_o^Z} \cdot K, \tag{2.1.7} $$

where D_o^Z is the number of deaths of children in the age group (0 to 1) and B_0^Z is the total number of births reported in the calendar year Z.

IMR is the chance (see Note 2.1) of dying of a newly born infant within a year of its life, under a given mortality condition. It is a highly sensitive index for evaluating the health status or the mortality standards of a community. Any change in health standard is reflected in IMR.

Remarks 2.2: The ASSDRand the IMR as defined above, no doubt, take into consideration of the chance of dying of a person belonging to a specific-age-group (inclusive of infant age-group). Strictly speaking, these rates, as defined above are not necessarily probability rates. This particularly applies to the case of IMR, as the same is considered to be a very sensitive index of health condition, in general. Now if we examine the definition of IMR, we can realize the point. For example, in case of an infant born in the month of November of the $(Z-1)$th calendar year and dying in February of Zth calendar year, the death is recorded in the numerator, whereas the birth is not recorded in the denominator. Similarly, in the case of an infant born in, say, July of the Zth calendar year and dying in January of the $(Z+1)$th calendar year, the birth is recorded in the denominator but the death is not recorded in the numerator. Therefore IMR defined above is not a probability rate. The same remark applies to age-specific mortality rates also. But the situation while constructing IMR is much more serious than that of age-specific mortality rates. Therefore, we present methods of adjustment of IMR converting the same to be a probability rate, in Section 2.2.

2.2 Methods of Adjustment of IMR

We shall discuss here two methods for the adjustment of IMR:

(A) This method is used by Actuaries. It runs as follows:

Here, we make two basic assumptions viz. (i) There is no secular trend in overall number of births in at least three consecutive years and (ii) There is no seasonal or cyclical changes in the number of births over years i.e. the number of births is uniform over the years.

We further assume that (iii) The available data consists of number of births ${}^tB_o^Z$ of infants in the tth quarter $(t = 1, 2, 3, 4)$ of the calendar year Z for a period of two consecutive years and (iv) The number of deaths ${}^tD_o^Z$ of infants in the tth quarter of their lives $(t = 1, 2, 3, 4)$ in the calendar year Z.

Then we can compute,

${}^tq_o^Z$ = Probability of dying for an infant in the tth quarter year of its life $(t = 1, 2, 3, 4)$ in the Zth calendar year.

Hence,

$$ {}^1q_o^Z = \frac{\text{No. of infants dying in the 1st quarter of their life in the } Z\text{th year}}{\text{Total No. of infants who could have died in the 1st quarter of their life}} \tag{2.2.1}$$

$$\text{Now,} \quad {}^1q_0^Z = \frac{{}^1D_o^Z}{\frac{1}{2}\cdot {}^4B_0^{Z-1} + {}^1B_0^Z + {}^2B_0^Z + {}^3B_o^Z + \frac{1}{2}\cdot {}^4B_0^Z} \tag{2.2.2}$$

Because some of the infants who could have died in the 1st quarter of their life in the calendar year Z might have born in the last quarter of $(Z-1)$th year. Assuming that the number of births in a quarter is uniformly distributed, half of the infants born in the 4th quarter of $(Z-1)$th year may be at the risk of dying in the 1st quarter of their life in the Zth year. Similarly, half of the infants born in the 4th quarter of Zth year will be at the risk of dying in the 1st quarter of their life in $(Z+1)$th year.

Similarly we can get,

$$ {}^2q_0^Z = \frac{{}^2D_0^Z}{\frac{1}{2}({}^3B_0^{Z-1}) + {}^4B_0^{z-1} + {}^1B_0^Z + {}^2B_0^Z + \frac{1}{2}({}^3B_0^Z)} \tag{2.2.3}$$

$$ {}^3q_0^Z = \frac{{}_3D_0^Z}{\frac{1}{2}({}^2B_0^{Z-1}) + {}^3B_0^{Z-1} + {}^4B_0^{Z-1} + {}^1B_0^Z + \frac{1}{2}({}^2B_0^Z)} \tag{2.2.4}$$

$$ {}^4q_0^z = \frac{{}_4D_0^Z}{\frac{1}{2}({}^1B_0^{Z-1}) + {}^2B_0^{z-1} + {}^3B_0^{z-1} + {}^4B_0^{Z-1} + \frac{1}{2}({}^1B_0^Z)} \tag{2.2.5}$$

$$\therefore \quad \text{Adjusted IMR} = \left(\sum_{t=1}^{4} {}^tq_0^Z\right)\cdot K. \tag{2.2.6}$$

2.2.1 An Alternative Method for Adjusting IMR (Mathen and Poti's method (1954))

The method is diagrammatically illustrated in Fig. 2.1

Notations:

(i) $B(t)\delta t =$ Number of births occurring between $(t, t+\delta t)$. (2.2.7)

(ii) $I(t) = I =$ True infant mortality rate per K live births per year holding for a period of two consecutive years. (2.2.8)

$\Rightarrow \dfrac{I}{K} =$ Probability that an infant will die within a year of life.

(iii) $f(x)dx =$ Conditional probability of dying for an infant between the ages $(x, x+dx)$ given that it will die within one year of life. (2.2.9)

(iv) $\displaystyle\int_0^t f(x)dx = P(t)$ (2.2.10)

(v) D_1 and D_2 represent the number of deaths in the $(Z+1)$th calender year of the births which took place in the Zth and $(Z+1)$th calender year respectively.

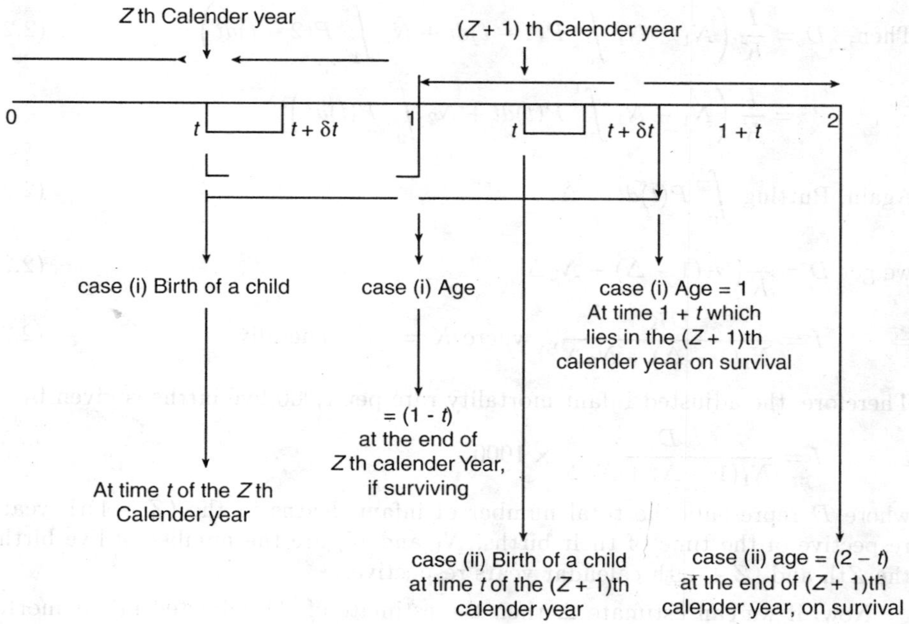

Fig. 2.1

Then from the case (i) in the above diagram, taking in to cognizance that the time of birth t lying in the Zth calender year i.e. $0 \le t \le 1$ and from the case (ii) in the diagram by similar reasoning, we have

$$D_1 = \frac{I}{K} \int_0^1 \left\{ B(t) \int_{1-t}^1 f(x)dx \right\} dt \tag{2.2.11}$$

and $$D_2 = \frac{I}{K} \int_1^2 \left\{ B(t) \int_0^{2-t} f(x)dx \right\} dt \tag{2.2.12}$$

$D_1 + D_2 =$ Number of all infant deaths which were reported in the $(Z+1)$th calender year irrespective of the time of birth. (2.2.13)

$$\therefore \; D_1 + D_2 = D = \frac{I}{K} \int_0^1 \left\{ B(t) \int_{1-t}^1 f(x)dx \right\} dt + \frac{I}{K} \int_1^2 \left\{ B(t) \int_0^{2-t} f(x)dx \right\} dt$$

$$\Rightarrow D = \frac{I}{K} \left\{ \int_0^1 B(t)[P(1) - P(1-t)]dt + \int_1^2 B(t)[P(2-t) - P(0)]dt \right\} \tag{2.2.14}$$

By definition, $P(0) = 0$ and $\int_0^1 f(x)dx = P(1) - P(0) = 1$

$$\therefore \; D = \frac{I}{K} \left\{ \int_0^1 B(t)[1 - P(1-t)]dt + \int_1^2 B(t)[P(2-t)] \right\}$$

$$= \frac{I}{K} \left[\int_0^1 B(t)dt - \int_0^1 B(t)P(1-t)dt + \int_1^2 B(t)[P(2-t)]dt \right] \tag{2.2.15}$$

Let $B(t) = N_1$, a constant in (0,1)

$= N_2$, a constant in (1,2)

Then, $D = \frac{I}{K}\left(N_1 - N_1\int_0^1 P(1-t)dt + N_2\int_1^2 P(2-t)dt\right)$ (2.2.16)

$$= \frac{I}{K}\left(N_1 - N_1\int_0^1 P(t)dt + N_2\int_0^1 P(t)dt\right)$$

Again, Putting $\int_0^1 P(t)dt = \Delta,$ (2.2.17)

we get $D = \frac{I}{K}[N_1(1-\Delta) + N_2\Delta]$ (2.2.18)

$\Rightarrow \quad I = \frac{DK}{N_1(1-\Delta)+N_2\Delta},$ where $K = 10^3$, generally. (2.2.19)

Therefore, the adjusted Infant mortality rate per 1,000 live births is given by

$$I = \frac{D}{N_1(1-\Delta)+N_2\Delta} \times 1000, \tag{2.2.20}$$

where D represents the total number of infant deaths in the $(Z+1)$th year, irrespective of the time of their births; N_1 and N_2 are the number of live births in the Zth and $(Z+1)$th calendar years respectively.

Now, if we can estimate Δ then the estimate of the adjusted infant mortality rate I can be obtained from (2.2.20). We illustrate the method of estimating Δ by the following example.

Example 2.1: The following data in Table 2.1, on infant deaths, were obtained from Kolkata City.

Table 2.1

Age at Death →	0 – 7 days	7 days to 1 month	1 – 2 months	2 – 3 months	3 – 6 months	6 – 12 months
Cumulative Percentage of Infant Deaths →	45.65 (2145)	58.63 (610)	65.19 (308)	73.83 (406)	84.15 (485)	100.00 (745)

Figures in the brackets, in the above table, are the actual frequencies in the respective age groups.

Now, from (2.2.17),

$$\Delta = \int_0^1 P(x)dx = \int_0^{7/365} P(x)dx + \int_{7/365}^{30/365} + \int_{30/365}^{90/365} P(x)dx$$
$$+ \int_{90/365}^{1/2} P(x)dx + \int_{1/2}^{1} P(x)dx$$

Using Trapezoidal (see (A.4.3)) and Simpson's $\frac{1}{3}$rd (see (A.4.2)) rules.

$$\Delta \cong \left[\frac{1}{2}\left\{P(0) + P\left(\frac{7}{365}\right) \times \frac{7}{365}\right\}\right] + \left[\frac{1}{2}\left\{P\left(\frac{7}{365}\right) + P\left(\frac{30}{365}\right)\right\} \times \frac{23}{365}\right]$$
$$+ \left[\frac{1}{3}\left\{P\left(\frac{30}{365}\right) + 4 \times P\left(\frac{60}{365}\right) + P\left(\frac{90}{365}\right)\right\} \times \frac{30}{365}\right]$$
$$+ \left[\frac{1}{2}\left\{P\left(\frac{90}{365}\right) + P\left(\frac{1}{2}\right)\right\} \times \frac{92}{730}\right] + \left[\frac{1}{2}\left\{P\left(\frac{1}{2}\right) + P(1)\right\} \times \frac{1}{2}\right]$$

Now, from definition (2.2.10),

$P(0) = 0,\ P(1) = 1$ and from the above table

$$P\left(\frac{7}{365}\right) = 0.4565,\ P\left(\frac{30}{365}\right) = 0.5863,\ P\left(\frac{60}{365}\right) = 0.6519,$$

$$P\left(\frac{90}{365}\right) = 0.7383 \quad \text{and} \quad P\left(\frac{1}{2}\right) = 0.8415.$$

Substituting these values in the above expression for Δ we get

$$\Delta \cong \left[\frac{1}{2}\{0 + 0.4565\}\right] + \left[\frac{1}{2}\{0.4565 + 0.5863\} \times \frac{23}{365}\right]$$

$$+ \left[\frac{1}{3}\{0.5863 + 4 \times 0.6519 + 0.7383\} \times \frac{30}{365}\right] + \left[\frac{1}{2}\{0.7383 + 0.8415\} + \frac{92}{730}\right]$$

$$+ \left[\frac{1}{2}\{0.8415 + 1\} \times \frac{1}{2}\right] = 0.9288$$

substituting this value of Δ in (2.2.18) when the values of N_1 and N_2 known, we can get the value of D and then putting the values of D, Δ, N_1 and N_2 in (2.2.20), we can get the required value of I.

2.2.2 Neonatal Mortality rate

While estimating infant mortality rate based on the data of the mortality of infants by months or by weeks during the first year of life it has been found that the number of infant deaths during the first to fourth week of life constitutes a major segment of the total number of infant deaths in a year. The mortality rate pertaining to this period of life i.e. upto 4 weeks (or the first month) is known as Neo-natal Mortality rate. (NMR).

Symbolically NMR is

$$\text{NMR} = \frac{D_N^{Z^-}}{B^Z} \cdot K \tag{2.2.21}$$

where $D_N^{Z^-}$ = Number of deaths of infants during the neo-natal period (0–1) month in the calendar year Z

B^Z = The total number of births during the calendar year Z

Again like the definition of infant mortality rate the present definition also subject to similar error. Because a child born in December in the $(Z-1)$th calendar year may die in January of the Zth calendar year. This makes an obvious overestimation of the NMR.

Hence an improved estimate of NMR is based on the following modified definition

$$D_N^Z = \left\{ \begin{aligned} &\frac{1}{2}\ \text{(Number of births in December } (Z-1))\\ &+ \frac{1}{2}\ \text{(Number of births in December } Z)\\ &+ \text{(Number of births from January to November } Z) \end{aligned} \right\} \tag{2.2.22}$$

Remark 2.3: However, if there is no significant variation in the number of births over months in two consecutive years then the above adjustment may be unnecessary; otherwise improvement in the estimated NMR by using the above adjustment can be made.

2.2.3 Perinatal Mortality Rate (PMR)

Many of the infants born with congenital malformations especially at the respiratory tract etc., prematurely die within the first week of life. These are called Perinatal deaths. We may define PMR as

$$\text{PMR} = \frac{{}^{P}D^{Z}}{B^{Z}} \cdot K \tag{2.2.23}$$

where ${}^{P}D^{Z}$ = number of deaths of infants during the first week of life in the calender year Z.

The deaths in this period are often not controllable by short term health plans. For the countries where IMR has fallen down considerably, it is found that the level of perinatal mortality rate, more of less, is held constant. Long term maternal and child health care programme however may lead to a decline in the perinatal mortality rate.

2.3 A Method of Calculation of IMR Based on in Incomplete Data

This method for calculating IMR is applicable when the data relating to the survival of infants are incomplete. The method is due tom Poti and Biswas (1963).

Let s_i be the number of infants surviving $(i = 1, 2, \ldots, 12)$ which we may takes at the end of ith month and let d_{ij} be the number of infants who died in jth month $(j = 1, 2 \ldots, i)$. Now each one among the s_i survivors contribute one unit of exposure for the ith month. Hence the total exposure contributed by the infants from the kth month to 12th month is given by

$$E_k = \sum_{i=k}^{12} s_i + \sum_{i=k}^{12}\sum_{j=k}^{i} d_{ij}, \; k = 1, 2, \ldots, 12,$$

and the total number of deaths from kth month to 12th month is given by

$$D_k = \sum_{i=k}^{12} d_{ik}, \; k = 1, 2, \ldots, 12.$$

Then the probability of survival for more than k months and less than 12 months is given by

$$S_k = \frac{E_k - D_k}{E_k}. \tag{2.3.1}$$

Calculating the survival probability in this manner, the probability of survival for the 1st year of the life is given as the product $\prod_{k=1}^{12} S_k$

IMR, therefore, is given by

$$\text{IMR} = \left(1 - \prod_{k=1}^{12} S_k\right) \cdot K \tag{2.3.2}$$

2.4 Standardization of Mortality Rates

The problem arise when we require to compare the mortality status of two communities of two different regions. Even if age specific mortality rates are same for both the regions (say A and B) the overall weighted crude mortality rates, defined later, (which may be considered as an overall index of mortality measure) may differ considerably on account of differing age distributions of two places. Thus, where really no differences in the mortality level exist, differences in age-sex composition brings out a picture of differences in mortality level, which is not true. Hence, our object is to obtain an overall mortality index for each of the two regions subject to the condition that the true mortality rates should remain unaffected by age distributions of the populations concerned. Symbolically, if M_x^A and M_x^B represent the mortality rates of A and B respectively at the age $[x \text{ to } x+1]$ and P_x^A and P_x^B are the respective age distributions (i.e. proportion of population in the age group $[x$ to $x+1]$ in respect of total population), then the overall mortality index (weighted crude mortality rates) for the regions A and B are, respectively,

$$M^A = \sum_x P_x^A M_x^A, \quad M^B = \sum_x P_x^B M_x^B \tag{2.4.1}$$

One may note that if $M_x^A \cong M_x^B \ \forall x$, M^A and M^B need not be the same due to different age distribution. Even if $M_x^A < M_A$, for almost all x we may have a peculiar situation leading to $M_B < M_A$, which is not at all true. This difficulty is removed by a technique known as 'Standardization technique' wherein we get the overall mortality indices, the indices independent of the age composition, although we use the entire set of age specific mortality rates.

There are two methods of Mortality standardization viz. (i) Direct Method and (ii) Indirect Method.

2.4.1 Direct Method

Direct standardization method is applicable where the age specific mortality rates M_x^A and M_x^B are known $\forall x$ for the region A and B and $P_x^S =$ proportion of population in the age sector $(x$ to $x+1)$ for some standard populations (which is comparable to both A and B), are also known. With this set up, the standardized mortality rates of A and B viz. SM^A and SM^B are obtained as follows.

$$SM^A = \sum_x P_x^S M_x^A, \quad SM^B = \sum_x P_x^S M_x^B \tag{2.4.2}$$

Note that SM^A and SM^B are unaffected by the age distribution since the age distributions of the standard population have been used as weights.

Sometime P_x^S is obtained as the mean of P_x^A and $P_x^B \ \forall x$ in absence of proper data concerning P_x^S.

2.4.2 Indirect Method of Standardization

To understand the technique of indirect standardization which is applied when

(i) Age specific mortality rates M_x^A and M_x^B are not known for each age groups but the overall mortality rates are known,

(ii) P_x^A, P_x^B and M_x^S (for standard population) are known $\forall\, x$.

We illustrate the indirect standardization technique as follows:
Different crude mortality rates for region A is given in the following table

P_x \ M_x	Given region (A)	Standard region (S)
Given region (A)	$\sum_x P_x^A M_x^A = (1)$	$\sum_x P_x^A M_x^S = (2)$
Standard region (S)	$\sum_x P_x^S M_x^A = (3)$	$\sum_x P_x^S M_x^S = (4)$

Similar table can be prepared for region B also.

Our object is to estimate the true standardized mortality measure for the region A (as well as of B) given by (3) in the above table where we do not have data pertaining to $M_x^A \, \forall \, x$.

For this, we assume (an empirically valid assumption)

$$\frac{(1)}{(2)} \cong \frac{(3)}{(4)} \quad \Rightarrow \quad (3) \cong \frac{(1) \times (4)}{(2)}$$

or the estimate of the standardized mortality rates of the region A is given by

$$\left\{ \sum_x P_x^A M_x^A \times \sum_x P_x^S M_x^S \div \sum_x P_x^A M_x^S \right\}$$
$$= \text{crude death rate of } A \times \sum P_x^S M_x^S \Big/ \sum_x P_x^A M_x^S \tag{2.4.3}$$

Similarly the standardized mortality rate of B is obtained as

$$= \text{crude death rate of } B \times \left(\sum P_x^S M_x^B \right) \Big/ \sum P_x^S M_x^S \tag{2.4.4}$$

2.5 Life Table

A Life Table is a mortality table which presents the survival experience of a hypothetical cohort of a number of new born infants, born at the same time, exposed to a particular type of mortality experience. A cohort, which is denoted by l_0, gradually diminishes by deaths. A Life Table tells us how the cohort experiencing changing mortality with advancing age will die out, and what would be the mortality rate or what is the probability that a person of specified age will survive a specified number of years etc. Life table population is a stationary population.

The Life Tables are constructed from census data or death registration data. These are constructed for various section of people which, supposedly, have different patterns of mortality. So there are Life Table for different sexes, occupational groups, regions, races etc.

In a Life table all the entries are for every year. A Life Table consists of the following columns:

(i) x = Exact age in year

(ii) l_x or $l(x)$ = No. of persons surviving at exact age x

(iii) d_x or $d(x)$ = Number of persons dying while passing from x to $(x+1)$

(iv) q_x or $q(x)$ = Probability of dying of a person within one year following the attainment of age x.

(v) m_x or $m(x)$ = central rate of mortality
= Probability of dying for a person whose exact age is between x to $(x+1)$ years

(vi) μ_x or $\mu(x)$ = Instantaneous force of mortality (or hazard rate) which means that $\mu_x dx$ is the conditional probability of dying between x to $x+dx$, given that the person is surviving at age x.

(vii) L_x or $L(x)$ = Person years lived by the cohort between x to $(x+1)$ years. $L(x)$ column provides age distribution of life table stationary population Note that Person years = Total number of years lived by the surviving person = Total number of living persons × (number of years lived)

(viii) T_x or $T(x)$ = Total person years lived by the cohort from x to ∞.

(ix) e°_x or $e^0(x)$ = complete expectation of life.

(x) ${}_{j-i}p_i$ = Probability of surviving from age i to j.

The above quantities l_x, q_x etc are called life table functions; they are inter-related.

The relation between them are

1. $$\left.\begin{aligned} & d_x = l_x - l_{x+1} \\ \text{and} \quad & \sum_{j=x}^{\infty} d_j = l_x \end{aligned}\right\} \qquad (2.5.1)$$

2. $q_x = d_x/l_x$ (2.5.2)

$= 1 - l_{x+1}/l_x = 1 - p_x$ (say) (2.5.3)

$\therefore\ p_x = l_{x+1}/l_x$ (2.5.4)

3. $m_x = d_x/l_{x+\frac{1}{2}}$ (2.5.5)

$= d_x/L_x$ (2.5.6)

Comparing (2.5.2) and (2.5.6) we note that $m_x > q_x\ \forall\ x$, since $L_x < l_x$ or $l_{x+\frac{1}{2}} < l_x$

4. $L_x = \int_o^l l(x+t)dt$ (2.5.7)

$\cong l_{x+\frac{1}{2}}$ (2.5.8)

5. $T_x = \sum_{t=0}^{\infty} L_{x+t}$ (2.5.9)

$= \sum_{t=x}^{\infty} L_t$ (2.5.10)

6. $e^0_x = T_x/l_x$ (2.5.11)

7. ${}_jp_i = l_{i+j}/l_i$ (2.5.12)

$p_x = {}_1p_x = l_{x+1}/l_x$ (2.5.13)

${}_xp_o = l_x/l_0$ (2.5.14)

8. $\dfrac{dL_x}{dx} = -d_x$ (2.5.15)

Because $L_x = \displaystyle\int_0^1 l_{x+t}dt$

Differentiating, w.r.t. x

$$\frac{dL_x}{dx} = \frac{d}{dx}\int_0^1 l_{x+t}dt = \int_0^1 \frac{dl_{x+t}}{dx}dt$$

$$= \int_0^1 \frac{d}{dt}l_{x+t}dt, \text{ since } l_{x+t} \text{ is symmetric in } x \text{ and } t.$$

$$= [l_{x+t}]_0^1 = l_{x+1} - l_x = -d_x, \text{ from (2.5.1).}$$

2.6 Relationships Between Life Table Functions

2.6.1 Relation between l_x and L_x

From (2.5.7), $L(x)$ is area bounded by the curve l_x, X-axis and and the ordinates at x and $x+1$ viz l_x and l_{x+1}, respectively (Fig. 2.2).

Assuming, linearity of l_x in the small interval $[x$ to $(x+1)]$,

$$L_x \cong \left(l_{x+\frac{1}{2}}\right)\cdot 1 = l_{x+\frac{1}{2}} \tag{2.6.1}$$

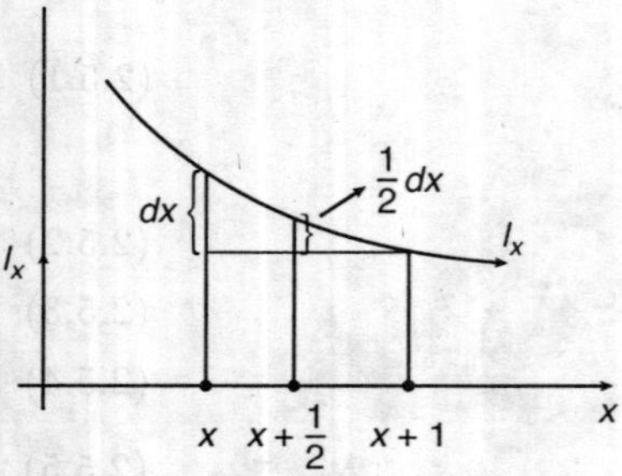

Fig. 2.2

Further, from the figure

$$l_x \cong l_{x+\frac{1}{2}} + \frac{1}{2}d_x \tag{2.6.2}$$

$$\Rightarrow l_{x+\frac{1}{2}} = l_x - \frac{1}{2}d_x$$

$$\Rightarrow l_{x+\frac{1}{2}} = l_x - \frac{1}{2}(l_x - l_{x+1}) = \frac{1}{2}(l_x + l_{x+1})$$

$$\therefore \quad L_x \cong \frac{1}{2}(l_x + l_{x+1}) \tag{2.6.3}$$

2.6.2 Relation between l_x, T_x

The person years lived by the cohort at age $(x+t)$ i.e. l_{x+t} for a period dt is given by $l_{x+t}dt$

Therefore $L_x = \displaystyle\int_0^1 l_{x+t}dt$ (2.6.4)

Hence the person years lived by the cohort from x to ∞ is given by

$$T_x = \sum_{t=0}^{\infty} L_{x+t} = \int_0^{\infty} l_{x+t}dt \tag{2.6.5}$$

2.6.3 Relation between Mortality Rates q_x and m_x

From the figure 2.1 the central rate of mortality, $m_x = \dfrac{d_x}{l_{x+\frac{1}{2}}}$.

From (2.5.2)

$$q_x = \frac{d_x}{l_x} \cong \frac{d_x}{l_{x+\frac{1}{2}} + \frac{1}{2}d_x}, \quad \text{from (2.6.2)}$$

$$= \frac{1}{\frac{l_{x+\frac{1}{2}}}{d_x} + \frac{1}{2}d_x} = \frac{1}{\frac{1}{m_x} + \frac{1}{2}} = \frac{2m_x}{2 + m_x}$$

$$\therefore \quad q_x = \frac{2m_x}{2 + m_x} \tag{2.6.6}$$

This is an approximates relation between q_x and m_x based on the assumption of linearity of l_x curve between $\{x \text{ to } (x+1)\}$. From (2.6.6) we get

$$m_x = \frac{2q_x}{2 - q_x} \tag{2.6.7}$$

2.6.4 Relation between q_x and μ_x

The instantaneous force of mortality μ_x is given by

$$\mu(x) = \frac{f(x)}{S(x)}, \quad \text{from (A.5.3)}$$

$$\Rightarrow \quad S(x) = \exp\left\{-\int_0^x \mu(t)dt\right\}, \quad \text{see (A.5.4)}$$

If we put

$$S(x) = \frac{l_x}{l_0} = P_r, \quad \text{(surviving at least upto age } x), \tag{2.6.8}$$

$$\text{then} \quad \frac{l_x}{l_0} = \exp\left\{-\int_0^x \mu(t)dt\right\}$$

$$\text{or,} \quad -\log\left(\frac{l_x}{l_0}\right) = \int_0^x \mu(t)dt$$

Differentiating both sides w.r.t. x we get

$$\mu_x = -\frac{1}{l_x}\frac{d\,l_x}{dx} = -\frac{d\log l_x}{dx}, \quad (l_0 \text{ is a constant and } \mu_0 = 0). \tag{2.6.9}$$

Further, the number of deaths between x to $(x + dx) = l_x - l_{x+dx}$

Then the relative decrease in the number of persons in the life table while passing from x to $x + dx$ is given by

$$\frac{l_x - l_{x+dx}}{l_x},$$

while the relative increase in age dx takes place. Therefore the elasticity of the decrease in the number of persons w.r.t. the age or age elasticity of the number of persons (analogous to the price elasticity of demand) is given by

$$\frac{1}{l_x}\left\{\frac{l_x - l_{x+dx}}{l_x}\right\} = \frac{-1}{l_x}\frac{d\,l_x}{dx} = \mu_x, \text{from (2.6.9)} \tag{2.6.10}$$

Thus μ_x represents the age elasticity of the decrease in the number of persons (at the age x).

From (2.6.9)

$$\mu_{x+t} = -\frac{1}{l_{x+t}}\frac{d\,l_{x+t}}{dx}$$

$$\Rightarrow \quad \int_0^1 \mu_x\, l_{x+t} dx = -\int_0^1 d\,l_{x+t}, \text{ since } x \text{ is fixed and } t \text{ is the variable in } (x+t).$$

$$\Rightarrow \quad \int_0^1 \mu(x+t) l_{x+t} dt = l_x - l_{x+1} = d_x, \text{ from (2.5.1)}$$

$$\Rightarrow \quad \frac{d_x}{l_x} = \frac{1}{l_x}\int_0^1 \mu(x+t) l_{x+t} dt$$

or, $$q_x = \frac{1}{l_x}\int_0^1 \mu(x+t) l_{x+t} dt \tag{2.6.11}$$

2.6.5 A Relation between e_x^0 and $\mu(x)$

Let X be a continuous r.v. denoting life time with distribution function $F(x)$ and force of mortality$\mu(x)$.

Then from (A.5.5)

$$E(X) = \int_0^\infty S(x)dx, \quad \text{when} \quad S(x) = 1 - F(x)$$

and $$S(x) = \exp\left\{-\int_0^x \mu(t)dt\right\}, \text{ from (A.5.4)} \tag{2.6.12}$$

$$\therefore \quad E(X) = \int_0^\infty \exp\left\{-\int_0^x \mu(t)dt\right\} dx. \tag{2.6.13}$$

Now e_x^0 is the complete expectation of life at the age x i.e. it represents the residual length of life the person that has already survived up to age x.

Let $S(x+y|x) = P$ [survival upto the age $(x+y)$ on the condition that it has survived upto x]

$$= \exp\left\{-\int_x^{x+y} \mu(t)dt\right\}, \text{ as in (2.6.12)} \tag{2.6.14}$$

Therefore, as in (2.6.13)

$$e_x^0 = \int_0^\infty \exp\left\{-\int_x^{x+y} \mu(t)dt\right\} dy \tag{2.6.15}$$

This is the required relation.

Further,

$f(x+y|x) = -\dfrac{\partial S(x+y|x)}{\partial y}$, from (2.6.14) which gives

$$f(x+y|x) = \exp\left\{-\int_x^{x+y} \mu(t)dt\right\}\left[\mu(x+y)\frac{d}{dy}(x+y)\right], \text{ from (A.1.15)}$$

$$= \mu(x+y)\exp\left\{-\int_x^{x+y} \mu(t)dt\right\} \tag{2.6.16}$$

$$\therefore \quad e_x^0 = \int_0^\infty y\, f(x+y|x)dy = \int_0^\infty y\,\mu(x+y)\exp\left\{-\int_x^{x+y} \mu(t)dt\right\} dy \tag{2.6.17}$$

2.6.6 A Relation between m_x and $\mu(x)$

From (2.5.15) we have

$$\frac{dL_x}{dx} = -d_x \Rightarrow -\frac{1}{L_x}\frac{dL_x}{dx} = \frac{d_x}{L_x} = m_x, \text{ from (2.5.6)} \tag{2.6.18}$$

Now $L_x \cong l_{x+\frac{1}{2}}$, from (2.5.8). Putting this in (2.6.18), we get

$$m_x = -\frac{1}{l_{x+\frac{1}{2}}}\frac{dl_{x+\frac{1}{2}}}{dx} = \mu\left(x+\frac{1}{2}\right), \text{ from (2.6.9).}$$

Thus the relation between m_x and $\mu(x)$ is

$$m_x = \mu\left(x+\frac{1}{2}\right). \tag{2.6.19}$$

2.6.7 A Relation between e_x^0 and e_{x+1}^0

We have from definition (2.5.11) $e_x^0 = \dfrac{T_x}{l_x}$.

Again from (2.5.9)

$$T_x = \sum_{t=x}^{\infty} L_t = L_x + \sum_{t=x+1}^{\infty} L_t = L_x + T_{x+1}$$

$$\therefore \quad e_x^0 = \frac{T_{x+1}+L_x}{l_x} = \frac{T_{x+1}}{l_x} + \frac{L_x}{l_x}$$

Now $L_x = \displaystyle\int_0^1 l_{x+t}dt = l_x \int_0^{\xi} dt + l_{x+1}\int_{\xi}^1 dt,$

$(0 < \xi < 1)$ from 2nd M.V. Theorem (A.1.12)

$$= l_x \times \xi + l_{x+1}(1-\xi)$$

$$\therefore \quad L_x = l_x f_x + l_{x+1}(1-fx), \text{ say} \tag{2.6.20}$$

$$\therefore \quad e_x^0 = \frac{T_{x+1}}{l_{x+1}}\cdot\frac{l_{x+1}}{l_x} + \frac{f_x l_x + (1-f_x)l_{x+1}}{l_x}$$

$$\text{or} \quad e_x^0 = e_{x+1}^0 p_x + f_x + (1-f_x)p_x \tag{2.6.21}$$

$$\because \quad p_x = 1 - q_x = 1 - \frac{d_x}{l_x} = \frac{l_x-(l_x-l_{x+1})}{l_x} = \frac{l_{x+1}}{l_x}$$

(2.6.21) is the required relation, where f_x is obtainable from (2.6.24) as

$$f_x = \frac{1}{m_x} - \frac{1}{q_x} + 1, \tag{2.6.22}$$

2.6.8 Another Relationship between e_x^0 and $\mu(x)$

We have $e_x^0 = \dfrac{T_x}{l_x} = \dfrac{\int_0^{\infty} l_{x+t}dt}{l_x}$, from (2.6.5).

Differentiating w.r.t. x we get

$$\frac{de_x^0}{dx} = \left(\int_0^{\infty} l_{x+t}dt\right)\frac{d}{dx}\left(\frac{1}{l_x}\right) + \frac{1}{l_x}\left(\int_0^{\infty}\frac{dl_{x+t}}{dx}\right)dt$$

$$= \left(\int_0^\infty l_{x+t}dt\right)\left(-\frac{1}{l_x^2}\frac{dl_x}{dx}\right) + \frac{1}{l_x}\int_0^\infty \frac{dl_{x+t}}{dt}dt$$

since l_{x+t} is a symmetric function of x and t.

$$= \left(-\frac{1}{l_x}\frac{dl_x}{dx}\right)\left(\frac{1}{l_x}\int_0^\infty l_{x+t}dt\right) + \frac{1}{l_x}[l_{x+t}]_0^\infty$$

$$= \mu(x)\frac{T_x}{l_x} + \frac{1}{l_x}\ [l_\infty - l_x], \text{ from (2.6.9) and (2.6.5)}$$

$$= \mu(x)e_x^0 - 1, \text{ from (2.5.11) and } \because l_\infty = 0.$$

$$\therefore \quad \mu(x) = \frac{1}{e_x^0}\left[\frac{de_x^0}{dx} + 1\right]. \qquad (2.6.23)$$

2.6.9 An Exact expression for L_x

We have $L_x = \int_0^1 l_{x+t}dt$, from (2.5.7)

and $\quad L_x = f_x l_x + (1 - f_x)l_{x+1}$,

where f_x is the weight function of l_x to be obtained.

or $\quad L_x = f_x(l_x - l_{x+1}) + l_{x+1} = f_x d_x + l_{x+1}$

$$\therefore \quad f_x = \frac{L_x - l_{x+1}}{d_x} = \frac{L_x}{d_x} - \frac{l_{x+1}}{d_x}$$

or $\quad f_x = \dfrac{1}{m_x} - \dfrac{l_x - d_x}{d_x}$, from (2.5.6) and (2.5.1)

$$= \frac{1}{m_x} - \frac{l_x}{d_x}\left(1 - \frac{d_x}{l_x}\right) = \frac{1}{m_x} - \frac{1}{q_x}(1 - q_x), \text{ from (2.5.2)}$$

$$\therefore \quad f_x = \frac{1}{m_x} - \frac{1}{q_x} + 1 \qquad (2.6.24)$$

Substituting the value of f_x from (2.6.24) in (2.6.20)

$$L_x = \left(1 + \frac{1}{m_x} - \frac{1}{q_x}\right)l_x + \left(\frac{1}{q_x} - \frac{1}{m_x}\right)l_{x+1} \qquad (2.6.25)$$

2.6.10 Curtailed (on Curtate) Expectation of life

The complete expectation of life at the age x is denoted e_x^0 (defined in (2.5.11)). The curtailed expectation of life at the age x is denoted as e_x, is defined as follows:

Let W (positive integer) be the last age of survival; which is obviously a.r.v.

Let ${}_xp_0 = P$ [surviving from the age 0 to x]

$$= S(x) = \exp\left[-\int_0^x \mu(t)dt\right], \text{ see (A.5.4)} \qquad (2.5.4)$$

In this notation $p_x = \dfrac{l_{x+1}}{l_x}$

$$\text{Hence} \quad {}_xp_0 = \frac{l_x}{l_0} \qquad (2.6.26)$$

∴ The probability of no person surviving upto age $(w+1)$ out of a cohort of l_0

$$= (1 - {}_{w+1}p_0)^{l_0} \tag{2.6.27}$$

Similarly, the probability that out of l_0 individuals no one is alive at w

$$= (1 - {}_{w}p_0)^{l_0} \tag{2.6.28}$$

Thus the probability 'p_w' of at least one person surviving, out of l_0 between w and $(w+1)$ by (2.6.27) and (2.6.28) is given by

$$p_w = (1 - {}_{w+1}p_0)^{l_0} - (1 - {}_{w}p_0)l_0. \tag{2.6.29}$$

This is the probability distribution of the r.v. W which takes a value w.

Then $E[W] = \sum_{w=0}^{\infty} w\, p_w = \sum_{w=0}^{\infty} w\{(1 - {}_{w+1}p_0)^{l_0} - (1 - {}_{w}p_0)^{l_0}\} = e_0$, say, (2.6.30)

which is called curtailed expectation of life at the age 0 (i.e., at birth) denoted by e_0. If we take $l_0 = 1$, then

$$E(w) = e_0 = \sum_{w=0}^{\infty} w({}_{w}p_0 - {}_{w+1}p_0) = \sum_{w=1}^{\infty} W p_0 \tag{2.6.31}$$

In the same way the curtailed expectation at age x is given by

$$e_x = \sum_{t=1}^{\infty} {}_{x+t}p_t. \tag{2.6.32}$$

e_x is the average number of years of life lived by each of the person of age x in the life table population viz. l_x.

2.6.11 Relationship between e_x^0 and e_x

From (2.5.11) and (2.6.5) the complete expectation of life at birth is

$$e_0^0 = \frac{T_0}{l_0} = \frac{1}{l_0}\int_0^{\infty} l_t dt \tag{2.6.33}$$

$$\simeq \frac{1}{l_0}\left[\frac{1}{2}l_0 + l_1 + l_2 + \ldots + l_n + \ldots\right], \text{ from Trapezoidal rule (A.4.3)}$$

$$= \frac{1}{2} + \frac{l_1}{l_0} + \frac{l_2}{l_0} + \ldots = \frac{1}{2} + {}_1p_0 + {}_2p_0 + \ldots, \quad \text{from (2.6.26)}$$

$$\therefore \quad e_0^0 \simeq \frac{1}{2} + \sum_{x=1}^{\infty} {}_xp_0 \tag{2.6.34}$$

or $e_0^0 \simeq \frac{1}{2} + e_0$, from (2.6.31) (2.6.35)

The relation (2.6.34) is true in general

$$\therefore \quad e_x^0 \simeq \frac{1}{2} + e_x, \tag{2.6.36}$$

which can be shown as follows:

Here $e_x^0 = \frac{1}{l_x}\int_0^{\infty} l_{x+t}dt$, from (2.5.11) and (2.6.5)

$$\simeq \frac{1}{2} + \sum_{t=1}^{\infty} {}_{x+t}p_x, \quad \text{as in (2.6.34)}$$

$$= \frac{1}{2} + e_x, \quad \text{from (2.6.35)} \tag{2.6.37}$$

Note 2.4: The death rate from the life table is given by $\frac{1}{e_0^0}$. Because $e_0^0 = \frac{T_0}{l_0}$ from (2.5.11).

$$\Rightarrow \quad \frac{1}{e_0^0} = \frac{l_0}{T_0} = \frac{\text{Total number of persons (Cohort)}}{\text{Total person-years lived by the cohort}} = \frac{d_0 + d_1 + d_2 + \ldots}{L_0 + L_1 + L_2 + \ldots}$$

Thus for a life table population of L_0 (whose age lives between 0 and 1) a number of d_0 person died within a course of a year. Similarly, out of L_1, d_1 died in a year and so on. So

$$\frac{1}{e_0^0} = \frac{\sum_{x=0}^{\infty} d_x}{\sum_{x=0}^{\infty} L_x} \cong D \text{ (say)} \Rightarrow \frac{l_0}{l_0} = D, \tag{2.6.38}$$

where D represents a measure of the crude death rate of stationary life table population.

2.7 Abridged Life Table

An Abridged life table is the abridged version of a complete Life Table. The abridgement is made w.r.t. two aspects viz (i) the age entries are in groups (often quinquennium age group) and (ii) number of columns are abridged in complete life table. The first Indian Abridged Life Table was constructed based on 1941 Census data collected on the basic of 1% sample, based on the design of Frank Yetes (1941), known as 1% Y-sample.

The following are the columns of an Abridge Life Table:

(i) x to $(x+n)$ = Age group for all x.

(ii) l_x or $l(x)$ = Number of persons surviving at the age x.

(iii) ${}_nd_x$ = Number of deaths while passing from x to $(x+n)$

$$= l_x - l_{x+n} \tag{2.7.1}$$

(iv) ${}_nq_x = \frac{{}_nd_x}{l_x}$, the yearly mortality rate while passing from x to $(x+n)$. (2.7.2)

and hence $${}_np_x = \frac{l_{x+n}}{l_x} \tag{2.7.3}$$

(v) ${}_nL_x = \int_0^n l_{x+t}\, dt$ = person years lived by the cohort l_x in x to $(x+n)$. (2.7.4)

Life table Population in the age group x to $(x+n)$

$$= \sum_{t=0}^{n-1} L_{x+t} = \sum_{t=x}^{x+n-1} L_t \tag{2.7.5}$$

(vi) $_nm_x = \dfrac{_nd_x}{_nL_x}$, the central rate of mortality in the age group x to $(x+n)$. (2.7.6)

$$\text{(vii)}\ T_x = \int_0^\infty l_{x+t} = \sum_{t=0}^{\infty} {}_nL_{x+t} \tag{2.7.7}$$

$$= \sum_{t=x}^{\infty} {}_nL_t = \sum_{t=x}^{\infty}\sum_{t=x}^{n} L_t, \text{ from (2.7.5)}$$

$$= \sum_{t=x}^{\infty} L_t \tag{2.7.8}$$

$$\text{(viii)}\ T_{x:\overline{n}|} = \int_0^n l_{x+t}dt = {}_nL_x \tag{2.7.9}$$

$$\text{(ix)}\ e_x^0 = \frac{T_x}{l_x} \tag{2.7.10}$$

In the Abridged life table the entries are for every nth year ($n = 5$ or 10, usually), not every year, that is why it is called Abridged life table.

2.7.1 Relations between Abridged Life Table Functions

(1) $_nd_x = l_x - l_{x+n}$. (2.7.11)

(2) We assume that l_x is linear in (x to $x+n$) i.e. the number of deaths $_nd_x$ in the interval (x to $x+n$) is uniform.

Hence $\dfrac{_nd_x}{n}$ is the number of deaths per year

$\Rightarrow$ $t\dfrac{_nd_x}{n}$ is the number of deaths in t years.

$$\therefore \quad l_{x+t} = l_x - t\cdot\frac{_nd_x}{n} \tag{2.7.12}$$

Hence, from (2.7.4)

$$_nL_x = \int_0^n \left\{ l_x - t\cdot\frac{_nd_x}{n}\right\} dt = {}_nl_x - \frac{n^2}{2}\frac{_nd_x}{n}$$

$$= n\left[1_x - \frac{1}{2}(l_x - l_{x+n})\right], \text{ from (2.7.11)} \tag{2.7.13}$$

$$\text{or} \quad _nL_x = \frac{n}{2}[l_x + l_{x+n}]. \tag{2.7.14}$$

For $n = 1$ we get (2.6.3).

Further, from (2.7.13) and (2.7.12) for $t = n$, we get

$$_nL_x = n\left[l_x - \frac{1}{2}{}_nd_x\right]$$

$$\text{or from (2.7.12) for } t = \frac{n}{2}, \quad _nL_x = nl_{x+\frac{n}{2}} \tag{2.7.15}$$

For $n = 1$, it corresponds to (2.6.1).

(3) $_nL_x = \int_0^n l_{x+t}dt = \int_0^1 l_{x+t}dt + \int_1^2 l_{x+t}dt + \ldots + \int_{n-1}^n l_{x+t}dt$

Now, putting $t' = t - 1$ the second integral on r.h.s becomes

$\int_0^l 1_{x+1+t'}dt' = L_{x+1}$, from (2.5.7) etc.

$$\therefore \quad _nL_x = L_x + L_{x+1} + \ldots + L_{x+n-1} = \sum_{t=0}^{n-1} L_{x+t} \tag{2.7.16}$$

$$= \sum_{t=x}^{x+n-1} L_t. \tag{2.7.17}$$

(4) A Recurrence relation for T_x: From (2.7.7)

$$T_x = \int_0^\infty l_{x+t}dt = \sum_{t=0}^{\infty} L_{x+t}, \quad \text{from (2.5.9)} \tag{2.7.18}$$

$$\Rightarrow \quad T_x = L_x + \sum_{t=1}^{\infty} L_{x+t} = L_x + \sum_{t=0}^{\infty} L_{x+1+t}$$

$$T_x = L_x + T_{x+1}. \tag{2.7.19}$$

(5) $T_x = \sum_{t=0}^{\infty} L_{x+t} \cong \sum_{t=0}^{\infty} \frac{1}{2}\{l_{x+t} + l_{x+t+1}\}$, from (2.6.3)

$$\Rightarrow \quad T_x = \frac{1}{2}l_x + \sum_{t=1}^{\infty} l_{x+t} \tag{2.7.20}$$

(6) $_np_x = \frac{l_{x+n}}{l_x} = \frac{l_{x+n}}{l_{x+n-1}} \frac{l_{x+n-1}}{l_{x+n-2}} \cdots \frac{l_{x+2}}{l_{x+1}} \frac{l_{x+1}}{l_x}$

$$\text{or} \quad _np_x = p_{x+n-1}p_{x+n-2} \cdots p_x = \prod_{i=0}^{n-1} p_{x+i}, \text{ from (2.5.4)} \tag{2.7.21}$$

(7) From (2.7.7)

$$\frac{dT_x}{dx} = \frac{d}{dx}\int_0^\infty l_{x+t}dt = \int_0^\infty \frac{d}{dx} l_{x+t}dt$$

$= \int_0^\infty \frac{d}{dt} l_{x+t}dt$, since l_{x+t} is a symmetric function in x and t.

$$= [l_{x+t}]_0^\infty = -l_x, \quad \because l_\infty = 0. \tag{2.7.22}$$

(8) From (2.7.4)

$$\frac{d(_nL_x)}{dx} = \frac{d}{dx}\int_0^n l_{x+t}dt = \int_0^n \frac{d}{dx} l_{x+t}dt$$

$= \int_0^n \frac{d}{dt} l_{x+t}dt$, as in (7) since l_{x+1} is a symmetrical function of x and t.

$$= [l_{x+t}]_0^n = l_{x+n} - l_x = -{_nd_x} \text{ from (2.7.1)} \tag{2.7.23}$$

Compare it with (2.5.15).

(9) On the assumption that l_x is linear in (x to $x+n$)

$$_nq_x = \frac{2n(_nm_x)}{2+n(_nm_x)} \tag{2.7.24}$$

This will be taken up in the next section.

2.8 Construction of Abridged Life Table

To tackle the problem of construction of Abridge Life Table we proceed as follows:

(i) We assume that the population $_nP_x$ and the number of deaths $_nD_x$, in the age group (x to $x+n$), are available, from vital registration or census.

(ii) We have $_nm_x = (_nd_x/_nL_x)$ and its estimate is given by $_n\hat{m}\hat{x} = (_nD_x/_nP_x)$.

(iii) We calculate $_nD_x/_nP_x(= {}_nm_x)$ for all x and n.

(iv) After having estimated $_nm_x$ the methods of construction of Abridged Life Table can be classified into two distinct groups:

(A) The methods which provide the estimates of $_nq_x$ directly from $_nm_x$, and the other columns of the life table can be constructed by employing the relationship between Abridge Life Table functions. We shall discuss three such methods here viz Abridged Greville's formula, Greville's formula and Reed and Merrell's formula.

(B) Actuarial methods, which enable us to obtain the population and number of deaths at central age points of the intervals and the life table is constructed on the basis of that. We shall discuss one such method viz King's method in the next section.

Actually, the three formulas mentioned in (A) are relations between $_nq_x$ and $_nm_x$; these relation can be used for the construction of Abridged Life Table whereas in King's method $_nD_x$ and $_nP_x$ are used to estimates the corresponding death and population parameters at the center age points using King's Interpolation formula.

2.8.1 Abridged Greville's Formula (Jaffe (1996))

This formula is obtained on the assumption that the curve for l_x is linear between (x to $x+n$).

We know from the definitions of abridged life table functions $_nq_x$ and $_nm_x$ that

$$_nq_x = {}_nd_x/l_x \quad \text{and} \quad _nm_x = {}_nd_x/nL_x \text{ from (2.7.2) and (2.7.6)}$$

$$\text{then} \quad _nd_x = {}_nm_{x\,n}L_x \quad \text{and so} \quad _nq_x = {}_nm_{x\,n}L_x/l_x \tag{2.8.1}$$

$$\text{or} \quad _nq_x = (_nm_x/l_x)\int_0^n l_{x+t}dt, \quad \text{from (2.7.4) and (2.7.6)}$$

$$\cong (_nm_x/l_x)\left[\frac{n}{2}(l_x + l_{x+n})\right], \text{ from Trapezoidal rule (A.4.3)}$$

$$= (_nm_x/l_x)\left[\frac{n}{2}(l_x + (l_x - {}_nd_x))\right], \quad \because l_x - l_{x+n} = {}_nd_x$$

$$= (_nm_x/l_x)\left[\frac{n}{2}(2\cdot l_x - {}_nd_x)\right] = n[_nm_x(1 - {}_nd_x/(2l_x))]$$

or $\quad {}_nq_x \cong n[{}_nm_x(1 - {}_nq_x/2)]$

$\therefore \quad {}_nq_x = [2n({}_nm_x)]/[2 + n({}_nm_x)]. \quad (2.8.2)$

2.8.2 Greville's Formula (Jaffe (1996))

Theorem 2.8.1: If the assumption of linearity of the curve l_x in (x to $x+n$), on which assumption (2.8.2) is obtained, does not hold good then we may assume that the curve for l_x is a third degree polynomial between x to $(x+n)$, and more general expression for ${}_nq_x$ in terms of ${}_nm_x$ is

$$ {}_nq_x = {}_nm_x / \left[\frac{1}{n} + ({}_nm_x)\left\{\frac{1}{2} + \frac{n}{12}({}_nm_x - k)\right\}\right], \quad (2.8.3)$$

where $k = \log C$, C is a parameter in the Gompertz curve viz.

$$ {}_nm_x \cong BC^x, \quad (2.8.4)$$

where B is another parameter.

(2.8.3) is called Greville's formula.

Note 2.5: Obviously (2.8.2) is a particular case of (2.8.3); the extra term here in the denominator is due to the assumption that l_x is a Cubic.

Proof: We have, $\frac{d}{dx}{}_nL_x = -{}_nd_x$, from (2.7.23).

Dividing both sides by ${}_nL_x$, we get

$$\frac{1}{({}_nL_x)}\frac{d({}_nL_x)}{dx} = -\frac{{}_nd_x}{{}_nL_x} \Longrightarrow \frac{d}{dx}\log{}_nL_x = -{}_nm_x, \text{ from (2.7.6)}$$

$$\therefore \quad \log_n L_x = -\int {}_nm_x dx + \log C \text{ (a constant)}.$$

$$\Rightarrow \quad {}_nL_x = C\exp(-\int {}_nm_x dx) \quad (2.8.5)$$

Now from Euler-Maclaurin's quadrature formula we have (since if $f(x)$ is a polynomial of 3rd order, higher than 3rd order derivatives will be zero) and derivatives upto there order will be constant.

$$\int_{x_0}^{x_0+mh} f(t)dt \simeq h\left[\sum_{i=0}^{m} f(x_0+ih) - \frac{1}{2}\{f(x_0) + f(x_0+mh)\}\right]$$
$$-\frac{h}{12}[f'(x_0+mh) - f'(x_0)] + \left(\frac{h^3}{720}\right)[f'''(x_0+mh) - f'''(x_0)] + \cdots$$

Put $x_0 = x$ and let $m \to \infty$, then

$$\frac{1}{h}\int_x^\infty f(t)dt \cong \sum_{i=0}^{\infty} f(x+ih) - \frac{1}{2}\{f(x) + f(\infty)\} - \frac{h}{12}[-f'(x) + f'(\infty)]$$

Put $f(x) = {}_nL_x$ and $h = n$ then $f(\infty) = \lim_{x\to\infty} {}_nL_x = 0$ and so $f'(\infty) = 0$, then we get

$$\frac{1}{n}\int_x^\infty {}_nL_t dt = \sum_{i=0}^{\infty} {}_nL_{x+in} - \frac{1}{2}{}_nL_x + \frac{h}{12}\left[\frac{d_nL_x}{dx}\right]$$

$$\text{or,}\quad \sum_{i=0}^{\infty} {}_nL_{x+in} = \frac{1}{n}\int_x^{\infty} {}_nL_t\,dt + \frac{1}{2}{}_nL_x - \frac{n}{12}\frac{d({}_nL_x)}{dx} \tag{2.8.6}$$

Putting the value of ${}_nL_x$ from (2.8.5) in the r.h.s. of (2.8.6) we get

$$\sum_{i=0}^{\infty} {}_nL_{x+i\cdot n} = \frac{1}{n}\int_x^{\infty} C\exp\left(-\int {}_nm_t dt\right) dt + \frac{1}{2}C\exp\left(-\int {}_nm_x dx\right)$$
$$+\frac{n}{12}C({}_nm_x)\exp\left(-\int {}_nm_x dx\right) \tag{2.8.7}$$

Differentiating both sides w.r.t. x we get (using (A.1.15) for the first term on r.h.s and noting that the integrand is independent of x).

$$\frac{d}{dx}\left[\sum_{i=0}^{\infty} {}_nL_{x+i\cdot n}\right] = -\frac{1}{n}C\exp\left(-\int {}_nm_x dx\right) - \frac{1}{2}({}_nm_x)C\cdot\exp\left(-\int {}_nm_x dx\right)$$
$$-\frac{n}{12}({}_nm_x)^2 C\exp\left(-\int {}_nm_x dx\right)$$
$$+\frac{n}{12}C\exp\left(-\int {}_nm_x dx\right)\frac{d}{dx}({}_nm_x) \tag{2.8.8}$$

$$\text{Now from (2.8.4),}\quad \frac{d}{dx}({}_nm_x) = BC^x\log C = kBC^x$$
$$= k\ {}_nm_x,\ k \text{ is a constant.} \tag{2.8.9}$$

Substituting in (2.8.8) from (2.8.9) we get

$$\frac{d}{dx}\left[\sum_{i=0}^{\infty} {}_nL_{x+in}\right] = -C\exp\left(-\int {}_nm_x dx\right)\left[\frac{1}{n} + \frac{1}{2}\,{}_nm_x + \frac{n}{12}({}_nm_x)^2 - \frac{n}{12}k({}_nm_x)\right]$$

$$\text{or}\quad -\frac{d}{dx}\left[\sum_{i=0}^{n} {}_nL_{x+in}\right] = {}_nL_x\left[\frac{1}{n} + \frac{{}_nm_x}{2}\{1+\frac{n}{6}({}_nm_x - k)\}\right] \tag{2.8.10}$$

$$\text{Now}\quad \frac{d}{dx}\left[\sum_{i=0}^{\infty} {}_xL_{x+in}\right] = \frac{d}{dx}[{}_nL_x + {}_nL_{x+n} + {}_nL_{x+2n} + \ldots]$$
$$= \frac{d}{dx}\left[\int_0^n l_{x+t}dt + \int_n^{2n} l_{x+t}dt + \ldots\right]$$
$$= \int_0^n \frac{d}{dt}l_{x+t}dt + \int_n^{2n}\frac{d}{dt}l_{x+t}dt + \ldots,$$

since l_{x+t} is symmetric in x and t

$$= [l_{x+n} - l_x] + [l_{x+2n} - l_{x+n}] + \ldots = -l_x, \quad \because l_\infty = 0.$$

Therefore (2.8.10) becomes

$$l_x = {}_nL_x\left[\frac{1}{n} + \frac{1}{2}({}_nm_x)\left\{1 + \frac{n}{6}({}_nm_x - k)\right\}\right]$$

$$\text{or}\quad \frac{1}{l_x} = \frac{1}{{}_nL_x\left[\frac{1}{n} + \frac{1}{2}({}_nm_x)\left\{1 + \frac{n}{6}({}_nm_x - k)\right\}\right]}$$

$$\text{or}\quad \frac{{}_nd_x}{l_x} = \frac{{}_nd_x/{}_nL_x}{\dfrac{1}{n} + {}_nm_x\left\{\dfrac{1}{2} + \dfrac{n}{12}({}_nm_x - k)\right\}}$$

$$\text{or}\quad {}_nq_x = \frac{{}_nm_x}{\frac{1}{n} + {}_nm_x\left\{\dfrac{1}{2} + \dfrac{n}{12}({}_nm_x - k)\right\}}, \tag{2.8.3}$$

from (2.7.2) and (2.7.6). //

2.8.3 Reed and Marrell's Formula (Jaffe (1966))

Theorem 2.8.2: Reed and Marrell's formula is obtained by ignoring the terms involving $({}_nm_x)^k$ for $k \geq 3$ and has therefore less accuracy as that of Greville's formula (2.8.3) but it has computational advantage over Greville's formula, of a quadrate formula.

This formula gives ${}_nq_x$ in terms of ${}_nm_x$ as

$${}_nq_x = 1 - \exp\left[-n({}_nm_x)\left\{1 + \frac{k}{12}n^2({}_nm_x)\right\}\right] \tag{2.8.11}$$

Proof: From Greville's formula (2.8.3.)

$${}_np_x = 1 - {}_nq_x = 1 - \frac{{}_nm_x}{\dfrac{1}{n} + {}_nm_x\left\{\dfrac{1}{2} + \dfrac{n}{12}({}_nm_x - k)\right\}}$$

$$= \frac{\dfrac{1}{n} + \dfrac{1}{2}({}_nm_x) + \dfrac{n}{12}\left\{({}_nm_x) - \dfrac{d}{dx}({}_nm_x)\right\} - {}_nm_x}{\dfrac{1}{n} + \dfrac{1}{2}({}_nm_x) + \dfrac{n}{12}\left\{({}_nm_x)^2 - \dfrac{d}{dx}({}_nm_x)\right\}},$$

putting the value of k from (2.8.9).

$$= \frac{1 + \dfrac{n^2}{12}\left\{({}_nm_x)^2 - \dfrac{d}{dx}({}_nm_x)\right\} - \dfrac{n}{2}({}_nm_x)}{1 + \dfrac{n^2}{12}\left\{({}_nm_x)^2 - \dfrac{d}{dx}({}_nm_x)\right\} + \dfrac{n}{2}({}_nm_x)} = \frac{1 + \alpha - \beta}{1 + \alpha + \beta}\ \text{(say)}$$

$$\text{when}\quad \alpha = \frac{n^2}{12}\left\{({}_nm_x)^2 - \frac{d}{dx}({}_nm_x)\right\} \quad\text{and}\quad \beta = \frac{n}{2}({}_nm_x) \tag{2.8.12}$$

Then $\log {}_np_x = \log(1 + \alpha - \beta) - \log(1 + \alpha + \beta)$

Expanding the l.h.s assuming that $|\alpha + \beta| < 1$ and $|\alpha - \beta| < 1$ and ignoring their powers higher than 2, we have

$$\log {}_np_x = \left\{(\alpha - \beta) - \frac{(\alpha-\beta)^2}{2} + O(\alpha-\beta)^3\right\} - \left\{(\alpha+\beta) - \frac{(\alpha+\beta)^2}{2} + O(\alpha+\beta)^3\right\}$$

$$\simeq -2\beta + 2\alpha\beta = -2\beta(1 - \alpha)$$

$$= -2 \cdot \frac{n}{2}({}_nm_x)\left[1 - \frac{n^2}{12}\left\{({}_nm_x)^2 - \frac{d}{dx}({}_nm_x)\right\}\right], \quad \text{from (2.8.12)}$$

$$= \left[-n({}_nm_x) + \frac{n^3}{12}({}_nm_x)^3 - \frac{n^3}{12}{}_nm_x\frac{d}{dx}({}_nm_x)\right]$$

$$= -n({}_nm_x) - \frac{n^3}{12}({}_nm_x)^2k,$$

from (2.8.9) and neglecting the terms involving $({}_nm_x)^3$.

$$\therefore \quad {}_np_x = \exp\left[-n({}_nm_x)\left\{1 + \frac{k}{12}n^2({}_nm_x)\right\}\right]$$

$$\therefore \quad {}_nq_x = 1 - \exp\left[-n({}_nm_x)\left\{1 + \frac{k}{12}n^2({}_nm_x)\right\}\right]. \quad // \tag{2.8.11}$$

2.8.4 Comparison of the Above Three Formulas

Greville's formula (2.8.3) is most accurate of the three formulas whereas the Greville's abridged formula (2.8.2) is least accurate; Reed and Merrell's formula (2.8.11) is in between, but fairly accurate. Through less accurate than Greville's formula but in terms of computational accessability Reed and Merrell's formula has its specific applications. However, through the Greville's abridged formula is least accurate but for quick construction of an abridged life table, often it is used. The life table prepared by this formula is not accurate for early age group as well as elderly age group because the assumption of linearity of the curve l_x, on which premise the formula (2.8.2) is derived, does not hold good for these age groups.

2.9 King's Method of Construction of an Abridged Life Table [Tetley (1950)]

As mentioned in Sec. 2.8, this is an acturial method.

Let ${}_nP_x$ = Population in the age group (x to $x+n$) and

${}_nD_x$ = Number of deaths in the age group (x to $x+n$).

Then we can write by employing King's Interpolation formula (Cf Whittikar and Robinson (1944),

$${}_nP_x = P_{x-\left[\frac{n-1}{2}\right]} + P_{x+1-\left[\frac{n-1}{2}\right]} + \ldots + P_{x+n-\left[\frac{n-1}{2}\right]} \tag{2.9.1}$$

$$\text{and} \quad {}_nD_x = D_{x-\left[\frac{n-1}{2}\right]} + D_{x+1-\left[\frac{n-1}{2}\right]} + \ldots + D_{x+n-\left[\frac{n-1}{2}\right]} \tag{2.9.2}$$

where P_i and D_i are respectively, the population and the number of deaths for the age i and $\left[\frac{n-1}{2}\right]$ stands for the greatest integer in $\left(\frac{n-1}{2}\right)$; $x > \left[\frac{n-1}{2}\right]$.

King obtained estimates of the population P_x^c and the deaths D_x^c for the central age point (middle point of x to $x+n$) in the age group (x to $x+n$) from the given values of ${}_nP_x$ and ${}_nD_x$ as

$$P_x^c = \frac{1}{n}({}_nP_x) - \frac{\frac{1}{n}\left(1 - \frac{1}{n^2}\right)}{24}\Delta^2{}_nP_{x-n} + \cdots \tag{2.9.3}$$

$$D_x^c = \frac{1}{n}({}_nD_x) - \frac{\frac{1}{n}\left(1 - \frac{1}{n^2}\right)}{24}\Delta^2{}_nD_{x-n} + \cdots \tag{2.9.4}$$

If $(x-n) < 0$, all the term involving $(x-n)$ will be taken as zero.

Note 2.5: ${}_nL_x$ is life table population, it is estimated by ${}_nP_x$.

Then $\hat{m}_x = \dfrac{D_x^c}{P_x^c}$ (for $x = x_0, x_0+h, x_0+2h, \cdots$) and $q_x = \left[\dfrac{2\hat{m}_x}{2+\hat{m}x}\right]$ from (2.6.6), and $p_x = 1 - q_x$.

Now to find the values of $l_{x+n}, l_{x+2n}, \ldots l_{x+kn}$ when l_0 (the cohort or radix of the life table) is given, we note that

$$_np_x = \frac{l_{x+n}}{l_x} \text{ (from (2.7.3))} = \frac{l_{x+n}}{l_{x+n-1}}\frac{l_{x+n-1}}{l_{x+n-2}}\cdots\frac{l_{x+1}}{l_x} = \prod_{i=0}^{n-1}\frac{l_{x+i+1}}{l_{x+i}}$$

$$\therefore \quad {}_np_x = p_x\,p_{x+1}\cdots p_{x+n-1}, \quad \text{from (2.5.4)} \tag{2.9.5}$$

$$\text{or} \quad \log {}_np_x = \sum_{i=0}^{n-1}\log p_{x+i} \tag{2.9.6}$$

Now we apply Everett's Central Difference formula which is given by

$$u_{x+h} = y\,u_{x+n} + u_x + \frac{y(y^2-1)}{3!}\Delta^2 u_x + \frac{z(z^2-1)}{3!}\Delta^2 u_{x-n} + \cdots \tag{2.9.7}$$

$$\text{where,} \quad 0 \le h \le n \text{ and } y = \frac{h}{n};\ z = 1 - y = 1 - \frac{h}{n} \text{ and } h = 0, 1, 2, \ldots, n \tag{2.9.8}$$

For $x < n$, all the term involving $(x-n)$ will be taken as zero. By putting $u_x = \log p_x$ in (2.9.7) and taking the terms only upto 2nd difference, we get:

For $h = 0, \log p_x = \log p_x$, using (2.9.8).

$$\text{For} \quad h = 1, \log p_{x+1} = \frac{1}{n}\log p_{x+n} + \left(1 - \frac{1}{n}\right)\log p_x + \frac{1}{3!}\frac{1}{n}\left(\frac{1}{n^2} - 1\right)\Delta^2\log p_x + \frac{1}{3!}\left(1 - \frac{1}{n}\right)\left\{\left(1 - \frac{1}{n}\right)^2 - 1\right\}\Delta^2\log p_{x-n}.$$

$$\text{For} \quad h = 2, \log p_{x+2} = \frac{2}{n}\log p_{x+n} + \left(1 - \frac{2}{n}\right)\log p_x + \frac{1}{3!}\frac{2}{n}\left(\frac{2^2}{n^2} - 1\right)\Delta^2\log p_x + \frac{1}{3!}\left(1 - \frac{2}{n}\right)\left\{\left(1 - \frac{2}{n}\right)^2 - 1\right\}\Delta^2\log p_{x-n}.$$

$$\cdots \quad \cdots \quad \cdots \quad \cdots \quad \cdots \quad \cdots \quad \cdots \quad \cdots$$

$$\text{For} \quad h = n-1, \log p_{x+n-1} = \frac{n-1}{n}\log p_{x+n} + \left(1 - \frac{n-1}{n}\right)\log p_x + \frac{1}{3!}\frac{n-1}{n}\left\{\left(\frac{n-1}{n}\right)^2 - 1\right\}\Delta^2\log p_{x-n} + \frac{1}{3!}\left(1 - \frac{n-1}{n}\right)\left\{\left(1 - \frac{n-1}{n}\right)^2 - 1\right\}\Delta^2\log p_{x-n}.$$

$$\text{Adding} \quad \sum_{h=0}^{n-1}\log p_{x+h} = \log_n p_x = \frac{n+1}{2}\log p_x + \frac{n-1}{2}\log p_{x+n} - \frac{n^2-1}{24n}(\Delta^2\log p_x + \Delta^2\log p_{x-n}). \tag{2.9.9}$$

So we are able to find out $_np_x$ for $x = x_0 + n$, and hence find $l_{x+n}, l_{n+2n}, \ldots$ If $x_0 = 0$ then $l_{x_0} = l_0$ and then

$$l_{x_0+n} = l_{x_0}\,{}_np_{x_0},\ l_{x_0+2n} = l_{x_0+n}\,{}_np_{x_0+n} \text{ etc.}$$

Next $T_{x:\overline{n}|} = {}_nL_x = \int_0^n l_{x+t}dt$, from (2.7.4)

$$\simeq \frac{1}{2}l_x + l_{x+1} + l_{x+z} + \ldots + l_{x+n-1} + \frac{1}{2}l_{x+n}$$

(by Trapezoidal rule (A.4.3)).

$$= \sum_{i=0}^{n-1} l_{x+i} - \left(\frac{1}{2}l_x - \frac{1}{2}l_{x+n}\right) = \sum_{i=0}^{n-1} l_{x+i} - \frac{l_x}{2}\left(1 - \frac{l_{x+n}}{l_x}\right)$$

$$= \sum_{i=0}^{n-1} l_{x+i} - \frac{1}{2}l_x(1 - {}_np_x), \text{ from (2.7.3)}$$

or $$T_{x:\overline{n}|} = \sum_{i=0}^{n-1} l_{x+i} - \frac{1}{2}l_x\ {}_nq_x \qquad (2.9.10)$$

Now as in (2.9.9)

$$\sum_{i=0}^{n-1} l_{x+i} = \frac{n+1}{2}l_x + \frac{n-1}{2}l_{x+n} - \frac{n^2-1}{24n}(\Delta^2 l_x + \Delta^2 l_{x-n})$$

$$= N_{x:\overline{n}|} \text{ (say)} \qquad (2.9.11)$$

Hence (2.9.10), becomes

$$T_{x:\overline{n}|} = N_{x:\overline{n}|} - \frac{1}{2}l_x{}_nq_x \qquad (2.9.12)$$

Finally, $e^0_{x:\overline{n}|} = \dfrac{T_{x:\overline{n}|}}{l_x}$, similar to (2.5.11)

or $$e^0_{x:\overline{n}|} = \frac{N_{x:\overline{n}|}}{l_x} - \frac{1}{2}\ {}_nq_x, \qquad (2.9.13),$$

can easily be obtained when $N_{x:\overline{n}|}$ and ${}_nq_x$ are known.

Finally $e^0_{x:\overline{n}|} = T_{x:\overline{n}|}/l_x = N_{x:\overline{n}|}/l_x - \frac{1}{2}{}_nq_x$

2.10 Sampling Distribution of Life Table Functions (Chiang (1968))

When we construct a life table from a sample, the problem arises (i) to test the reliability of the life table and (ii) to obtain the sampling variance associated with the estimates of different life table functions.

Basically, the parameters in the life table (which are to be estimated from the sample) are the survival probabilities (or intrinsic risk of mortality). We have from (2.5.4)

$${}_xp_0 = \exp\left[-\int_0^x \mu(t)dt\right] \quad \text{and} \quad p_x = \left[-\int_x^{x+1} \mu(t)dt\right] \qquad (2.10.1)$$

or $${}_xp_0 = \exp\left[-\left\{\int_0^1 \mu(t)dt + \int_1^2 \mu(t)dt + \ldots + \int_{x-1}^x \mu(t)dt\right\}\right] = p_0\ p_1 \ldots p_{x-1}$$

or $${}_xp_0 = \prod_{i=0}^{x-1} p_i = {}_{x-1}p_0\ p_{x-1}. \qquad (2.10.2)$$

Here $\mu(x)$, ${}_xp_0$ and p_x are parameters of the life table. The sequence $\{l_x\}$, $\{d_x\}$, $\{L_x\}$, $\{T_x\}$ and $\{e_x^0\}$, $x = 0, 1, 2, \ldots$ are sequences of r.v.'s is in the life table.

2.10.1 Distribution of l_x

The distribution (p.m.f.) of l_x is

$$p(l_x) = \binom{l_0}{l_x} ({}_xp_0)^{l_x} (1 - {}_xp_0)^{l_0 - l_x}$$

i.e. $l_x \sim B(l_0, {}_xp_0)$ (A.2.1(b)(ii)).

2.10.2 Joint Probability Distribution of Number of Deaths

To obtain the joint probability distribution of $d_0, d_1, \ldots, d_w$, where w is the last age of survival, we note that

$$\left.\begin{aligned} &d_0 + d_1 + \cdots + d_w = l_0 \\ \text{and} \quad &{}_0p_0\, q_0 + {}_1p_0\, q_1 + {}_2p_0 q_2 + \cdots + {}_wp_0\, q_w = 1, \quad ({}_0p_0 = 1) \end{aligned}\right\} \tag{2.10.4}$$

which show that the joint probability by distribution of $d_0, d_1, \ldots, d_w$ is multinomial i.e. their p.m.f is given by

$$p(d_0, d_1, \ldots, d_w) = \frac{l_0!}{d_0! d_1! \ldots, d_w!} ({}_0p_0\, q_0)^{d_0} ({}_1p_0\, q_1)^{d_1} \cdots ({}_wp_0\, q_w)^{d_w} \tag{2.10.5}$$

$$\text{where } E(d_i) = l_0\ {}_ip_0\ q_i, \ \mathrm{Var}(d_i) = l_0 {}_ip_0 \quad q_i(1 - {}_ip_0\ q_i) \tag{2.10.6}$$

$$\text{and} \quad \mathrm{Cov}(d_i, d_j) = -l_0\ {}_ip_0\ q_i\ {}_jp_0\ q_j, \quad i \cdot j = 1, 2, \ldots, w.$$

2.10.3 Distribution of the Observed Residual Expectation of Life

Let l_x be the number of survivors at age x. The future life times of l_x survivors may be regarded as a sample of l_x identically and independently distributed random variables (i.i.d.r.v's) Y_k, $k = 1, 2, \ldots, l_x$.

The pdf of Y_k is given by (from (2.6.16))

$$f(y) = \mu(x + y) \exp\left[-\int_x^{x+y} \mu(t)dt\right], y \geq 0$$

$$\therefore \quad e_y^0 = E[Y_k] = \int_0^\infty y\mu(x + y) \exp\left[-\int_x^{x+y} \mu(t)dt\right] dy \tag{2.10.7}$$

$$\left.\begin{aligned} \text{or} \quad &E[Y_k] = \int_0^\infty \exp\left[-\int_x^{x+y} \mu(t)dt\right] dy, \text{ from (2.6.17) and (2.6.15)} \\ \text{and} \quad &\mathrm{Var}[Y_k] = \int_0^\infty (y - e_y^0)^2 f(y)dy = \sigma_y^2 \text{ (say)}, \end{aligned}\right\} \tag{2.10.8}$$

where e_y^0 is the complete expectation of life at y; e_y^0 and $f(y)$ are given above.

According to Central Limit Theorem (A.2.2(b)) the distribution of the sample mean $\overline{y} = \frac{1}{l_{\hat{x}}} \sum_{k=1}^{l_{\hat{x}}} Y_k$ is asymptotically normal as $N(e_y^0, \sigma_y^2/l_x)$ (A.2.1(b)(i)). It follows that the sample mean length of life, $\overline{y}$, is Consistently Asymptotic Normal (CAN) estimator (A.2.5).

2.11 Probability Generating Functions (p.g.f.) of l_x's

The p.g.f. of the non-negative integer valued r.v. X, with pmf p_x, is given by {see (A.2.4)}

$$P(s) = E[s^X] = \sum_{x=0}^{\infty} s^x p_x.$$

The number of survivors at age x viz l_x is a non-negative integer valued r.v.

2.11.1 The p.g.f. of a single l_x for a fixed l_0

From (2.10.3) $l_x \sim B(l_0, {}_xp_0)$ (A.2.1(b)(ii)). Hence, the p.g.f. of l_x is

$$P(s) = E[s^{l_x}|l_0] = \sum_{l_x=0}^{l_0} s^{l_x} \binom{l_0}{l_x} ({}_xp_0)^{l_x}(1-{}_xp_0)^{l_0-l_x}$$

$$= \sum_{l_x=0}^{l_0} \binom{l_0}{l_x} (s\ {}_xp_0)^{l_x}(1-{}_xp_0)^{l_0-l_x}$$

or $P(s) = \{1 - {}_xp_0(1-s)\}^{l_0}, |s| < 1.$ (2.11.1)

From (A.2.15), $E(l_x) = \left.\frac{P(s)}{ds}\right|_{s=1} = l_0\ {}_xp_0.$ (2.11.2)

Similarly, the p.g.f. for the conditional distribution of l_j given $l_i, i < j$, is

$$P(s_j) = E[s_j^{l_j}/l_i] = \sum_{l_j=0}^{l_i} s_j^{l_j} \binom{l_i}{l_j} ({}_{j-i}p_i)^{l_j}(1-{}_{j-i}p_i)^{l_i-l_j}$$

or $P(s_j) = [1 - {}_{j-i}p_i(1-s_j)]^{l_i}.$ (2.11.3)

From (A.2.15), $\mu'_{(1)} = \mu'_1 = E[l_j] = \left.\frac{dP(s_j)}{ds_j}\right|_{s_j=1}$

$\therefore$ From (2.11.3), $E(l_j|l_i) = l_i[1-{}_{j-i}p_i(1-s_j)]^{l_i-1}({}_{j-i}p_i)\Big|_{s_j=1}$

$$= l_i\ {}_{j-i}p_i. \quad (2.11.4)$$

Similarly, $\mu'_{(2)} = E[l_j(l_j-1)/l_i] = \left.\frac{d^2P(s_j)}{ds_j^2}\right|_{s_j=1}$

or $E[l_j^2|l_i] = l_i(l_i-1)[1-{}_{j-i}p_i(1-s_j)]^{l_i-2}{}_{j-i}p_i^2\Big|_{s_j=1} + E(l_j/l_i)$

$$= l_i(l_i-1)_{j-i}p_i^2 + l_{ij-i}p_i \quad (2.11.5)$$

$\therefore$ $\text{Var}(l_j|l_i) = l_i^2{}_{j-1}p_i^2 - l_i\ {}_{j-i}p_i^2 + l_i\ {}_{j-i}p_i - l_i^2\ {}_{j-i}p_i^2$

$$= l_i\ {}_{j-i}p_i(1-{}_{j-i}p_i). \quad (2.11.6)$$

2.11.2 Multivariate p.g.f. of $l_1, l_2, \ldots, l_w$ for a fixed l_0

Theorem 2.11.1: The pgf of $l_1, l_2, \ldots, l_w$ for a fixed l_0 is given by

$$P(\underset{\sim}{s}) = [1 - \{{}_1p_0(1-s_1) + {}_2p_0s_1(1-s_2) + \ldots$$
$$+ {}_{w-1}p_0s_1s_2\ldots s_{w-2}(1-s_{w-1}) + {}_wp_0s_1s_2\ldots s_{w-1}(1-s_w)]^{l_0} \quad (2.11.7)$$

where $\underset{\sim}{s} = (s_1, s_2, \ldots, s_w)$.

Proof: By definition (A.2.14)

$$P(\underset{\sim}{s}) = E[s_1^{l_1} s_2^{l_2} \dots s_w^{l_w} | l_0]$$
$$= E[s_1^{l_1} s_2^{l_2} \dots s_{w-1}^{l_{w-1}} E(s_w^{l_w} | l_{w-1}) | l_0], \text{ from (A.2.5)}$$
$$= E[s_1^{l_1} s_2^{l_2} \dots s_{w-1}^{l_{w-1}} \sum_{l_w=0}^{l_{w-1}} s_w^{l_w} \binom{l_{w-1}}{l_w} p_{w-1}(1-p_{w-1})^{l_{w-1}-l_w} | l_0]$$

Note that ${}_1p_{w-1} = p_{w-1}$

$$= E[s_1^{l_1} s_2^{l_2} \dots s_{w-1}^{l_{w-1}} (1 - p_{w-1} + s_w\, p_{w-1})^{l_{w-1}} | l_0], \text{ from (2.11.1).} \tag{2.11.8}$$

Let us put

$$s_{w-1}\{1 - p_{w-1}(1 - s_w)\} = t_{w-1} \tag{2.11.9}$$

in (2.11.8), then

$$P(\underset{\sim}{s}) = E[s_1^{l_1} s_2^{l_2} \dots s_{w-2}^{l_{w-2}} t_{w-1}^{l_{w-1}} | l_0] \tag{2.11.10}$$

We shall prove the result (2.11.7) by mathematical induction.

For $w = 1$, (2.11.10) gives

$$P(s_1) = E[s_1^{l_1} | l_0] = [1 - {}_1p_0 + s_1 {}_1p_0]^{l_0} \text{ from (2.11.1)} \tag{2.11.11}$$

Hence, the relation (2.11.7) is true for $w = 1$.

Assuming that the result is true for $w - 1$ r.v.'s viz $l_1, l_2, \dots, l_{w-1}$ we may get (replacing the arbitrary constant s_{w-1} by another arbitrary constant t_{w-1}) from (2.11.7)

$$P(s_1 s_2, \dots, s_{w-1}, t_{w-1}) = E[s_1^{l_1} s_2^{l_2} \dots s_{w-2}^{l_{w-2}} t_{w-1}^{l_{w-1}} | l_0]$$
$$= [1 - \{{}_1p_0(1 - s_1) + {}_2p_0 s_1(1 - s_2) + \dots$$
$$+ {}_{w-1}p_0 s_1 s_2 \dots s_{w-2}(1 - t_{w-1})\}]^{l_0} \tag{2.11.12}$$

Now, from (2.11.9)

$$1 - t_{w-1} = 1 - s_{w-1} + p_{w-1} s_{w-1}(1 - s_w).$$

Substituting this in (2.11.12) we get

$$P(s_1, s_2, \dots, s_{w-2}, s_{w-1}, s_w)$$
$$= [1 - [{}_1p_0(1 - s_1) + {}_2p_0 s_1(1 - s_2)$$
$$+ \dots + {}_{w-1}p_0 s_1 s_2 \dots s_{w-2}\{(1 - s_{w-1}) + p_{w-1} s_{w-1}(1 - s_w)\}]^{l_0}]$$
$$= [1 - \{{}_1p_0(1 - s_1) + {}_2p_0 s_1(1 - s_2) + \dots + {}_{w-1}p_0 s_1 s_2 \dots s_{w-2}(1 - s_{w-1})$$
$$+ {}_{w-1}p_0 s_1 s_2 \dots s_{w-2} p_{w-1} s_{w-1}(1 - s_w)\}]^{l_0}$$

$$\therefore \quad P(\underset{\sim}{s}) = [1 - \{{}_1p_0(1 - s_1) + {}_2p_0 s_1(1 - s_2) + \dots + {}_{w-1}p_0 s_1 s_2 \dots s_{w-2}(1 - S_{w-1})$$
$$+ {}_wp_0 s_1 s_2 \dots s_{w-1}(1 - s_w)\}]^{l_0} \tag{2.11.7}$$

$\because \ {}_{w-1}p_0 p_{w-1} = {}_wp_0,$ from (2.10.2),

which shows that the result is true for $l_1, l_2, \dots, l_w$.

Hence, the result is true for any finite number of r.v.'s. //

2.11.3 An Application of the above Result

From (2.11.7) we can get mean and variance of l_i's, covariance between l_i and l_j etc. as follows:

Differentiating (2.11.7) w.r.t. s_1 we get

$$\frac{\partial P(\underset{\sim}{s})}{\partial s_1} = l_0[1 - \{{}_1p_0(1-s_1) + {}_2p_0 s_1(1-s_2) + \cdots$$

$$+ {}_wp_0 s_1 s_2 \cdots s_{w-1}(1-s_2)\}]^{l_0-1}[{}_1p_0 - {}_2p_0(1-s_2) - \cdots - {}_wp_0\ s_2 \cdots s_{w-1}(1-s_w)]$$

Then $$\left.\frac{\partial P(\underset{\sim}{s})}{\partial s_1}\right]_{s_i=1\forall i} = l_0\ {}_1p_0 = E[l_1|l_0] = E(l_i). \tag{2.11.13}$$

In multinomial distribution $E(X_i) = n \cdot p_i$

Similarly $$\left.\frac{\partial P(\underset{\sim}{s})}{\partial s_i}\right]_{s_i=1\ \forall i} = l_0\ {}_ip_0 = E(l_i/l_0) = E(l_i) \tag{2.11.14}$$

and $$\left.\frac{\partial^2 P(\underset{\sim}{s})}{\partial s_i^2}\right]_{s_i=1\forall i} = l_0(l_0-1)\{{}_ip_0 - {}_ip_0^2\}$$

$$\therefore \quad \text{Var}(l_i) = l_0\ {}_ip_0(1-{}_ip_0) \tag{2.11.15}$$

$$= l_0(l_0-1){}_1p_0\ {}_2p_0 + l_0\ {}_2p_0.$$

Further $$\left.\frac{\partial^2 P(\underset{\sim}{s})}{\partial s_1 \partial s_2}\right]_{s_i=1\forall i} = E[l_1 l_2/l_0]$$

$$\therefore \quad \text{Cov}(l_1, l_2) = l_0\ {}_2p_0(1-{}_1p_0). \tag{2.11.16}$$

Similarly $$\text{Cov}(l_i, l_j) = l_0\ {}_jp_0(1-{}_ip_0), \quad \text{for } i<j \tag{2.11.17}$$

$$\text{Cor}(l_i, l_j) = \sqrt{{}_jp_0(1-{}_ip_0)}\Big/\sqrt{{}_ip_0(1-{}_jp_0)} \tag{2.11.18}$$

2.12 Estimation of Survival Probability p_x by the Method of MLE [Chiang (1968)]

Let ϵ_{xi} be a r.v. such that

$$\left.\begin{aligned} \epsilon_{xi} &= 1 \text{ if } i\text{th person dies between the year } x \text{ to } x+1 \\ &= 0, \quad \text{otherwise} \\ i &= 1, 2, \ldots, l_0 \text{ and } x = 0, 1, 2 \ldots \end{aligned}\right\} \tag{2.12.1}$$

Thus the likelihood function of the sample is given by

$$L = \prod_{x=0}^{\infty} \prod_{i=1}^{l_0} [{}_xp_0(1-p_x)]^{\epsilon_{xi}}, \quad \text{see (2.10.1)} \tag{2.12.2}$$

or $$L = \prod_{x=0}^{\infty} [{}_xp_0(1-p_x)]^{\sum_{i=1}^{l_0} \epsilon_{xi}} \tag{2.12.3}$$

Now $$\sum_{i=1}^{l_0} \epsilon_{xi} = d_x = \text{Number of deaths in } x \text{ to } x+1$$

$$\therefore \quad L = \prod_{x=0}^{\infty} [{}_xp_0(1-p_x)]^{d_x} \tag{2.12.4}$$

$$\Rightarrow \quad \log L = \sum_{x=0}^{\infty} dx[\log {}_xp_0 + \log(1-p_x)]$$

$$= d_0[\log {}_0p_0 + \log(1-p_0)] + d_1[\log {}_1p_0 + \log(1-p_1)]$$

$$+ \ldots + d_x[\log {}_xp_0 + \log(1-p_x)] + d_{x+1}[\log_{x+1} p_0 + \log(1-p_x)] + \ldots$$

Now, from (2.10.2) ${}_xp_0 = p_0p_1 \ldots p_{x-1}$ (2.12.5)

$$\Rightarrow \quad \log {}_xp_0 = \log p_0 + \log p_1 + \ldots + \log p_{x-1} +$$

$$\Rightarrow \quad \frac{\partial \log {}_xp_0}{\partial p_j} = \frac{1}{p_j}, \quad \text{for } x > j. \tag{2.12.6}$$

From (2.12.5), we further note that ${}_xp_0$ does not contain p_x but all ${}_jp_0$ for $j = x+1, x+2, \ldots$ will contain p_x. Hence the Maximum likelihood equation

$$\frac{\partial \log L}{\partial p_x} = 0 \Rightarrow -\frac{dx}{1-p_x} + \frac{d_{x+1}}{p_x} + \frac{d_{x+2}}{p_x} + \ldots = 0, \text{ using (2.12.6)} \tag{2.12.7}$$

$$\text{or} \quad \frac{1}{p_x}\sum_{j=x+1}^{\infty} d_j = \frac{dx}{1-p_x} \Rightarrow \frac{d_x}{\sum_{j=x+1}^{\infty} d_j} = \frac{1-p_x}{p_x}$$

$$\Rightarrow \quad \frac{dx}{\sum_{j=x+1}^{\infty} d_j} + 1 = \frac{1}{p_x} \Rightarrow \frac{\sum_{j=x}^{\infty} d_j}{\sum_{j=x+1}^{\infty} d_j} = \frac{1}{p_x}$$

$$\Rightarrow \quad \hat{p}_x = \frac{\sum_{j=x+1}^{\infty} d_j}{\sum_{j=x}^{\infty} d_j} = j = x+1 \tag{2.12.8}$$

$$\text{or} \quad \hat{p}_x = \frac{l_{x+1}}{l_x}, \quad \because \sum_{j=x}^{\infty} d_j = l_x \text{ from (2.5.1)} \tag{2.12.9}$$

which is the Maximum Likelihood Estimator (MLE) of p_x.

2.12.1 The Properties of MLE $\hat{p}_x$

1. **$\hat{p}_x$ is unbiased for p_x**: From (2.12.9) we have

$$E[\hat{p}_x] = E\left[\frac{l_{x+1}}{l_x}\right] = E\left[\frac{1}{l_x}E(l_{x+1}|l_x)\right], \text{ see (A.2.2)}$$

$$= E\left[\frac{1}{l_x}l_xp_x\right], \text{ from (2.11.4)}$$

$$= p_x, \text{ remember } p_x \text{ is a parameter i.e. a constant.} \tag{2.12.10}$$

2. **Variance of $\hat{p}_x$**: Again from (2.12.9) we have

$$E[\hat{p}_x^2] = E\left[\left(\frac{l_{x+1}}{l_x}\right)^2\right] = E\left[\frac{1}{l_x^2}E(l_{x+1}^2|l_x)\right]$$

$$= E\left[\frac{1}{l_x^2}\{l_x(l_x-1)p_x^2 + l_x p_x\}\right], \text{ from (2.11.5)}$$

$$= E\left[\frac{1}{l_x}p_x(1-p_x) + p_x^2\right] = p_x^2 + p_x(1-p_x)E\left(\frac{1}{l_x}\right) \quad (2.12.11)$$

Now, from (2.11.4), (2.11.5) and (2.11.6) we have

$$E[l_{x+1}|l_x] = l_x p_x, E[l_{x+1}^2|l_x] = l_x p_x(1-p_x) + l_x^2 p_x^2$$

and $\text{Var}[l_{x+1}|l_x] = l_x p_x(1-p_x)$

$$\therefore \quad \text{Var}(\hat{p}_x) = E(\hat{p}_x^2) - [E(\hat{p}_x)]^2$$

$$= p_x^2 + p_x(1-p_x)E\left(\frac{1}{l_x}\right) - p_x^2, \text{ from (2.12.10) and (2.12.11)}$$

or $$\text{Var}(\hat{p}_x) = p_x(1-p_x)E\left(\frac{1}{l_x}\right) \quad (2.12.12)$$

3. Cramer-Rao Bound for Var($\hat{p}_x$): We can obtain the Cramer-Rao lower bound (see (A.2.29)) for the variance $\hat{p}_x$ as follows:

From (2.12.7)

$$\frac{\partial \log L}{\partial p_x} = -\frac{d_x}{1-p_x} + \frac{\sum_{j=x+1}^{\infty} d_j}{p_x} \quad \Rightarrow \quad \frac{\partial^2 \log L}{\partial p_x^2} = -\frac{d_x}{(1-p_x)^2} - \frac{\sum_{j=x+1}^{\infty} d_j}{p_x^2}$$

$$\Rightarrow \quad -\frac{\partial^2 \log L}{\partial p_x^2} = \frac{l_x - l_{x+1}}{(1-p_x)^2} + \frac{l_{x+1}}{p_x^2} \quad \because l_x = \sum_{j=x}^{\infty} d_j$$

$$\therefore \quad E\left[-\frac{\partial^2 \log L}{\partial p_x^2}\right] = \frac{E(l_x) - E(l_{x+1})}{(1-p_x)^2} + \frac{E(l_{x+1})}{p_x^2}$$

$$= \frac{l_0\, {}_xp_0 - l_0\, {}_xp_0\, p_x}{(1-p_x)^2} + \frac{l_0\, {}_xp_0\, p_x}{p_x^2}, \quad \because E(l_x) = l_0\, {}_xp_0 \text{ from (2.11.2)}$$

$$= \frac{l_0\, {}_xp_0}{1-p_x} + \frac{l_0\, {}_xp_0}{p_x} = \frac{l_0\, {}_xp_0}{p_x(1-p_x)} \quad (2.12.13)$$

Hence, C-R lower bound for Var($\hat{p}_x$) is given by (see (A.2.29))

$$-\frac{1}{E\left(\dfrac{\partial^2 \log L}{\partial p_x^2}\right)} = \frac{p_x(1-p_x)}{l_0\, {}_xp_0} = \frac{p_x(1-p_x)}{E[l_x]} \quad (2.12.14)$$

Now, $\text{Var}(\hat{p}_x) = p_x(1-p_x)E\left(\frac{1}{l_x}\right)$ (2.12.12)

But, $E\left(\frac{1}{l_x}\right) \geq \frac{1}{E(l_x)} \quad \because \text{Harmonic mean} \geq \frac{1}{\text{Arithmetic mean}}$ (2.12.15)

Hence, from (2.12.12), (2.12.14) and (2.12.15)

$$\text{Var}(\hat{p}_x) \geq \frac{p_x(1-p_x)}{E(l_x)} \quad (2.12.16)$$

Variance of MLE $\hat{p}_x \geq$ C.R. lower bound (L.B.)

$$\therefore \quad \text{Efficiency of } (\hat{p}_x) = \frac{E(l_x)}{E\left(\frac{1}{l_x}\right)}, \text{ from (2.12.12) and (2.12.14).} \tag{2.12.17}$$

Next we shall show that MLE is MVUE and C.R. L.B. is not attained.

4. Sufficiency of $\hat{p}_x$:

Theorem 2.12.1: Show that $\hat{p}_x = \dfrac{l_{x+1}}{l_x}$ given by (2.12.9) is sufficient for p_x or (l_x, l_{x+1}) are jointly sufficient for p_x and further, $\hat{p}_x$ is a MVUE of p_x.

Proof: From (2.12.4) the likelihood of the sample is

$$L = \prod_{x=0}^{\infty} \{{}_xp_0(1-p_x)\}^{d_x} \tag{2.12.18}$$

$$\begin{aligned}
\therefore \quad \prod_{x=0}^{\infty} ({}_xp_0)^{d_x} &= ({}_0p_0)^{d_0}({}_1p_0)^{d_1}({}_2p_0)^{d_2}({}_3p_0)^{d_3}\ldots \\
&= (p_0)^{d_1}(p_0p_1)^{d_2}(p_0p_1p_2)^{d_3}\ldots, \quad \text{assuming } {}_0p_0 = 1 \\
&= p_0^{\sum_{x=1}^{\infty} d_x} \; p_1^{\sum_{x=2}^{\infty} d_x} \; p_2^{\sum_{x=1}^{\infty} d_x} \; \ldots \\
&= p_0^{l_1} p_1^{l_2} p_2^{l_3} \ldots, \quad \because \sum_{i=x}^{\infty} d_i = l_x \quad \text{and} \quad d_x = l_x - l_{x+1} \\
&= \prod_{x=0}^{\infty} p_x^{l_{x+1}}
\end{aligned}$$

$$L = \prod_{x=0}^{\infty} p_x^{l_{x+1}}(1-p_x)^{l_x - l_{x+1}} \tag{2.12.19}$$

5. $\hat{p}_x$ is a Minimum variance unbiased estimator (MVUE)

We reconsider the likelihood function of the sample form (2.2.19)

$$L = \prod_{x=0}^{\infty} p_x^{l_{x+1}}(1-p_x)^{1_x - 1_{x+1}} = \prod_{x=0}^{\infty} p_x^{1_{x+1}}(1-p_x)^{d_x} \quad (\because d_x = 1_x - 1_{x+1})$$

Note that l_x 's are random variables and p_x 'sare set of parameters in the joint distribution of probability function.

Therefore, (2.12.9) can be written as

$$\begin{aligned}
&\prod_{x=0}^{\infty} \frac{1_x!}{d_x!(1_{x+1})!} p_x^{1_{x+1}}(1-p_x)^{1_x - 1_{x+1}} \prod_{i=0}^{\infty} \frac{d_x! l_{x+1}!}{1_x!} \\
&= \prod_{x=0}^{\infty} \frac{1_x p_x^{1_{x+1}}(1-p_x)^{1_x - 1_{x+1}}}{(1_x - 1_{x+1})!(1_{x+1})!} \prod_{i=0}^{\infty} \frac{(\in_{x\beta})!(1_x - \sum_{\beta=1}^{\infty} \in_{x\beta})!}{1_x!}
\end{aligned}$$

$$\left(\because d_x = 1_x = 1 + x - 1_{x+1} = \sum_{\beta=0}^{\infty}(\in_{x\beta})\right) \tag{2.12.20}$$

Therefore, L is capable of being factorized into two distinct products. The first being a function of the parameters p_x and (l_x, l_{x+1}) and the second being independent of px (i.e. they are function of r.v.s only). By Neyman Factorization criterion (1.3.64) (l_x and l_{x+1}) are jointly sufficient for p_x

Also if $\widetilde{p}_x$ be any other estimate of p_x then $(\widetilde{p}_x|1_x, l_{x+1})$ must be independent for p_x.

Let $E(\widetilde{p}_x|1_x, 1_{x+1}) = f(1_x, 1_{x+1})$, say. (2.12.21)

Let $\widetilde{p}_x$ be unbiased for p_x, i.e. $E(\widetilde{p}_x) = p_x$ (2.12.22)

Next we shall compare the efficiency of $\widetilde{p}_x$ with the maximum likelihood extima or of p_x i.e. $\hat{p}_x$.

Since $E(\hat{p}_x) = p_x$

then $E(\widetilde{p}_x) = E[E(\widetilde{p}|1_x, 1_{x+1})] = E[f(1_x, 1_{x+1})]$ (2.12.23)

To obtain $E[f(1_x, 1_{x+1})]$ we require the joint probability density function of 1_x and 1_{x+1} subject to the condition $1_x \neq 0$ is given by

$$p(1_x, 1_{x+1}) = \frac{\binom{1_0}{1_x}({}_xP_0)^{1_x}(1 - {}_xP_0)^{1_0 - 1_x}\binom{1_x}{1_{x-1}}p_x^{1_{x+1}}(1 - p_x)^{1_x - 1_{x+1}}}{[1 - (1 - {}_xP_0)^{1_0}]} \tag{2.12.24}$$

so form (2.12.23) and (2.12.24), it follows that

$p_x = E(\widetilde{p}_x)$

$$= \frac{\sum_{1_x=0}^{1_0}\sum_{1_{x+1}=0}^{1_0} f(1_x, 1_{x+1})\binom{1_0}{1_x}({}_xp_0)^{1_x}(1 - {}_xp_0)^{1_0 - 1_x}\binom{1_x}{1_{x_1}}p_x^{1_{z+1}}(1 - p_x)^{1_x - 1_{x+1}}}{[1 - (1 - {}_xP_0)^{1_0}]} \tag{2.12.25}$$

$$\sum_{1_x=0}^{1_0}\sum_{1_{x+1}=0}^{1_0} f(1_x, 1_{x+1})p(1_x, 1_{x+1}) = p_x \tag{2.12.25}$$

Also we have shown in (2.12.10)

$$E(\hat{p}_x = E\left(\frac{1_{x+1}}{1_x}\right) = p_x \Rightarrow \sum_{1_x=0}^{1_0}\sum_{1_{x+1}=0}^{1_0}\left(\frac{1_{x+1}}{1_x}\right)p(1_x, 1_{x+1}) = p_x$$

(2.12.25) is an identity in p_x by comparing (2.12.10) and (2.12.25) we have

$$F(1_x, 1_{x+1}) = \frac{1_{x+1}}{1_x} = \hat{p}_x \tag{2.12.26}$$

$\Rightarrow \quad E(\widetilde{p}_x|1_x, 1_{x+1}) = \hat{p}_x$, from (2.12.21)

Hence by Rao-Blackwell Theorem (Appendix 1.3.5) the m.l.e. $\hat{p}_x = \dfrac{1_{x+1}}{1_x}$ is MVUE though we have seen that it is not MVBE since Cramer-Rao lower bound is not attained.

Alternative Method

We can prove the above result alternatively as follows:

We have $\text{Var}(\widetilde{p}_j) = E(\widetilde{p}_j - p_j)^2 = E(\widetilde{p}_j - \hat{p}_j + \hat{p}_j - p_j)^2$

$$= E(\widetilde{p}_j - \hat{p}_j)^2 + \text{Var}(\hat{p}_j) \tag{2.12.27}$$

The product term vanishes, because

$$E[(\widetilde{p}_j - \hat{p}_j)(\hat{p}_j - p_j)] = E[E(\widetilde{p}_j - \hat{p}_j)(\hat{p}_j - p_j)|l_j, l_{j+1}]$$
$$= E[E\{(\widetilde{p}_j - \hat{p}_j)|l_{j_1} l_{j+1}\}(\hat{p}_j - p_j)]$$

Since $\hat{p}_j = l_{j+1}/l_j$, from (2.12.9), hence given l_j and l_{j+1}, $\hat{p}_j$ is a constant.

Now $\because E[\widetilde{p}_j|l_j, l_{j+1}] = \hat{p}_j$ from (2.12.26)

$\therefore$ $E[(\widetilde{p}_j - \hat{p}_j)|l_j, l_{j+1}] = 0$ as $\hat{p}_j$ is a constant here

$\therefore$ $E[(\hat{p}_j - \hat{p}_j)(\hat{p}_j - p_j)] = 0.$

Hence, from (2.12.27) $\text{Var}(\widetilde{p}_j) \geq \text{Var}(\hat{p}_j)$ (2.12.28)

Since $\widetilde{p}_j$ is an arbitrary unbiased estimator, hence there exist no unbiased estimator whose variance is lower than that of $\hat{p}_j$. Hence $\hat{p}_j$ is MVUE of p_j.

2.13 Large Sample S.E. of Estimated Life Table Functions

In this section we shall derive large sample S.E. of e_x^0, $\hat{q}_i$ and ${}_{j-i}\hat{p}_i$. But first of all we shall prove the following lemma.

2.13.1 Lemma 2.1

Show that ${}_nq_x = \dfrac{n({}_nm_x)}{1 + (n - {}_na_x)_n m_x}$, (2.13.1)

where ${}_nq_x$ and ${}_nm_x$ are defined by (2.7.2) and (2.7.6) respectively, and ${}_na_x$ = average number of years lived from x to $(x+1)$ by those who die during this period.

Proof: Now the person years lived by the cohort l_x in the interval x to $(x+n)$ i.e. ${}_nL_x$ is equal to the person years lived by the cohort l_{x+n} who survived after the interval, each one of which lived n person years, so in total $n \cdot l_{x+n}$ person years and in addition the person years lived by the person who died in the interval i.e. ${}_nd_x$ and each one of these lived, on the average, ${}_na_x$ person years, so in total ${}_na_x\, {}_nd_x$ person years. Hence, we can write

$${}_nL_x = n \cdot l_{x+n} + {}_na_x\, {}_nd_x. \tag{2.13.2}$$

or $1 = n \cdot \dfrac{l_{x+n}}{{}_nL_x} + {}_na_x \dfrac{{}_nd_x}{{}_nL_x} = n\dfrac{l_{x+n}}{{}_nL_x} + {}_na_x\, {}_nm_x$, from (2.7.6)

or $1 - {}_na_x\, {}_nm_x = n\dfrac{(l_x - {}_nd_x)}{{}_nL_x} = n\left(\dfrac{l_x}{{}_nd_x} - 1\right) \Big/ \dfrac{{}_nL_x}{{}_nd_x}$

$= n \cdot \left(\dfrac{1}{{}_nq_x} - 1\right) \Big/ \dfrac{1}{{}_nm_x}$, from (2.7.2) and (2.7.6)

$= n \cdot \dfrac{{}_nm_x}{{}_nq_x} - n \cdot {}_nm_x$

or ${}_nq_x[1 - {}_na_x\, {}_nm_x + n \cdot {}_nm_x] = n \cdot {}_nm_x$, which gives (2.13.1) //

2.13.2 Large Sample S.E. of e_x^0

From (2.7.10) we have

$$e_x^0 = \frac{T_x}{l_x} = \frac{T_x}{l_x} - \frac{\sum_{t=y}^{\infty} L_t}{l_x} + \frac{\sum_{t=y+n}^{\infty} L_t}{l_x} + \frac{{}_nL_y}{l_x} \text{ (identically)}$$

$$\therefore \quad \sum_{t=y}^{\infty} L_t = \sum_{t=y}^{y+n-1} L_t + \sum_{t=y+n}^{\infty} L_t$$

$$\text{or} \quad \sum_{t=y}^{\infty} L_t - \sum_{t=y+n}^{\infty} L_t = {}_nL_y, \text{ from (2.7.5)}$$

$$\text{or} \quad e_x^0 = \frac{T_x - T_y}{l_x} + \sum_{t=y+n}^{\infty} \frac{L_t}{l_x} + \frac{{}_nL_y}{l_x}, \text{ from (2.7.8)} \tag{2.13.3}$$

Also, from (2.13.2)

$${}_nL_y = n \cdot l_{y+n} + {}_na_y \, {}_nd_y \tag{2.13.4}$$

$$\Rightarrow \quad \frac{{}_nL_y}{l_y} = n \cdot \frac{l_{y+n}}{l_y} + {}_na_y \frac{{}_nd_y}{l_y} = n \cdot (1 - {}_nq_y) + {}_na_y \, {}_nq_y, \text{ from (2.7.2)}$$

$$\therefore \quad \frac{{}_nL_y}{l_x} = \frac{{}_nL_y}{l_y} \cdot \frac{l_y}{l_x} = \frac{\{n \cdot (1 - {}_nq_y) + {}_na_y \, {}_nq_y\} l_y}{l_x} \tag{2.13.5}$$

$$\text{Also} \quad \frac{\sum_{t=y+n}^{\infty} L_t}{l_x} = \frac{T_{y+n}}{l_x} \text{ (from (2.7.8))} = \frac{T_{y+n}}{l_{y+n}} \cdot \frac{l_{y+n}}{l_y} \frac{l_y}{l_x}$$

$$= e_{y+n}^0 (1 - {}_nq_y) \frac{l_y}{l_x}, \text{ from (2.7.10) and (2.7.2)} \tag{2.13.6}$$

Substituting from (2.13.5), (2.13.6) in (2.13.3) we get

$$e_x^0 = \frac{T_x - T_y}{l_x} + \frac{l_y}{l_x} [e_{y+n}^0 (1 - {}_nq_y) + n \cdot (1 - {}_nq_y) + {}_na_y \, {}_nq_y] \tag{2.13.7}$$

Note that $(T_x - T_y)/l_x$ and e_{y+n}^0 are independent of ${}_nq_y$ and we assume that ${}_na_y$ is also independent of ${}_nq_y$. Then

$$\frac{\partial e_x^0}{\partial {}_nq_y} = -\frac{l_y}{l_x} [e_{y+n}^0 + n - {}_na_y] \tag{2.13.8}$$

By the approximate S.E. formula for large sample (A.2.18) we have

$$\text{Var}(e_x^0) \simeq \sum_{y \geq x} \left(\frac{\partial e_x^0}{\partial {}_nq_y} \right)^2 \text{Var}({}_n\hat{q}_y), \tag{2.13.9}$$

noting that e_x^0 depends on all ${}_nq_y$ for $y \geq x \, \forall \, x$ and since ${}_nq_x$ are independent among themselves and so variance of the sum = sum of variances. Now

$$\text{Var}({}_nq_y) = \frac{{}_nq_y (1 - {}_nq_y)}{l_0 \; {}_yp_0}, \quad \text{see (2.12.14)} \tag{2.13.10}$$

$$\because \quad \text{Var}(\hat{p}) = \text{Var}(1 - \hat{p}) = \text{Var}(\hat{q}).$$

Substituting from (2.13.8) and (2.13.10) in (2.13.9), the large sample S.E. of e_x^0 is given by

$$\text{S.E.}(e_x^0) \simeq \sqrt{\sum_{y \geq x} \left(\frac{l_y}{l_x} \right)^2 (e_{y+n}^0 + n - {}_na_y)^2 \frac{{}_nq_y (1 - {}_nq_y)}{l_0 \; {}_yp_0}} \tag{2.13.11}$$

2.13.3 Large Sample Variances of $\hat{q}_i$ and ${}_{j-i}\hat{p}_i$

We shall derive large sample variances of $\hat{q}_i$ and ${}_{j-i}\hat{p}_i$ for (i) current life table and (ii) Generation (Cohort) life table. But first we shall derive the following relation

$$\hat{q}_i = n_i M_i/[1 + (1 - a_i) n_i M_i], \tag{2.13.12}$$

where n_i is the length of the age interval (x_i to $x_i + n_i$), M_i is the central rate of mortality (for natural population) and a_i is the average number of years lived in the interval (x_i to $x_i + 1$) for those who die in it.

Note 2.6: Current life table is life table for natural population where as the Generation life table is for the hypothetical population (cohort).

We may note that the variances of $\hat{q}_i$ and $\hat{p}_i$ are equal i.e.

$$\sigma^2_{q_i} = \text{Var}(\hat{q}_i) = \text{Var}(1 - \hat{p}_i) = \text{Var}(\hat{p}_i) = \sigma^2_{\hat{p}_i} \tag{2.13.13}$$

and consequently their sample variances are also equal i.e.

$$S^2_{\hat{q}_i} = S^2_{\hat{p}x}. \tag{2.13.14}$$

(i) Current Life Table

For a current life table

$$\hat{q}_i = D_i/N_i, \tag{2.13.15}$$

where D_i is the number of deaths is (x_i, x_{i+1}) given that N_i persons are alive at age x_i.

If we assume that all N_i individuals in the population have the same probability of dying is the interval then D_i is a binomial variate and $\hat{q}_i$ is the estimated binomial proportion. Hence,

$$\hat{\text{V}}\text{ar}(D_i) = S^2_{D_i} = N_i \hat{q}_i (1 - \hat{q}_i)$$

and $$\text{Var}(\hat{q}_i) = S\hat{q}_i = \text{Var}\left(\frac{D_i}{N_i}\right) = \frac{N_i \hat{q}_i (1 - \hat{q}_i)}{N_i^2}$$

$$= \frac{\hat{q}_i(1 - \hat{q}_i)}{N_i} = \frac{\hat{q}_i^2(1 - \hat{q}_i)}{D_i},$$

$$\because \quad N_i = \frac{D_i}{\hat{q}_i}, \quad \text{from (2.13.15).} \tag{2.13.16}$$

The Central rate of mortality is defined as

$$m_x = \frac{d_x}{L_x}, \quad \text{(see (2.5.6))}.$$

Now here person years lived by the population N_i is the period

$$(x_i \text{ to } x_i + n_i) = (N_i - D_i) n_i + a_i n_i D_i = P_i \text{ (say)}$$

$$\therefore \quad M_i = \frac{D_i}{P_i} = \frac{D_i}{(N_i - D_i) n_i + a_i n_i D_i} \tag{2.13.17}$$

Substituting the value of D_i from (2.13.15) in (2.13.17) and rearranging it we get

$$\hat{q}_i = \frac{n_i M_i}{1 + (1 - a_i) n_i M_i},$$

which is the required relation (2.13.12).

Now substituting the values of $\hat{q}_i$ from (2.13.18) in (2.13.16)

$$\text{Var}(\hat{q}_i) = \frac{n_i^2 M_i^2}{\{1+(1-a_i)n_iM_i\}^2} \frac{\left\{1 - \dfrac{n_iM_i}{1+(1-a_i)n_iM_i}\right\}}{D_i}$$

$$= \frac{n_iM_i^2(1-a_in_iM_i)}{\{1+(1-a_i)n_iM_i\}^3 D_i} = \frac{n_iM_i(1-a_in_iM_i)}{\{1+(1-a_i)n_iM_i\}^3\left(\dfrac{D_i}{M_i}\right)}$$

$$\text{Var}(\hat{q}_i) = \frac{n_iM_i(1-a_in_iM_i)}{\{1+(1-a_i)n_iM_i\}^3 P_i}, \quad \text{from (2.13.17).} \tag{2.13.18}$$

If $n_i = 1$, thus replacing i by x is (2.13.18) to get the formula for single year of age

$$\text{Var}(\hat{q}_x) = \frac{M_x(1-a_xM_x)}{P_x\{1+(1-a_x)M_x\}^3}. \tag{2.13.19}$$

Taking $a_x = \dfrac{1}{2}$ we have

$$\text{S.E. } (\hat{q}_x) = \sqrt{\text{Var}(\hat{q}_x)} = \sqrt{\frac{4M_x(2-M_x)}{P_x(2+M_x)^3}}. \tag{2.13.20}$$

The sampling Variance Var $({}_{j-i}p_i)$

We have ${}_{j-i}\hat{p}_i = \hat{p}_x\hat{p}_{x+1}\ldots\hat{p}_{j-1}$, $j > i, (i, j = 0, 1, 2, \ldots)$, from (2.10.2) (2.13.21)

where the sample proportions $\hat{p}_i = 1 - \hat{q}_i$, $i = 0, 1, 2, \ldots$ (2.13.22)

are based on the estimates of the age specific death rates and $\hat{p}_i$ and $\hat{p}_k (i \neq k)$ are independent. Therefore by the formula (A.2.18) of large sample variance of a function of an estimator (remembering that ${}_j\hat{p}_i$ is a function of $\hat{p}_1, \hat{p}_2, \ldots, \hat{p}_{j-1}$)

$$\text{Var}({}_j\hat{p}_i) \simeq \sum_{k=1}^{j-1} \left(\frac{\partial_j\hat{p}_i}{\partial\hat{p}_k}\right)^2_{{}_j\hat{p}_i = {}_jp_i} \text{Var}(\hat{p}_k). \tag{2.13.23}$$

Now $\log_j \hat{p}_i = \log\hat{p}_1 + \log\hat{p}_2 + \ldots + \log\hat{p}_{j-1}$

$$\therefore \quad \frac{\partial \log_j \hat{p}_i}{\partial\hat{p}_k} = \frac{1}{\hat{p}_k} \quad \text{or} \quad \frac{1}{{}_j\hat{p}_i}\frac{\partial_j\hat{p}_i}{\partial\hat{p}_k} = \frac{1}{\hat{p}_k} \quad \text{or} \quad \frac{\partial_j\hat{p}_i}{\partial\hat{p}_k} = {}_j\hat{p}_i \cdot \frac{1}{\hat{p}_k},$$

$$\therefore \quad \left(\frac{\partial_j\hat{p}_i}{\partial\hat{p}_k}\right)^2_{{}_j\hat{p}_i = {}_jp_i} = ({}_jp_i)^2\left(\frac{1}{p_k}\right)^2. \tag{2.13.24}$$

Substituting in (2.13.23) from (2.13.24) we have

$$\text{Var}({}_j\hat{p}_i) = ({}_jp_i)^2 \sum_{k=1}^{j-1}(p_k)^{-2}\text{Var}(\hat{p}_k) \tag{2.13.25}$$

$$\therefore \quad \text{S.E. } ({}_j\hat{p}_i) = {}_jp_i\sqrt{\sum_{k=1}^{j-1}(p_k)^{-2}\text{Var}(\hat{p}_k)}. \tag{2.13.26}$$

(ii) Cohort (Generation) Life Table

In generation life table, $\hat{q}_i = \dfrac{d_i}{l_i}$ and $\hat{p}_i = \dfrac{l_{i+1}}{l_i}$.

These are ordinary binomial proportions with sampling variances

$$\operatorname{Var}(\hat{q}_i) = \operatorname{Var}(\hat{p}_i) = \frac{\hat{p}_i \hat{q}_i}{l_i} \tag{2.13.27}$$

Note that ${}_j\hat{p}_i$ is a general form of $\hat{p}_i$, as ${}_{i+1}\hat{p}_i = \hat{p}_i$. Hence,

$$\operatorname{Var}({}_j\hat{p}_i) = \frac{1}{l_i}\, {}_jp_i \cdot (1 - {}_j\hat{p}_i), i < j-1, \; (i, j = 0, 1, 2,) \tag{2.13.28}$$

Further, as in (2.13.25)

$$\operatorname{Var}({}_j\hat{p}_i) = ({}_jp_i)^2 \sum_{k=1}^{j-1} (p_k)^{-2} \operatorname{Var}(\hat{p}_k) \tag{2.13.29}$$

$$= ({}_jp_i)^2 \sum_{k=1}^{j-1} p_k^{-2} \frac{1}{l_k} \hat{p}_k (1 - \hat{p}_k), \text{ from (2.13.27).} \tag{2.13.30}$$

Hence, from (2.13.28) and (2.13.30)

$$\frac{1}{l_i}\, {}_j\hat{p}_i (1 - {}_j\hat{p}_i) = {}_jp_i^2 \sum_{k=1}^{j-1} p_k^{-2} \frac{1}{l_k} \hat{p}_k (1 - \hat{p}_k). \tag{2.13.29}$$

Thus from (2.13.25) and (2.13.28) we see that the result of cohort life table and current life table is same.

2.14 Mathematical Models in Mortality—Graduation of Mortality Curves

If we calculate the mortality rates q_x or m_x from ordinary complete enumeration of data then one notable feature of the q_x or m_x curves plotted against x that may be noted is that the curves are unlikely to be smooth. The tendency of the curves, so drawn, will be to move or fluctuate around some smooth pattern of curves. The purpose of mortality graduation is to obtain the smooth pattern. In other words we are required to obtain smooth mathematical functions of x for m_x (or q_x) which correspond to the observed values of m_x (or q_x) fairly closely; so that the same mathematical functions may be used, safely, for prediction purposes.

We consider here two models (or functions) viz. Makeham's model and Gompertz's model.

2.14.1 Makeham's Model

Makeham (1860), for the development of this model, assumed that the mortality is caused due to either of the two factors viz (a) Accident and (b) Age. He further assumed that the risk of mortality due to accident is independent of the age of the individual. His argument was that although young persons are less liable to accidents than their elderly counterparts; but it is also true that young persons take more risks. Hence, it is not unjustified that the risk due to accident remains more or less invariant over age. The second factor viz the effect of aging is to increase the level of mortality.

Hence, we may write the deterministic model of Makeham, known as Makeham's graduation law as

$$\mu(x) = A + \frac{B}{f(x)}, \tag{2.14.1}$$

where $\mu(x)$ is the instantaneous force of mortality or hazard function at the age x and $f(x)$ represents the resistance function (on resistivity) at age x, which decreases with x; A and B are age independent constants.

Since $f(x)$ is a decreasing function of x,

$$\therefore \quad \frac{df(x)}{dx} = -a\,f(x), \quad \text{where } a(>0) \text{ is a constant}$$

$$\Rightarrow \quad \frac{d\log f(x)}{dx} = -a \Rightarrow \log f(x) = -ax + \log b, \text{ where } b \text{ is an integration constant.}$$

$$\Rightarrow \quad f(x) = be^{-ax}.$$

$$\text{Hence, } \mu(x) = A + \frac{B}{b}e^{ax}, \text{ from (2.14.1)} \tag{2.14.2}$$

$$\text{or From (2.6.9)} \; -\frac{1}{l(x)}\frac{d\,l(x)}{dx} = -\frac{d\log l(x)}{dx} = A + \frac{B}{b}e^{ax},$$

$$\Rightarrow \quad \log l(x) = -Ax - \frac{B}{b}\frac{e^{ax}}{a} + \log K, \quad \text{where } K \text{ is an integration constant.}$$

$$\Rightarrow \quad l(x) = KS^x g^{C^x} \tag{2.14.3}$$

$$\text{where } e^{-A} = S,\; g = \exp\left(-\frac{B}{ba}\right) \quad \text{and} \quad e^a = C$$

$$\text{Now} \quad q_x = 1 - p_x = 1 - \frac{l(x+1)}{l(x)}, \text{ from (2.5.4)}$$

$$= 1 - KS^{x+1}g^{C^{x+1}}/KS^x g^{C^x} = 1 - Sg^{C^x(C-1)} \tag{2.14.4}$$

$$\text{Now,} \quad m_x = \frac{2q_x}{2-q_x}, \text{ from (2.6.7)}$$

$$\Rightarrow \quad m_x = 2[1 - Sg^{C^x(C-1)}]/[1 + Sg^{C^x(C-1)}] \tag{2.14.5}$$

2.14.2 Gompertz' Model

If we neglect the accident factor 'A' in the Makeham's model (2.14.1), we get Gompertz' model as

$$\mu(x) = B/f(x) = \frac{B}{b}e^{-ax}, \text{ from (2.14.2)} \tag{2.14.6}$$

$$\text{Then,} \quad l(x) = Kg^{C^x}, \text{ from (2.14.3)}, \; \because A = 0 \Rightarrow S = e^{-A} = 1, \tag{2.14.7}$$

$$q_x = 1 - g^{C^x(C-1)}, \text{ from (2.14.4)}, \; \because S = 1 \tag{2.14.8}$$

$$\text{and} \quad m_x = 2[1 - g^{C^x(C-1)}]/[1 + g^{C^x(C-1)}]. \tag{2.14.9}$$

2.14.3 Test For Justification of Graduation

Let $x, x+n, x+2n, \ldots, x+kn$ be $(k+1)$ central ages each at period n apart and the corresponding values of $l(\cdot)$ are given as $l(x), l(x+n), \ldots, l(x+kn)$, respectively. Using these data we require to judge whether the graduation by a particular model (Makeham or Gompertz) is justified or not.

1. Makeham's Model

Here, we have $l(x) = KS^x g^{C^x}$ (2.14.3)

Then for the first four observations we have

$$\left.\begin{aligned} &\log l(x) = \log K + x\log S + C^x \log g, \\ &\log l(x+n) = \log K + (x+n)\log S + C^{x+n}\log g, \\ &\log l(x+2n) = \log K + (x+2n)\log S + C^{x+2n}\log g \\ \text{and}\quad &\log l(x+3n) = \log K + (x+3n)\log S + C^{x+3n}\log g \end{aligned}\right\} \tag{2.14.10}$$

Then, $\log l(x+n) - \log l(x) = \Delta \log l(x) = n\log S + C^x(C^n - 1)\log g$ (2.14.11)

Similarly

$$\Delta \log l(x+n) = n\log S + C^{x+n}(C^n - 1)\log g$$

$$\Delta \log l(x+2n) = n\log S + C^{x+2n}(C^n - 1)\log g$$

$$\therefore \quad \Delta \log(x+n) - \Delta\log l(x) = \Delta^2 \log l(x) = C^x(C^n-1)^2 \log g \tag{2.14.12}$$

$$\text{and}\quad \Delta^2 \log l(x+n) = C^{x+n}(C^n - 1)^2 \log g$$

$$\Rightarrow \quad \frac{\Delta^2 \log l(x+n)}{\Delta^2 \log l(x)} = \frac{C^{x+n}}{C^x} = C^n \tag{2.14.13}$$

Thus, for four consecutive values of $(x, l(x))$ we get one value of C^n (and hence C). Hence, for $(k+1)$ values of $(x, l(X))$ we get $(k-3)$ values of C^n. Now, we test the graduation based on agreement of all the $(k-3)$ values of C^n (or C). This may justify the graduation of mortality rates by Makeham model.

2. Gompertz' Model

Here, $l(x) = Kg^{C^x}$ (2.14.7)

For the first three observations from (2.14.7) we get

$$\log l(x) = \log K + C^x \log g,$$

$$\log l(x+n) = \log K + C^{x+n}\log g$$

$$\text{and}\quad \log l(x+2n) = \log K + C^{x+2n}\log g$$

$$\therefore \quad \Delta \log l(x) = C^x(C-1)\log g$$

$$\text{and}\quad \Delta \log l(x+n) = C^{x+n}(C-1)\log g$$

$$\therefore \quad \frac{\Delta \log l(x+n)}{\Delta \log l(x)} = C^n \tag{2.14.14}$$

Thus, from three values of $(x, l(x))$ we get one value of C^n (or C). Hence, from $(k+1)$ values of $(x, l(x))$ we shall get $(k-2)$ values of C^n (or C). If these values are close to each other, then graduation by Gompretz's law is justified.

2.14.4 Fitting of Makeham and Gompertz Curves

The approach of fitting of Makeham an Gompertz curves used by Actuarists can be broadly be classified into two categories viz.

(a) Estimating the parameters of Makeham (or Gompertz) curve by choosing number of equidistant pivotal points equal to the number of parameters of the curve and then solve the equations containing the parameters as unknowns equated, respectively, to the observable $l(x)$'s, and their differences of various orders like $\Delta l(x)$, $\Delta^2 l(x), \ldots, \Delta^{k-1} l(x)$, where k is the number of parameters of the graduation formula which is equal to the number of equations.

(b) The method of summation applied on $d(x)$ and $P(x)$, where $d(x)$ and $P(x)$ are the observed number of deaths and population size in the age group x to $(x+1)$, respectively, which are used to estimate the central rate of mortality $m(x) = \dfrac{d(x)}{P(x)}$ and thus the parameters of the curve are estimated by using the relation $m(x) \simeq \mu\left(x+\frac{1}{2}\right)$.

Different methods based on each approach are given, for two curves, below:

1. Fitting of Makeham's Curve

(i) Approach (a): Method I

The Makeham's curve $l(x) = KS^x g^{C^x}$ has four parameters viz K, S, g and C. So four estimating equations, given in (2.14.10), would be necessary to get their estimates.

From (2.14.13) we shall get a value of C; and in all $(k-3)$ values of C. If they are fairly close to each other then we estimate C from their mean C (say). Putting this value of C in (2.14.12) we get the estimate of g. Similarly substituting estimates of C and g in (2.14.11) we get estimate of S and then from first among (2.14.10), the estimate of K.

Remarks 2.4: The method is defective in the sense that entire data is not used, only a part of the data is used. Thus, although C is estimated using all the values $l(x), l(x+n), l(x+2n), l(x+3n)$ but g is estimated from only $l(x)$, $l(x+n)$, $l(x+2n)$ and S from $l(x)$, $l(x+n)$ whereas K is estimated from $l(x)$ only.

(ii) Approach (a): Method II

This method is an improvement over Method I. Here, C is estimated exactly as in Method I. But to obtain the estimates of the other parameters we group the observations into three distinct classes; for this number of observations must be a multiple of 3, say $3r$ (if necessary we may leave one or two extreme observations for this).

Let $$S_1 = \sum_{i=0}^{r-1} w_i \log l(x+i), \tag{2.14.15}$$

where $w_1, w_2, \ldots, w_{r-1}$ be a set of arbitrary weights subject to the restriction $\sum_{i-0}^{r-1} w_i = 1$.

Similarly, $S_2 = \sum_{i=r}^{2r-1} w_{i-r} \log l(x+i)$ and $S_3 = \sum_{i=2r}^{3r-1} w_{i-2r} \log(x+i)$

On substituting the values of $l(x+i)$ from (2.14.3) viz

$$\log l(x+i) = \log K + (x+i)\log S + C^{x+i}\log g \tag{2.14.16}$$

we have $$S_1 = \log K + B'C^x \log g + A' \log S,$$
$$S_2 = \log K + B'C^{x+r} \log g + (r + A') \log S,$$
and $$S_3 = \log K + B'C^{x+2r} \log g + (2r + A') \log S,$$

where $$A' = \sum_{i=0}^{r-1}(x+i)w_i \quad \text{and} \quad B' = \sum_{i=0}^{r-1} C^i w_i. \tag{2.14.17}$$

A' and B' can be easily obtained from (2.14.17) given C's which has been obtained from the entire range of data.

Now, $S_2 - S_1 = \Delta S_1 = C^x(C^r - 1)B' \log g + r \log S$ (2.14.18)

$$\Delta S_2 = C^{x+r}(C^r - 1)B' \log g + r \log S$$

and $\therefore \quad \Delta S_2 - \Delta S_1 = \Delta^2 S_1 = C^x(C^r - 1)^2 B' \log g$ (2.14.19)

Given C we can obtained g from (2.14.19) and then S from (2.14.18).

Further, from (2.14.16)

$$\sum_{i=0}^{3r-1} \log l(x+i) = 3r \log K + \sum_{i=0}^{3r-1} C^{x+i} \log g + \sum_{i=0}^{3r-1} (x+i) \log S. \tag{2.14.20}$$

Knowing the values of C, g and S we can get K from (2.14.20)

As regard the choice of the weights $w_0, w_i, \ldots, w_{r-1}$ we may take the coefficients of the binomial expansion as a basis of the weights; the weights $w_0, w_1, \ldots, w_{r-1}$ may be taken as

$$w_0 = mC_0/2^m, w_1 = mC_1/2^m, \ldots, w_{r-1} = mC_m/2^m \tag{2.14.21}$$

$$\because \quad \sum_{i=0}^{m} \binom{m}{i} = 2^m, \quad \text{where } m = r - 1.$$

(iii) Approach (b): Method of Summation

We have $m(x) = \mu\left(x + \frac{1}{2}\right)$ from (2.6.19) where $m(x) = \dfrac{d(x)}{P(x)}$ = the central rate of mortality at the age x, where

$d(x)$ = Number of deaths between the x to $(x+1)$ and

$P(x)$ = Population in the age sector x to $(x+1)$.

N.B. 2.7 L_x is the life table population and is estimated by $P(x)$.

Thus $\dfrac{d(x)}{P(x)} = \mu\left(x + \dfrac{1}{2}\right) = A + BC^{x+\frac{1}{2}}$, as in (2.14.2) under Makeham's law (2.14.22)

$$\Rightarrow \quad d(x) = AP(x) + BP(x)C^{x+\frac{1}{2}} \tag{2.14.23}$$

$$\Rightarrow \quad \sum_x d(x) = A \sum_x P(x) + BC^{\frac{1}{2}} \sum_x C^x P(x) \tag{2.14.24}$$

Multiplying (2.14.23) by C^x and then summing we get

$$\sum C^x d(x) = A \sum_x C^x P(x) + BC^{\frac{1}{2}} \sum_x C^{2x} P(x) \tag{2.14.25}$$

We first estimate C as in Approach (a) Method I. Substituting this value of C in (2.14.24) and (2.14.25) and solving them for A and B we get the estimates of A and B. This is known as the method summation.

2. Fitting of Gompertz Curve

Approach (a): Method I

We have $l(x) = Kg^{C^x}$ (2.14.7)

$$\Rightarrow \quad \log l(x) = \log K + C^x \log g$$

As in (2.14.14), $C^n = \{\Delta \log l(x+n)\}/\{\Delta \log l(x)\}$. Then we estimate C^n by the mean of $(k-2)$ values of C^n as

$$C^n = \frac{1}{k-2}\sum_x \{\Delta \log l(x+n)\}/\{\Delta \log l(x)\} \tag{2.14.26}$$

and then using $\Delta \log l(x) = C^x(C^n - 1)\log g$, '$g$' is estimated.

Finally, from $\sum_{i=0}^{k-1} \log l(x+i) = k \log K + \log g \sum_{i=0}^{k-1} C^{x+i}$, '$K$' is estimated.

Approach (a): Method II

As in Method I we obtain the estimate of C^n (i.e. C) from (2.14.26).

Next we divide the data into two groups. Following the same notations as per Approach (a) Method II for Makeham's curve, we have

$$S_1 = \log K + C^x B' \log g \tag{2.14.27}$$

$$S_2 = \log K + C^{x+r} B' \log g,$$

$$\Rightarrow \ \Delta S_1 = B' C^x (C^r - 1)\log g \tag{2.14.28}$$

where B' is given by, $B' = \sum_{i=0}^{r-1} C^i w_i$ see (2.14.17).

Knowing C and B', g is given by (2.14.28) and K is obtained from (2.14.27).

Approach (b): Method of Summation

As before

$$m(x) = \frac{d(x)}{P(x)} \quad \text{and} \quad m(x) \simeq \mu\left(x + \frac{1}{2}\right) = BC^{x+\frac{1}{2}}, \text{ see (2.14.6)}$$

$$\therefore \quad \frac{d(x)}{P(x)} = BC^{x+\frac{1}{2}}$$

$$\Rightarrow \quad \sum_x d(x) = BC^{\frac{1}{2}} \sum_x C^x P(x) \tag{2.14.29}$$

Knowing C^n from (2.14.26) we can obtain the estimate of B directly from (2.14.29).

Theorem 2.14.1 A necessary and sufficient condition (NASC) for

$$\mu(x) = A + BC^x \tag{2.14.30}$$

where A, B, C are constants, is that

$$\text{Colog}\, p_x = \alpha + \beta C^x \tag{2.14.31}$$

where α, β are constants and Colog $z = -\log z$

Proof: The Condition is Necessary.

We have $\mu(x) = A + BC^x$

$$\Rightarrow \quad -\frac{d\log(x)}{dx} = A + BC^x, \text{ from (2.6.9)}$$

$$\Rightarrow \quad -\log(x) = Ax + \frac{BC^x}{\log C} + K, \ K \text{ is a constant of integration}$$

$\Rightarrow \quad -\log l(x) = Ax + B'C^x + K, \quad \text{where} \quad B' = B/\log C$

$\Rightarrow \quad -\log l(x+n) = A(x+n) + B'C^{x+n} + K$

$\Rightarrow \quad -[\log l(x+n) - \log l(x)] = -\log p_x = An + B'C^x(C^n - 1)$

or $\quad \text{Colog}\, p_x = An + B'C^x(C^n - 1)$

For $\quad n = 1, \text{Colog}\, p_x = A + B'(C-1)C^x = \alpha + \beta C^x,$

where $\quad \alpha = A \quad \text{and} \quad \beta = B'(C-1).$

The Condition is sufficient

We have, $\dfrac{l(x)}{l(0)} = \prod_{j=1}^{x} \dfrac{l(j)}{l(j-1)} = \prod_{j=1}^{x} p_{j-1}, \quad \text{from (2.5.4)}$

$$\Rightarrow \quad \log l(x) = \log l(0) - \sum_{j=1}^{x} \text{Colog}\, p_{j-1} \tag{2.14.32}$$

Now since (2.14.31) is assumed to hold i.e.

$$\text{Colog}\, p_{j-1} = \alpha + \beta C^{j-1}$$

$$\Rightarrow \quad \sum_{j=1}^{x} \text{Colog}\, p_{j-1} = \alpha x + \beta \frac{(C^x - 1)}{(C-1)} \tag{2.14.33}$$

Substituting from (2.14.33) is (2.14.32) we get

$$\log l(x) = \log l(0) + \alpha x + \beta(C^x - 1)/(C-1)$$

$$\Rightarrow \quad \frac{d \log l(x)}{dx} = \alpha + \beta C^x \log C/(C-1)$$

$$\Rightarrow \quad \mu(x) = A + BC^x \quad \text{where} \quad A = -\alpha, B = \beta \log C/(1-C). \quad //$$

Corollary: We can get a relation of type (2.14.31) for ${}_xp_0$ and ${}_np_x$ also.

We have $\quad {}_xp_0 = \exp\left\{-\int_0^x \mu(u)du\right\},$

$$\Rightarrow \quad \log {}_xp_0 = -\int_0^x \mu(u)du = -\int_0^x [A + BC^u]du \text{ from (2.14.30)}$$

$$= -A \cdot x - B\frac{C^x}{\log C} + \frac{B}{\log C}$$

$$\Rightarrow \quad \text{Colog}\, {}_xp_0 = \alpha \cdot x + \beta(C^x - 1) \tag{2.14.34}$$

where $\quad \alpha = A, \beta = B/\log C$

Similarly, $\quad {}_np_x = \exp\left\{-\int_x^{x+n} \mu(u)du\right\}$

$$\Rightarrow \quad \text{Colog}\, {}_np_x = \alpha \cdot n + \beta C^x(C^n - 1) \tag{2.14.35}$$

Further $\quad \text{Colog}\, {}_np_x = -\log(1 - {}_nq_x) \simeq {}_nq_x + 0({}_nq_x^2)$

$$\text{Then} \quad {}_nq_x \simeq \text{Colog}_n p_x = \alpha \cdot n + \frac{B}{\log C} C^{x(C^n - 1)} \text{ from (2.14.3)} \tag{2.14.36}$$

References

1. Berclay, G.W. (1968): Techniques of Population Analysis; John Wiley and Sons, New York.

2. Chiang, C.L. (1968): Introduction to Stochastic Processes in Biostatistics. John Wiley and Sons, New York.

3. Cox, D.R. (1976): Demography, Cambridge University Press.

4. Jaffee, A.J. (1966): Handbook of Statistical Methods for Demographers; U.S. Govt. Printing press. Washington.

5. Keyfitz Nathan (1966): A life table that agrees with the data I,—Journal of American Statistical Associates, Vol. 61, pp. 305–12.

6. Keyfitz Nathan (1968): A life table that agrees with the data II—Journal of American Statistical association, Vol. 63, pp. 1253–68.

7. Keyfitz N. (1977): Introduction to the Mathematics of Population—with Revisions, Addition-Wesley, London.

8. Logan, W.P. (1953): The measurement of Infant mortality. Population Bulletin of United Nations, No. 3, pp. 30–55, New York.

9. Mathen, K.K. and Poti, S.J. (1954): An adjustment for the effect of changing birth rates on Infant mortality rates—Sankhya Vol. 15, Part 4, pp. 417–422.

10. Makeham, W.H. (1866): On the law of mortality and Construction of Annuity Tables. Journal of Institute of Actuaries, Vol. 8.

11. Poti, S.J. and Biswas, S. (1963)—A study of child health during the first year of life, Sankhya, Vol. 26, Series I and II, pp. 35–120.

12. Rogers, A. (1975): Introduction to multiregional Demography—John Wiley and Sons, New York.

13. Spiegelman, M. (1980): Introduction to Demography (Revised edition), Harvard University press, Cambridge.

14. Telly, H. (1950): Acturial Statistics, vol 3, Statistics and graduation. The Institute of Actuaries and the faculty of Actuaries, Cambridge.

15. Wolfenden, H.H. (1954): Population Statistics and their Compilation, The University of Chicago Press, Chicago.

16. United Nations (1955): A Sex Patterns of Mortality Model. Life Tables for Under Developed Countries, New York.

17. Whittikar, E.T. and Robinson, G. (1944): Calculus of Observations, D. Van Nostrand Company, New York.

3

Techniques of Demographic Analysis: Fertility

3.0 Introduction

Fertility corresponds to the additive component of the growth of a population. The inherent biological capacity to give births is known as "fecundity"; whereas the realized level of producing births under a given socio-economic and cultural set up is known as 'Fertility'. Unlike mortality, fertility measures are estimable mostly through Female Population who are exposed to the risk of child bearing. Again certain measures of fertility as 'Net Reproduction Rate' are associated with the mortality condition in the sense that to what extent mortality in the child bearing period affects total fertility performance of the mothers. Below we present a few measures of fertility, most of which, can be taken up independent of mortality situation.

3.1 Indices of Fertility Measures

The following fertility rates are indices of fertility measures which provide the measures of the rate of growth of population.

(i) Crude Birth Rate (GBR)
(ii) General Fertility Rate (GFR)
(iii) Age-Sex-Specific Fertility Rate (ASSFR)
(iv) Total Fertility Rate (TFR)
(v) Pearl's vital Index (PVI)
(vi) Gross reproduction Rate (GRR)
(vii) Net Reproduction Rate (NRR)

3.1.1 Crude Birth Rate

In a particular locality in the period Z (calender year), let the total number of live births be B^Z and the mean population measured in the middle of the calender year Z be P^Z. Then this fertility index is given by

$$\text{CBR} = \frac{B^Z}{P^Z} \times K, \text{ where } K = 10^3, \text{ usually.} \tag{3.1.1}$$

Remarks 3.1: (1) Z will always refer to the calender year Z. See Note 2.3.

(2) It is a very crude measure of fertility but easily understandable index and easy to calculate. It is not necessary to have the births in different sectors of the population.

(3) It is not a probability rate since whole population is not exposed to the risk of giving births. The whole of male population and the part of the female population not exposed to the risk of child bearing should have been excluded.

(4) CBR may not be a suitable index for comparing the fertility status of two communities. For example, two populations may have the same CBR. This does not mean that the two populations have the same fertility status. The age distribution of the female population or the distortion or imbalance in the sex ratios in the population may mask the actual fertility picture presented in the form of CBR.

3.1.2 General Fertility Rate

It is defined as

$$\text{GFR} = \frac{B^Z}{{}_rW^Z} \times K, \tag{3.1.2}$$

where B^Z = Total number of live births in the calender year Z

W^Z = Female population in the reproductive age group, (generally, 15 to 45 years); denotes the reproductive span.

Remarks 3.2: (1) It is a probability rate because the denominator corresponds to the female population who are exposed to the risk of child bearing.

(2) It does not enable us to study the variation of fertility by ages as it overlooks the age composition of the female population.

(3) it has also a defect of non-comparability in respect of time and place. It may so happen that for two places 'A' and 'B', general fertility rates are the same. Even in that case we can hardly say that the two places have the same fertility status; for the proportion of young females (who are under greater risk of giving birth) for 'A' may be different than from B.

3.1.3 Age-Specific Fertility Rate

Age-specific fertility rate between the age-group x to $(x+n)$ is given by

$$\text{ASFR} = \frac{{}_nB_x^Z}{{}_nW_x^Z} \times K, \tag{3.1.3}$$

where ${}_nB_x^Z$ = number of live births to women of age x to $x+n$ in the calender year Z

${}_nW_x^Z$ = number of women of age x to $x+n$ in the calender year Z

n = 5 (usually for the data collected), i.e., the total range is divided into intervals to 5 years.

Remarks 3.3: (1) It is a probability rate and is not a heterogeneous figure like CBR.

(2) It is seen that, for any community, generally, it starts at a low point, peaks somewhere between 20 to 29 years of age and after that decreases steadily. Thus, ASFR gives a highly positively spewed curve.

(3) It enable us to compare the fertility status of two different populations corresponding to the same age group.

(4) The fertility status of two populations cannot be compared on the basis of two sets of age specific fertility rates because it may be higher for certain age-groups and lower for other age-groups in one populations than in the other. As remarked above, ASFR cannot be used for over all comparison of two communities. For this the ASF Rates are to be combined in a single quantity; TFR is such a quantity.

3.1.4 Total Fertility Rate

It is define as,

$$\text{TFR} = K \sum_{x=w_1}^{w_2} n \,\{\text{age specific fertility rate between ages } x \text{ to } (x+n)\},$$

where w_1 and w_2 are the lower and upper bounds of the child bearing age groups

$$= K \sum_{x=w_1}^{w_2} \left[\frac{{}_nB_x^Z}{{}_nW_x^Z}\right] \cdot n \tag{3.1.3}$$

$$= K \sum_{x=w_1}^{w_2} n \; {}_nf_x, \quad \text{where} \quad {}_nf_x = \frac{{}_nB_x^Z}{{}_nW_x^Z}. \tag{3.1.4}$$

The Physical interpretation of total fertility rate

Suppose 1,000 females are born at the same time and none of them die before reaching the end of the child bearing age interval. Then the total fertility rate (TFR) can be interpreted as the number of babies born to the cohort of 1000 women (all born in the same period) assuming that in each group (in the child bearing ages) they are subjected to the fertility conditions given by the observed age specific fertility rates.

3.1.5 Pearl's Vital Index

Pearl's Vital Index (PVI) is defined by

$$\text{PVI} = \frac{\text{No. of births in a given period of time}}{\text{No. of Deaths in a given period of time}} \times K \tag{3.1.5}$$

From the above, it is possible to employ PVI to find out whether the population is increasing or decreasing. As PVI is very simple to calculate Pearl has suggested the index for measuring Populations growth. But as may be found in the following lines it has some serious defects.

(1) PVI, no doubt, gives a measure, whether births exceed deaths or not. But our object is not merely to know this. We may, as well, want to know whether a Population has a tendency to increase or decrease and to measure the trend in Population growth. But this index certainly fails to provide such measures.

(2) For comparative purpose PVI is defective. Suppose for two countries I and II births B I $= B$ II holds but D I $> D$ II deaths. Then PVI I $<$ PVI II but this does not imply that the country II has a tendency to grow faster. The snag is obvious for we do not consider the age composition of either of the two countries.

(3) In Pearl's Vital Index we take into consideration the whole Population; but population can increase through females only, so that we should consider only female populations and their births and deaths.

The last point given in (3) is employed to construct two important fertility growth indices viz. (i) Gross Reproduction Rate (GRR) and (ii) Net Reproduction Rate (NRR).

3.1.6 Gross Reproduction Rate

The gross reproduction rate is defined as

$$\text{GRR} = \left\{ \sum n \frac{{}_nB_x^f}{{}_nW_x} \cdot \right\} K, \tag{3.1.6}$$

where (x to $x+n$) represents the age groups of female population.

${}_nW_x$ = number of females in the age group (x to $x+n$)

${}_nB_x^f$ = number of female births to mothers in the age group (x to $x+n$) in a year (calender year).

Physical interpretation of GRR

GRR gives the number of female babies that would be born to K women at same age when none of these women die before reaching the end of child bearing age interval and if all of them are subjected to the risk of giving births under the given fertility schedule. A value of GRR $= K' \Rightarrow$ under the existing conditions of fertility, K women would replace themselves in the next generation by K' women so that the GRR can be taken as the measure of population growth under the hypothesis of no mortality during the entire fertility span.

3.1.7 Net Reproduction Rate

Net reproduction rate (NRR) improves Gross reproduction rate (GRR) not only by considering the extent of replacement of one female cohort of newly born babies to another cohort formed of the female children born to these female cohort of the newly born babies (who were potential mothers); but also by considering the ability to produce children subject to the condition of survival of the mothers during the fertility span. More precisely, if

$f(x) = P$ [of a female aged x to produce a female child during the course of a calender year Z under the given fertility condition]

and $p(x) = P$ [that a newly born female baby will survive up to the age x under the given mortality schedule],

then

$$\text{NRR per women} = \int_{w_1}^{w_2} p(x)f(x)dx, \tag{3.1.7}$$

where w_1 and w_2 are the lower and the upper bounds of the fertility span.

If we require to estimate NRR from grouped data, then

$$\text{NRR per } K \text{ woman} = \sum_{x=w_1}^{w_2} \left[\frac{{}^Z_xB_x^f}{{}_nW_x} \right] \left(\frac{{}_nL_x}{l_0} \right) \tag{3.1.8}$$

where ${}^{Z}_{n}B^{f}_{x}$ = number of female born to women in the age group (x to $x+n$) in the calender year Z and

${}_{n}W_{x}$ = number of women in the age group (x to $x+n$).

Further, $\frac{{}_{n}L_{x}}{l_0}$ will represent the average number of years lived by the Cohort of newly born potential mother in the age group (x to $x+n$) over which the probability of having a female baby, $\frac{{}^{Z}_{n}B^{f}_{x}}{{}_{n}W_{x}}$, per year will be applied for a period of $\frac{{}_{n}L_{n}}{n}$ for K mothers will give rise to (3.1.8).

The Physical Interpretation of the NRR:

The NRR per K women will imply the total number of female babies born by K number of newly born female infants during their fertility span subject to their survival at the present mortality schedule throughout the fertility span.

Thus NRR can be taken as a growth Index. If NRR $> K$, we can conclude that the population is increasing and if NRR $< K$ we may conclude that the population is decreasing. In other words, K number of potential mothers will fail to produce sufficient number of future mothers to maintain the same level.

Hence, for rough work NRR may be used as the measure of replacement index of the population; replacement index is the quantitative measurement of one generation replaced by other generation.

Note that the following assumptions have been made for the estimation of NRR:

(1) The same survival factor (i.e. mortality) as given in the current life table valid for the previous decade will be applicable for the entire newly born babies throughout their life time i.e.

$$\left(\frac{l_x}{l_0}\right) = \frac{l(x,t)}{l(0,t)} = \text{constant i.e. independent of } t \tag{3.1.9}$$

(2) Same fertility pattern as available in the latest fertilely schedule. will be applicable for the newly born babies throughout their life time.

(3) the sex ratio $\left(= \frac{\text{number of males}}{\text{total population}}\right)$ should remain the same over all the years to come.

But the above assumptions are not always justified. If the mortality rate has a decreasing trend then the actual NRR $\geq$ estimated NRR on the basis of the above assumption and if the mortality is increasing then a reverse conclusion holds good.

3.2 A Relationship Between Crude Birth Rate, General Fertility Rate and Total Fertility Rate

Let, $c(x;t)$ = Observed proportion of females in the age group (x to $x+1$) at time t, (3.2.1)

and $f(x;t)$ = Observed proportion of females giving birth to female children in the age group (x to $x+1$) at time t (3.2.2)

Then, $\int_{\alpha}^{\beta} f(x;t)dx$ = Estimated total fertility rate at time $t = \hat{T}_f(t)$, say (3.2.3)

$$\text{and} \quad \int_\alpha^\beta c(x;t)f(x;t)dx = \text{Estimated female birth rate at time } t$$
$$= \hat{B}_f(t), \text{ say,} \tag{3.2.4}$$

where α and β represent the lower and the upper bound of the child bearing age. The correlation between $c(x;t)$ and $f(x;t)$, at a given t, is given by

$$\text{Cor }(c(x;t), f(x;t)|t) = \frac{E(c(x;t)f(x:t)|t) - E(c(x;t)|t)E(f(x;t)|t)}{\sqrt{\text{Var}(c(x;t)|t)}\sqrt{\text{Var}(f(x;t)|t)}}$$
$$= r_{c,f|t}, \text{ say} \tag{3.2.5}$$

We have, $E(c(x;t)f(x;t)|t) = \int_\alpha^\beta [c(x;t)f(x;t)|t]\phi_x dx,$

where ϕ_x is the probability distribution of the r.v. X, the child bearing age.

Also, $E(c(x;t)|t) = \int_\alpha^\beta [c(x;t)|t]\phi_x dx,$

$$E(f(x;t)|t) = \int_\alpha^\beta [f(x;t)|t]\phi_x dx,$$

$$\text{Var}(c(x;t)|t) = \int_\alpha^\beta [c^2(x;t)|t]\phi_x dx - [E(c(x;t)|t)]^2 = \sigma^2_{c/t},$$

$$\text{Var}[f(x;t)|t] = \int_\alpha^\beta [f^2(x;t)|t]\phi_x dx - [E(f(x;t)|t)]^2 = \sigma^2_{f/t}.$$

Let us assume that X is uniformly distributed in (α, β) i.e.

$$\phi_x = \frac{1}{\beta - \alpha}; \ \alpha \le x \le \beta = 0 \text{ otherwise} \tag{3.2.6}$$

Given $\hat{r}_{c,f|t} = \text{Cor}(c(x;t), f(x;t)|t) = \left[\frac{1}{\beta-\alpha}\int_\alpha^\beta [c(x;t)f(x;t)|t]dx\right.$

$$\left. - \frac{1}{\beta-\alpha}\int_\alpha^\beta [c(x;t)|t]dx \frac{1}{\beta-\alpha}\int_\alpha^\beta [f(x;t)|t]dx\right] \Big/ \sigma_{c|t}\sigma_{f|t}$$

$$\text{where } \sigma_{c|t} = \sqrt{\left[\left\{\frac{1}{\beta-\alpha}\int_\alpha^\beta [c(x,t)|t\right]^2 dx - \frac{1}{\beta-\alpha}\left[\int_\alpha^\beta [c(x,t)|t]dx\right]^2}$$

$$\text{and} \quad \sigma_{f|t} = \sqrt{\left[\left\{\frac{1}{\beta-\alpha}\int_\alpha^\beta [f(x,t)|t\right]^2 - \frac{1}{\beta-\alpha}\left[\int_\alpha^\beta [f(x,t)|t]dx\right]^2}$$

$$\Rightarrow \quad r_{c,f|t\sigma_{c|t}\sigma_{f|t}} = \left\{\frac{1}{\beta-\alpha}\int_\alpha^\beta [c(x;t)f(x;t)|t]dx\right.$$
$$- \frac{1}{(\beta-\alpha)^2}\int_\alpha^\beta [c(x;t)|t]dx \int_\alpha^\beta [f(x,t)|t]dx$$
$$= \frac{1}{\beta-\alpha}\hat{B}_f(t) - \frac{\hat{T}_t(t)}{(\beta-\alpha)^2}$$

where $\hat{B}_f(t) = \int_\alpha^\beta [c(x,t)f(x,t)|t]dx$

and $\hat{T}_f(t) = \int_\alpha^\beta [c(x,t)|t] \int_\alpha^\beta [f(x,t)|t]dx$

Again, estimated General fertility rate (GFR) is given by

$$\hat{G}_F(t) = \frac{N\int_\alpha^\beta [f(x;t)|t]dx}{N\int_\alpha^\beta [c(x;t)|t]dx} \text{ from (3.23) and (3.24)} \tag{3.2.7}$$

$$\Rightarrow \int_\alpha^\beta [c(x;t)|t]dx = \frac{\hat{T}_f(t)}{\hat{G}_f(t)} \tag{3.2.8}$$

Putting (3.5) in (3.4)

$$\Rightarrow [\hat{r}_{c,f|t}]\hat{\sigma}_{c|t}\hat{\sigma}_{f|t} = \frac{\hat{B}_f(t)}{\beta-\alpha} - \frac{\hat{T}_f(t)}{(\beta-\alpha)^2}\frac{\hat{T}_f(t)}{\hat{G}_F(t)} \tag{3.2.9}$$

$$\Rightarrow (\beta-\alpha)\hat{r}_{c,f|t}\hat{\sigma}_{c|t}\hat{\sigma}_{f|t} = \hat{B}_f(t) - \frac{(\hat{T}_f(t))^2}{\beta-\alpha}\frac{1}{\hat{G}_F(t)} \tag{3.2.10}$$

$$\Rightarrow \hat{B}_f(t) = \left[\hat{r}_{c,f|t}\hat{\sigma}_{c|t}\hat{\sigma}_{f|t}(\beta-\alpha) + \frac{(\hat{T}_f(t))^2}{(\alpha-\beta)}\frac{1}{\hat{G}_F(t)}\right] \tag{3.2.11}$$

which established a relation between birth rate, total fertility rate and general fertility rate.

A special case

$$\text{if} \quad \hat{r}_{c,f|t} = 0 \Rightarrow \hat{B}_f(t) = \frac{(\hat{T}_f(t))^2}{\beta-\alpha}\frac{1}{\hat{G}_F(t)} \tag{3.2.12}$$

3.3 Mathematical Models on Fertility

The origin of the systematic development of Stochastic models in India, representing the fertility behaviour under uncontrolled condition of fertility or a description of the Human Reproductive Process, perhaps has its foundation is the work of Dandekar (1955). Dandekar while formulating an appropriate probability model of the number of births in a given marital exposure of duration t developed special interruped probability distributions known as 'Modified Binomial' 'Modified and Poisson' distributions.

3.3.1 Dandekar's Modified Binomial Distribution

Dandekar's Modified Binomial distribution is based on the following assumptions:

(i) Probability of a conception (Vis-a-vis a birth, assuming a one-to-one correspondence between a conception and a birth) is p in every trial (every trial is assumed to be of duration of one month approximately which is the interval between two consecutive ovulatory cycles).

(ii) Given that there is a success (a conception leading to a birth), the probability of a further success in another π number of trials (π is an integer) inclusive of the trial in which a success took place is zero i.e. in π number of trials then can be almost one success.

Let (i) X = number of successes (conceptions or births); a.r.v.

(ii) n = number of trials

(iii) p = Probability of a success in a trial; $q = 1 - p$

(iv) $P(x;n)$ = Probability of exactly x successes in n trials (3.3.1)

(v) $F(x,n)$ = Probability of not more than x successes in n trials,

i.e., $F(x,n) = P[X \leq x \mid n]$ (3.3.2)

Development of the Model

Consider a sequence of n trials. If there is no success in $(n - x\pi)$ number of trials then obviously in the remaining $n-(n-x\pi) = x\pi$ number of trials at most x number of successes can occur. Probability of more than x successes in the remaining $x\pi$ number of trials given that no successes occurred upto the first $(n - x\pi)$ number of trials is zero. In other words, given that the $(n-x\pi)$ leads to failures the probability of at most x number of successes in the remaining $x\pi$ number of trials is one. Again when the first success occurs after S number of consecutive failure $(S < n-x\pi-1)$, then in the remaining $(n - \pi - S)$ trials at most $(x - 1)$ successes can occur with probability $F(x-1;\ n-\pi-S)$. Thus,

$$F(x;n) = F(0;n - x\pi) + \sum_{S=0}^{n-x\pi-1} pq^S F(x-1;n-\pi-S), \tag{3.3.3}$$

where x and π are integers; $x = 0, 1, 2, \ldots, \dfrac{n}{\pi}$.

Obviously, $F(0;\ n) = P(0,n) = q^n$, from (3.3.1) and (3.3.2) (3.3.4)

Putting $x = 1$ in (3.3.3) we get,

$$F(1;n) = F(0;n-\pi) + \sum_{S=0}^{n-\pi-1} pq^S F(0:n-\pi-S)$$

$$= q^{n-\pi} + \sum_{S=0}^{n-\pi-1} pq^S q^{n-\pi-S}, \text{ form (3.3.4)}$$

$$= q^{n-\pi} + pq^{n-\pi}(n-\pi) = q^{n-\pi}[1 + p(n-\pi)]$$

$$\therefore\ F(1,n) = q^{n-\pi}[1 + (n-\pi)p], \tag{3.3.5}$$

which is generalised Binomial distribution.

It can be shown that for $\pi = 0$ it gives Binomial distribution.

For $x = 2$, from (3.3.3) we get,

$$F(2,n) = F(0, n-2\pi) + \sum_{S=0}^{n-2\pi-1} q^S p\ F(1, n-\pi-S)$$

$$= q^{n-2\pi} + \sum_{S=0}^{n-2\pi-1} q^S p\ q^{(n-\pi-S)-\pi}[1+((n-\pi-S)-\pi)p], \text{ (from(3.3.5))}$$

$$= q^{n-2\pi} + \sum_{S=0}^{n-2\pi+1} q^S pq^{n-\pi-S-\pi} + \sum_{S=0}^{n-2\pi-1} q^S pq^{n-\pi-s-\pi}(n-\pi-S-\pi)p$$

$$= q^{n-2\pi} + (n-2\pi)pq^{n-2\pi} + p^2q^{n-2\pi}\left(\sum_{S=0}^{n-2\pi-1}(n-2\pi-S)\right)$$

$$= q^{n-2\pi} + (n-2\pi)pq^{n-2\pi} + p^2q^{n-2\pi}(n-2\pi)^2 - \frac{p^2q^{n-2\pi}(n-2\pi)(n-2\pi-1)}{1\cdot 2}$$

$$= q^{n-2\pi}\left\{1+(n-2\pi)p+p^2(n-2\pi)^2 - \frac{p^2(n-2\pi)(n-2\pi-1}{2}\right\}$$

$$= q^{n-2\pi}\left[1+(n-2\pi)p+\frac{(n-2\pi)(n-2\pi+1)}{2}p^2\right] \tag{3.3.6}$$

Proceeding in this way

$$F(3,n) = q^{n-3\pi}\left[1+(n-3\pi)p+\frac{(n-3\pi)(n-3\pi+1)}{2}p^2 + \frac{(n-3\pi)(n-3\pi+1)(n-3\pi+2)}{1\cdot 2\cdot 3}p^3\right]$$

$\cdots \quad \cdots \quad \cdots \quad \cdots \quad \cdots \quad \cdots$

$$F(x,n) = q^{n-x\pi}\left[1+p(n-x\pi)+\frac{(n-x\pi)(n-x\pi+1)}{2\,!}p^2 + \cdots + \frac{(n-x\pi)(n-x\pi+1)\cdots(n-x\pi+x-1)}{x\,!}p^x\right] \tag{3.3.7}$$

which are the first $(x+1)$ terms in the expansion of

$q^{n-x\pi}(1-p)^{-(n-x\pi)}$.

The relation (3.3.7) is true for all integral values of x for which

$n - x\pi \geq 0$ i.e. $\frac{n}{\pi} \geq x$.

For large values of x for which $n - x\pi < 0$, clearly $F(x;\ n) = 1$

From this the required probability $P(x,n)$ is given by

$$P(x;\ n) = F(x;\ n) - F(x-1:\ n). \tag{3.3.8}$$

3.3.2 Dandekar's Modified Poisson Distribution

Dandekar (1955) took the limiting case of his modified Binomial distribution and obtained the modified Poisson distribution as follows:

We can write (3.3.7) as

$$F(x,n) = (1-p)^{n(1-\frac{\pi}{n}x)}\left[1+np\left(1-\frac{\pi}{n}x\right)+n^2p^2\frac{\left(1-\frac{\pi}{n}x\right)\left(1-\frac{\pi}{n}x+\frac{1}{n}\right)}{2!} + \cdots + \frac{(np)^x}{x\,!}\left(1-\frac{\pi}{n}x\right)\left(1-\frac{\pi}{n}x+\frac{1}{n}\right)\cdots\left(1-\frac{\pi}{n}x+\frac{x-1}{n}\right)\right]$$

Putting $np = \lambda t; \ \dfrac{\pi}{n} = \dfrac{\theta}{t},$ (3.3.9)

$$\text{We get, } F(x,n) = \left\{\left(1-\frac{\lambda t}{n}\right)^n\right\}^{\left(1-\frac{\theta x}{t}\right)}\left[1+\lambda t\left(1-\frac{\theta x}{t}\right)+\frac{(\lambda t)^2}{2\,!}\left(1-\frac{\theta}{t}x\right)\left(1-\frac{\theta}{t}x+\frac{1}{n}\right)\right.$$

$$\left.+\cdots+\frac{(\lambda t)^x}{x\,!}\left(1-\frac{\theta}{t}x\right)\left(1-\frac{\theta}{t}x+\frac{1}{n}\right)\cdots\left(1-\frac{\theta}{t}x+\frac{x}{n}-\frac{1}{n}\right)\right]$$

As $n \to \infty$, the above reduces to

$$\lim_{n\to\infty} F(x,n) = F(x), \text{ say } = \exp(-\lambda(t-\theta x))\sum_{r=0}^{x}\frac{(\lambda t)^r}{r\,!}\,\frac{(t-\theta x)^r}{t^r}$$

$$\text{or} \quad F(x) = \exp(-\lambda(t-\theta x))\sum_{r=0}^{x}\frac{[\lambda(t-\theta x)]^r}{r\,!}. \qquad (3.3.10)$$

This is the modified Poisson Process of Dandekar which gives the probability distribution of the number of conceptions or births in time $(0, t)$ subject to the condition that each conception with intensity λ is followed by an infecundable exposure of duration θ; θ and t are continuous analog of $\pi$and n, respectively.

From (3.3.10), we have

$$\begin{aligned} P(x) = P(X = x) &= F(x) - F(x-1) \\ &= \exp(-\lambda(t-\theta x))\sum_{r=0}^{x}\frac{(\lambda(t-\theta x))^r}{r\,!} \\ &\quad - \exp(-\lambda(t-(x-1)\theta))\sum_{r=0}^{x-1}\frac{[\lambda(t-(x-1)\theta]^r}{r\,!} \end{aligned} \qquad (3.3.11)$$

However, the modified Poisson Process derived using difference equation (3.3.11) seems to lack mathematical rigour, precisely due to switching from discrete process to continuous by taking $n \to \infty$. Here π and n both being positive integers, $\dfrac{\pi}{n}$ is a rational number which cannot essentially be equal to $\dfrac{\theta}{t}$, where both θ and t are continuous variables admitting irrational values also.

3.3.3 William Brass Model

William Brass (1958) generalised the model given by (3.3.11) further by assuming 'λ' to conform to a probability distribution of the form

$$f(\lambda) = \frac{a^k e^{-a\lambda}\lambda^{k-1}}{\Gamma(k)}; \ 0 \le \lambda < \infty, a, k > 0, \qquad (3.3.12)$$

where λ = fecundability parameter in fertility analysis or intensity of sickness in morbidity studies. This assumptions is a fairly plausible one in representing the differential risk in human fertility or differential susceptibility in communicable diseases or accident proneness.

Writing, (3.3.10) as

$$\left.\begin{aligned} F(x,t|\lambda) &= \sum_{r=0}^{x} e^{-\lambda(t-\theta x)}\frac{[\lambda(t-\theta x)]^r}{r!} \text{ if } t > \theta \\ &= 0 \quad \text{otherwise.} \end{aligned}\right\} \qquad (3.3.13)$$

Therefore the interrupted Poisson Process in view of (3.3.13) is thus given by

$$\begin{aligned}
F(x,t) &= \int_0^\infty F(x,t|\lambda)f(\lambda)d\lambda \\
&= \int_0^\infty \sum_{r=0}^{x} e^{-\lambda(t-\theta x)} \frac{[\lambda(t-\theta x)]^r}{r!} \cdot \frac{a^k e^{-a\lambda}\lambda^{k-1}}{\Gamma(k)} d\lambda \\
&= \sum_{r=0}^{x} \frac{a^k}{\Gamma(k)} \cdot \frac{(t-\theta x)^r}{r!} \int_0^\infty e^{-\lambda(a+t-\theta x)}\lambda^{r+k-1} d\lambda \\
&= \sum_{r=0}^{x} \frac{a^k}{\Gamma(k)} \cdot \frac{(t-\theta x)^r}{r!} \cdot \frac{\Gamma(r+k)}{(a+t-\theta x)^{r+k}} \qquad (3.3.14)
\end{aligned}$$

Now, $P(x,t) = F(x,t) - F(x-1,t)$

$$\begin{aligned}
&= \sum_{r=0}^{x} \frac{a^k}{\Gamma(k)} \frac{(t-\theta x)^r}{r!} \cdot \frac{\Gamma(r+k)}{(a+t-\theta x)^{r+k}} \\
&\quad - \sum_{r=0}^{x-1} \frac{a^k}{\Gamma(k)} \frac{(t-(x-1)^r\theta)\Gamma(r+k)}{r!(a+t-(x-1)^r\theta)^{r+k}}, \text{ from (3.3.14)} \qquad (3.3.15) \\
&= \frac{a^k}{\Gamma(k)} \frac{(t-\theta x)^x}{x!} \frac{\Gamma(x+k)}{(a+t-\theta x)^{x+k}} \qquad (3.3.16)
\end{aligned}$$

The result (3.3.14) and (3.3.16) are the cumulative distribution function (c.d.f.) and probability mass function, respectively, of the model of William Brass which he used for obtaining the probability distribution of births in $(0,t)$ (in a given marriage duration).

We can see that the probability distribution of the number of births in time $(0,t)$ given by $P(t,x)$ in (3.3.16) for an integer k and $x(x>0)$ is a Negative Binomial Distribution with a, $k(a>0, k>0)$ and θ as parameters:

$$\begin{aligned}
P(t;x) &= \frac{\Gamma(x+k)}{\Gamma(k)x!}\left(1-\frac{a}{a+t-\theta x}\right)^x \left(\frac{a}{a+t-\theta x}\right)^k \qquad (3.3.17) \\
&= \binom{x+k-1}{k-1}\left(\frac{a}{a+t-\theta x}\right)^{k-1} \left(1-\frac{a}{a+t-\theta x}\right)^x \cdot \left(\frac{a}{a+t-\theta x}\right)
\end{aligned}$$

Putting $p' = \dfrac{a}{a+t-\theta x} \geq 0$ and $p' \leq 1$ when $t \geq \theta x \Rightarrow x \leq \left[\dfrac{t}{\theta}\right]$

where $\left[\dfrac{t}{a}\right]$ is the greatest integer contained in $\dfrac{t}{a}$.

$\therefore q' = 1 - \dfrac{a}{t+a-\theta x}$, where $p'+q'=1$

It follows that

$$P(x;t) = \binom{x+k-1}{k-1}(p')^{k-1}(q')^{x\cdot p'} \qquad (3.3.18)$$

represents the probability of precisely k number of successes in $(x+k)$ trials subject to the condition that the last trial leads to a success, i.e. a negative binomial distribution.

Brass fitted the distribution to the data of of births to women in the United States with completed marital span while ignoring θ, the period of infecundable exposure corresponding to every birth. This gave rise to the following model

$$P(x,t) = \frac{\Gamma(x+k)}{\Gamma(k)x!}\left(\frac{t}{a+t}\right)^x\left(\frac{a}{a+t}\right)^k, \text{ putting } \theta = 0 \text{ in (3.3.16)}, (x = 0, 1, 2 \ldots), \tag{3.3.19}$$

As the fit was not found to be good, especially for the mothers with no births in the entire marital span which could have been due to a heterogeneous combination of two different groups of women one group of mothers having Biological sterility, the other group of mothers prefering to have no children. Brass therefore, excluded mothers with zero number of births and considered a zero truncated model of (3.3.19) given by the probability distribution,

$$P(x;t) = \frac{\Gamma(x+k)}{\Gamma(k)x!}\left(\frac{t}{a+t}\right)^x\left(\frac{a}{a+t}\right)^k \Big/ \left\{1 - \left(\frac{a}{a+t}\right)^k\right\}, \quad x = 1, 2, 3, \ldots \tag{3.3.20}$$

Under the above assumption, Brass obtained a good fit of (3.3.20) to the data.

The raw moments of the distribution (3.3.20) can easily be obtained, as

$$\mu_1' = \frac{kt}{a} \Big/ \left\{1 - \left(\frac{a}{a+t}\right)^k\right\} \tag{3.3.21}$$

$$\mu_2' = \frac{kt}{a} + \frac{k(k+1)t^2}{a^2} \Big/ \left[1 - \left(\frac{a}{a+t}\right)^k\right] \tag{3.3.22}$$

$$\mu_3' = \frac{kt}{a} + \frac{3k(k+1)t^2}{a^2} + \frac{k(k+1)(k+2)t^3}{a^3} \Big/ \left[1 - \left(\frac{a}{a+t}\right)^k\right] \tag{3.3.23}$$

For fitting the distribution (3.3.20) to the data using the above three equations, the parameters a, k and t can be estimated. However, Brass model suffered from an inherent defect by ignoring the infecundable exposure corresponding to every live birth as pointed out by Biswas (1973).

Biswas (1973) generalized Brass model (3.3.19) by (i) introducing a factor $(1-\alpha)$, representing the proportion of biologically sterile mothers and (ii) assuming that following each live birth there is an infecundable period of constant length θ so that for x number of births, $\left(\frac{x\theta}{t}\right)^n = O(1)\,\forall n \geq 2$, subjected to $\max x = 6$ and t being the total marital exposure; n is the number of trials and $n \geq x$.

Starting from (3.3.11) under assumption of (i) and (ii), he got

$$f(t - x\theta \mid \lambda) = \alpha\, f\left(t\left(1 - \frac{x\theta}{t}\right) \mid \lambda\right) = P[x;t]$$

$$= \alpha\left[e^{-\lambda t}\frac{(\lambda t)^x}{x!} - x\theta\lambda\left\{e^{-\lambda t}\frac{(\lambda t)^{x-1}}{(x-1)!} - e^{-\lambda t}\frac{(\lambda t)^x}{x!}\right\}\right.$$

$$\left. + \frac{x\theta}{t}e^{-\lambda t}\frac{(\lambda t)^x}{x!}\right] + 0\left(\frac{x\theta\lambda}{t}\right) \tag{3.3.24}$$

By using Taylor series expansion about t

$$P[X = x] = f(t - x\theta \mid \lambda) = \sum_{r=0}^{x} e^{-\lambda(t-\theta x)} \frac{\lambda(t-\theta x)^r}{r!}$$

$$- \sum_{r=0}^{x-1} e^{-\lambda(t-(x-1)\theta)} \frac{\lambda(t-(x-1)\theta)^r}{r!} \quad (3.3.25)$$

the probability of zero number of births in $(0, t)$ as given by (3.3.24) is $\alpha e^{-\lambda t}$, where α is the proportion of biologically non-sterile mothers in a population and $(1 - \alpha)$ is the proportion of sterile mothers in the population.

The probability of x number of births in $(0, t)$ is given by

$$\alpha f\left(t\left(1 - \frac{x\theta}{t}\right) \mid \lambda\right), \quad (3.3.26)$$

under the condition that fecundability is λ.

Further, assuming the probability distribution of λ (i.e., variation of fecundability) as in (3.3.12),

$$\phi(\lambda) = \frac{ak}{\Gamma(k)} e^{-a\lambda} \lambda^{k-1}; \; 0 \leq \lambda < \infty, a, k > 0. \quad (3.3.27)$$

The unconditional probability of a biologically fecund mother to have x number of births $(x = 1, 2, 3 \ldots)$ is given by

$$(1-\alpha)\int_{\cup}^{\infty} f\left[t\left(1 - \frac{x\theta}{t}\right) \mid \lambda\right] \phi(\lambda) d\lambda = (1-\alpha)\left[\left(\frac{a}{a+t}\right)^k \frac{\Gamma(x = k)}{\Gamma(k)x!} \left(\frac{t}{a+t}\right)^x\right.$$

$$+ \frac{k(k+1)\theta t}{a^2}\left(\frac{a}{a+t}\right)^{k+2} \frac{\Gamma(k+2+x-1)}{\Gamma(k+2)(x-1)!}\left(\frac{t}{a+t}\right)^{x-1}$$

$$\left. \frac{k(k+1)\theta t}{a^2}\left(\frac{a}{a+t}\right)^{k+2} \frac{\Gamma(k+2+x-2)}{\Gamma(k+2)(x-2)!}\left(\frac{t}{a+t}\right)^{x-2}\right] \quad (3.3.28)$$

and that the probability of zero number of births is given by

$(1-\alpha) + \alpha\left(\dfrac{a}{a+t}\right)^k$, from (3.3.19) for $x = 0$ and assumption (i) above.

Remarks 3.1: Now the probability of having x number of births in $(0, t)$ given that (i) the probability of a conception (or a birth) in an infinitesimal interval of length δt is $\lambda\delta t + 0(\delta t)$ where λ is a constant and (ii) given that a renewal (conception or birth) taking place at time t for a fixed time π following t, there cannot occur another success, (conception or birth). It is really a problem of Geiger Muller Counter model type I with a fixed dead time π. Hence the Dandekar's model derived in (3.3.13) can be derived in a much more rigorous way using Counter theory as in Sec. 1.6.1. One may compare (3.3.25) with that of (1.6.15) for $n = x$ and $\theta = \pi$. Again starting form (1.6.15) one can derive William Brass model given in (3.3.28).

Using (3.3.14), we have on replacing x by n

$$F(t, n) = \sum_{r=0}^{n} \frac{a^k}{\Gamma(k)} \frac{(t-\theta n)^r}{r!} \frac{\Gamma(r+k)}{(a+t-\theta n)^{r+k}}$$

$$= P[X \leq n \mid t] \Rightarrow 1 - F(t, n) = P[X > n \mid t]$$

$$\Rightarrow \qquad 1 - F(t,n) = P[T_n \le t], \tag{3.3.29}$$

where T_n represents the waiting time for the nth renewal (i.e. $(n+1)$th conception or birth).

Therefore, $P[t \le T_n \le t + \delta t \mid t] = \dfrac{d}{dt}[t - F(t,n)]$ (3.3.30)

$$\Rightarrow \qquad f_n(t_n) = \frac{d}{dt}\left[1 - \sum_{r=0}^{n} \frac{a^k}{\Gamma(k)} \frac{(t-\theta n)^r}{r!} \frac{\Gamma(r+k)}{(a+t-\theta n)^{r+k}}\right]_{t=t_n}. \tag{3.3.31}$$

Putting $n = 0$ we have the waiting time distribution for the first conception or first birth as

$$f_0(t_0) = \frac{d}{dt}\left[1 - \frac{a^k}{\Gamma(k)} \frac{\Gamma(k)}{(a+t)^k}\right]_{t=t_0} = \frac{ka^k}{(a+t_0)^{k+1}}. \tag{3.3.32}$$

This is the model used by Singh (1964 a) for obtaining the waiting time distribution of the first birth from marriage.

3.3.4 A Modification of the Singh's (1964 a) Result

The model for the waiting time of first conception given in (3.3.32) is based on the premises (i) that the waiting time distribution from marriage to first conception is given by

$$f(t \mid \lambda) = \lambda e^{-\lambda t}, \;\; 0 \le t < \infty, \lambda > 0,$$

where the conception rate (i.e., fecundity) remains fixed at λ. (ii) However, given that the fecundity varies from woman to woman and the distribution of λ follows a Gamma distribution (3.3.27) viz.

$$\phi(\lambda) = \frac{a^k}{\Gamma(k)} e^{-a\lambda}\lambda^{k-1}; \; \lambda > 0, \; a, k > 0,$$

the unconditional waiting time distribution for first conception is given by

$$\int_0^\infty f(t \mid \lambda)\, \phi\,(\lambda)\; d\lambda = \frac{ka^k}{(a+t)^{k+1}}. \tag{3.3.33}$$

One major objection in the model (3.3.33) is that the marital exposure for the first conception need not be infinite. Suppose,

$$f(t_1 \mid \lambda) = \frac{\lambda \exp(-\lambda t_1)}{1 - \exp(-\lambda T')}, \; \lambda > 0, \; 0 \le t_1 < T',$$

which is a negative exponential right truncated at T', it may be taken as the distribution of reasonable exposure for the first conception given λ.

Then the unconditional waiting time distribution for first conception is given by

$$f(t_1) = \int_0^\infty f(t_1 \mid \lambda)\, \phi\,(\lambda)\; d\lambda = \int_0^\infty \frac{\lambda \exp(-\lambda t_1)}{1 - \exp(-\lambda T')} \frac{a^k}{\Gamma(k)} e^{-a\lambda}\lambda^{k-1} d\lambda$$

$$= \frac{a^k}{\Gamma(k)} \int_0^\infty \lambda^k \exp(-\lambda(t_1 + a))\; (1 - \exp(-\lambda T'))^{-1} d\lambda.$$

Expansion of the 3rd term gives

$$= \frac{a^k}{\Gamma(k)} \int_0^\infty [\exp(-\lambda(t+a)) + \exp(-\lambda(t_1 + a + T')) + \cdots]\lambda^k d\lambda$$

$$= \frac{a^k}{\Gamma(k)} \int_0^\infty \sum_{i=0}^{\infty} \exp[-\lambda(t_1 + a + i\,T')]\lambda^k d\lambda$$

$$= \frac{a^k}{\Gamma(k)} \sum_{i=0}^{\infty} \frac{\Gamma(k+1)}{(t_1 + a + iT')^{k+1}} = \sum_{i=0}^{\infty} \frac{ka^k}{(t_1 + a + iT')^{k+1}}, \tag{3.3.34}$$

which is the modified distribution for the waiting time for first conception due to Biswas and Shrestha (1986).

Replacing $\sum_{i=0}^{\infty} \frac{ka^k}{(t_1 + a + i\,T')^{k+1}}$ by $\int_0^\infty \frac{ka^k dT'}{(t_1 + a + i\,T')^{k+1}}$ it can be shown that

$$E(T_1) \cong \frac{a}{k-1}\left[1 - \left(\frac{a}{a+T'}\right)^{k-1}\right]. \tag{3.3.35}$$

Similarly, after some routine calculations, we get

$$\text{Var}\,(T_1) = \frac{a}{(k-1)^2}\left[\frac{ka}{k-2} - \left(\frac{a}{a+T'}\right)^{k-1}\right]\left\{\left(\frac{kT'(k-1)+2a}{k-2}\right) + a\left(\frac{a}{a+T'}\right)^{k-1}\right\} \tag{3.3.36}$$

The parameters of the distribution (3.3.3) are estimable by the method of moments by using (3.3.5) and (3.3.6).

3.3.5 A Model for the Waiting Time of Conception of Various Orders

Putting $n = 1$ in (3.3.1), we have

$$f_1(t_1) = \frac{d}{dt}\left[1 - \sum_{r=0}^{1} \frac{a^k}{\Gamma(k)} \frac{(t-\theta)^r}{r!} \cdot \frac{\Gamma(k+r)}{(a+t-\theta)^{r+k}}\right]$$

$$= \frac{d}{dt}[1 - F, (t; 1)]_{t=t_1} = \frac{d}{dt}(P[T_1 \le t])_{t=t_1}$$

$$\Rightarrow \quad P[t_1 \le T_1 \le t_1 + dt_1] = f_1(t_1)dt_1 = \frac{ka^k(k+1)(t_1-\theta)}{(a+t_1-\theta)^{k+2}};\ \theta \le t_1 < \infty, \tag{3.3.37}$$

which is the waiting time distribution of the second conception or birth from the marriage.

Putting $n = 2$ in (3.3.1),

$$f_2(t_2) = \frac{d}{dt}\left[1 - \sum_{r=0}^{2} \frac{a^k}{\Gamma(k)} \frac{(t-2\theta)^r}{r!} \frac{\Gamma(k+r)}{(a+t-2\theta)^{r+k}}\right]_{t=t_2}$$

$$= \frac{a^k k(k+1)(k+2)(t_2-2\theta)^2}{2!(a+t_2-2\theta)^{k+3}};\ 2\theta \le t_2 < \infty. \tag{3.3.38}$$

which is the waiting time distribution of the third conception (or birth) from marriage.

Finally, the waiting time distribution of the nth conception (or birth) from marriage is given by

$$f_{n-1}(t_{n-1}) = \frac{k(k+1)\cdots(k+n-1)}{(n-1)!} \frac{a^k(t_{n-1}-(n-1)\theta)^{n-1}}{(a+t_{n-1}-(n-1)\theta)^{k+r}};$$

$$(n-1)\theta \le t_{n-1} < \infty.$$

It is seen that $(t_{n-1} - (n-1)\theta)$ conforms to a Beta distribution of Type II. The results are due to Biswas and Nauhria (1980).

3.3.6 Another Generalization of Dandekar's Model

This generalization is due to Biswas (1980). Here, it is assumed that infecundable exposure period θ, following every conception, is fixed but the conceptions occur with gradually decreasing intensities with differential fecundities $\lambda_1, \lambda_2, \ldots, \lambda_n$ (instead of one conception rate λ) given by

$$\lambda_i = \lambda \exp[-(i-1)\delta],\ \delta \text{ (a constant) } > 0,\ i = 1, 2, \ldots, n, \tag{3.3.39}$$

for the ith conception.

As earlier, it is assumed that λ_i are Poisson intensities leading to ith registration of conception/birth during a fecundable time t_i after which follows an infecundable or dead period θ. Thus t_i's follow exponential distribution given by

$$g_i(t_i) = \lambda_1 \exp(-\lambda_i t_i),\ i = 1, 2, \ldots, n. \tag{3.3.40}$$

Here t_i's are assumed to be independently but not identically distributed, because t_1 is preceded by no dead time whereas $t_2, t_3, \ldots, t_n$ are preceded by dead time of length θ. Let

$$S_n = t_1 + t_2 + \cdots + t_n. \tag{3.3.41}$$

Then, S_n represents the waiting time for nth conception. We are to find the distribution of S_n for given $\lambda_1, \lambda_2, \ldots, \lambda_n$. The density of S_n is given by the convolution of the densities of $t_1, t_2, \ldots, t_n$. We proceed to find it step by step.

Now, $S_2 = t_1 + t_2$, then for positive r,v.'s (see (A.2.4)),

$$\begin{aligned} f_2(S_2|\lambda_1, \lambda_2) &= \int_0^{S_2} g_1(t_1) g_2(S_2 - t_1) dt_1 \\ &= \int_0^{S_2} \lambda_1 \exp(-\lambda_1 t_1) \lambda_2 \exp[-\lambda_2 (S_2 - t_1) dt_1, \text{ from (3.3.40)} \\ &= \frac{\lambda_1 \lambda_2}{\lambda_1 - \lambda_2} [\exp(-\lambda_2 S_2) - \exp(-\lambda_1 S_2)] \\ &= \sum_{i=1}^{2} \lambda_i \exp(-\lambda_i S_2) \prod_{i \neq j = 1}^{2} \frac{\lambda_j}{\lambda_i - \lambda_j}. \end{aligned} \tag{3.3.42}$$

Similarly, $f_3(S_3|\lambda_1, \lambda_2, \lambda_3) = \int_0^{S_3} g_3(t_3) f_2(S_3 - t_3) dt_3.$

Using (3.3.40) and (3.3.42), after some straightforward calculations, we get

$$\begin{aligned} f_3(S_3|\lambda_1, \lambda_2, \lambda_3) &= \frac{\lambda_1\lambda_2\lambda_3 \exp(-\lambda_1 S_3)}{(\lambda_1 - \lambda_2)(\lambda_1 - \lambda_3)} + \frac{\lambda_1\lambda_2\lambda_3 \exp(-\lambda_2 S_3)}{(\lambda_2 - \lambda_1)(\lambda_2 - \lambda_3)} \\ &\quad + \frac{\lambda_1\lambda_2\lambda_3 \exp(-\lambda_3 S_3)}{(\lambda_1 - \lambda_2)(\lambda_2 - \lambda_3)} + \frac{\lambda_1\lambda_2\lambda_3 \exp(-\lambda_3 S_3)}{(\lambda_1 - \lambda_2)(\lambda_3 - \lambda_1)}. \end{aligned}$$

Adding the last two terms together, we get

$$f_3(S_3|\lambda_1, \lambda_2, \lambda_3) = \frac{\lambda_1\lambda_2\lambda_3 \exp(-\lambda_1 S_3)}{(\lambda_1 - \lambda_2)(\lambda_1 - \lambda_3)} + \frac{\lambda_1\lambda_2\lambda_3 \exp(-\lambda_2 S_3)}{(\lambda_2 - \lambda_1)(\lambda_2 - \lambda_3)} + \frac{\lambda_1\lambda_2\lambda_3 \exp(-\lambda_3 S_3)}{(\lambda_3 - \lambda_1)(\lambda_3 - \lambda_2)}$$

$$= \sum_{i=1}^{3} \lambda_i \exp(-\lambda_i S_3) \prod_{i\neq j=1}^{3} \frac{\lambda_i}{(\lambda_i - \lambda_j)} \tag{3.3.43}$$

Proceeding similarly or by mathematical induction, we can get

$$f_n(S_n|\lambda_1, \lambda_2, \ldots, \lambda_n) = \sum_{i=1}^{n} \lambda_i \exp(-\lambda_i S_n) \prod_{i\neq j=1}^{n} \frac{\lambda_i}{(\lambda_i - \lambda_j)}. \tag{3.3.44}$$

Further, using (3.3.39) we get,

$$f_n(S_n;\delta|\lambda) = \sum_{i=1}^{n} \lambda e^{-(i-1)\delta} \exp\left[-\left\{\lambda e^{-(i-1)\delta}\right\} S_n\right] \prod_{i\neq j=1}^{n} \frac{\lambda_i}{(\lambda_i - \lambda_j)}. \tag{3.3.45}$$

Following Brass (1958), we assume that λ is a gamma variate (3.3.27), then the unconditional density of S_n is given by

$$f_n(S_n;\delta) = \int_0^\infty f_n(S_n;\delta|\lambda)\phi(\lambda)d\lambda$$

$$= \int_0^\infty \sum_{i=1}^{n} \lambda \exp\left[-\left\{\lambda e^{-(i-1)\delta}\right\} S_n\right] \prod_{i\neq j=1}^{n} \frac{e^{-(i+j-2)\delta}}{e^{-(i-1)\delta} - e^{-(j-1)\delta}} \frac{a^k e^{-a\lambda}\lambda^{k-1}}{\Gamma(k)} d\lambda$$

$$= \frac{a^k}{\Gamma(k)} \sum_{i=1}^{n} \prod_{i\neq j=1}^{n} L(i,j;\delta) \int_0^\infty \lambda^k \exp\left[-\left\{\lambda e^{-(i-1)\delta}\right\} S_n + a\right] d\lambda,$$

$$\text{where} \quad L(i,j;\delta) = \frac{e^{-(i+j-2)\delta}}{e^{-(i-1)\delta} - e^{-(j-1)\delta}}.$$

$$= \frac{a^k}{\Gamma(k)} \sum_{i=1}^{n} \prod_{i\neq j=1}^{n} L(i,j;\delta) \frac{1}{b^{k+1}} \int_0^\infty b^{k+1} \lambda^k e^{-\lambda\cdot b} d\lambda,$$

$$\text{where} \quad b = \left\{e^{-(i-1)\delta}\right\} S_n + a$$

$$= \frac{a^k}{\Gamma(k)} \sum_{i=1}^{n} \prod_{i\neq j=1}^{n} L(i,j;\delta) \frac{1}{b^{k+1}} \Gamma(k+1)$$

$$\text{or,} \quad f_n(S_n;\delta) = ka^k \sum_{i=1}^{n} \frac{1}{b^{k+1}} \prod_{i\neq j=1}^{n} L(i,j;\delta). \tag{3.3.46}$$

Finally, we incorporate the dead period θ, then the time for the nth registration of conception (or birth) is

$$\xi_n = S_n + (n-1)\theta. \tag{3.3.47}$$

Replacing S_n, in (3.3.46), by ξ_n and restoring the value of b, we get the density function of ξ_n as

$$f_n(\xi_n;\delta) = ka^k \sum_{i=1}^{n} \frac{1}{\left[a + e^{-(i-1)\delta}\{\xi_n - (n-1)\}\theta\right]^{k+1}} \prod_{i\neq j=1}^{n} L(i,j;\delta), \tag{3.3.48}$$

which is also the distribution of waiting time of the nth arrival in a Counter Model I (see Sec. 1.6.1).

Moments of S_n: The rth raw moment of S_n is given by

$$\mu_r' = E[S_n^r] = ka^k \sum_{i=1}^{n} L(i,j;\delta) \int_0^\infty \frac{S_n^r}{\left[e^{-(i-1)\delta} \cdot S_n + a\right]^{k+1}} dS_n.$$

After some simple calculations we get,

$$\mu'_r = ka^{r-k} \sum_{i=1}^{n} L(i,j;\delta)\, e^{[(i-1)\delta(r+1)]}\, B(r+1, k-r), \tag{3.3.49}$$

where, $B(.,.)$, is a Beta function of second kin α. μ'_r is the rth moment of the uninterrupted waiting time S_n, for the nth conception (or birth).
The corresponding moment of the interrupted time ξ_n is given by

$$\nu'_r = E[\xi_n^r] = E[S_n + (n-1)\theta]^r, \quad \text{from (3.3.47)}$$

$$\text{or} \quad \nu'_r = \mu'_r + \binom{r}{1}\mu'_{r-1}(n-1)\theta + \binom{r}{2}\mu'_{r-2}(n-1)^2\theta^2 + \cdots + (n-1)^r\theta^r, \tag{3.3.50}$$

since $\mu'_0 = 1$ and where $r = 1, 2, \ldots$
We may note that to estimate a, k, δ in (3.3.48), it is sufficient to consider the estimating equations of ν'_1, ν'_2 and ν'_3.

A Special case

Let $\lambda_1 = \lambda_2 = \cdots = \lambda_n = \lambda$.
In this case,

$$L\{f_n(S_n|\lambda)\} = [L\{g(t_i/\lambda)\}]^n, \ L \text{ stands for } L.T. \tag{3.3.51}$$

Since, $S_n = t_1 + t_2 + \cdots + t_n$ and t_i's are i.i.d. r.v.'s.

$$L\{f_n(S_n|\lambda)\} = E\left[e^{-s(t_1+t_2+\cdots+t_n)}\right] = \left[\lambda \int_0^\infty e^{-s.t} e^{-\lambda.t} dt\right]^n$$

$$= \frac{\lambda^n}{(\lambda+s)^n}.$$

Note that $S_1, S_2, \ldots, S_n$ are r.v.'s whereas s is a parameter.
Then, taking the Inverse L.T. and using the table of L.T. (see Appendix Table A-1) we get

$$f_n(S_n|\lambda) = L^{-1}\left\{\frac{\lambda^n}{(\lambda+s)^n}\right\} = \frac{\lambda^n e^{-\lambda.S_n} S_n^{n-1}}{\Gamma(n)} \tag{3.3.52}$$

If λ follows the gamma distribution (3.3.27), then

$$f_n(S_n) = \int_0^\infty f_n(S_n|\lambda)\ \phi\ (\lambda)\ d\lambda$$

$$= \frac{a^k S_n^{n-1}}{\Gamma(n)\Gamma(k)} \int_0^\infty \exp[-\lambda(a+S_n)]\lambda^{n+k-1} d\lambda$$

Putting $\lambda(a+S_n) = z \Rightarrow d\lambda = \dfrac{dz}{a+S_n}$, we get

$$f_n(S_n) = \frac{a^k S_n^{n-1}}{\Gamma(n)\Gamma(k)} \int_0^\infty e^{-z}\left(\frac{z}{a+S_n}\right)^{n+k-1} \frac{1}{a+S_n} dz$$

$$= \frac{a^k S_n^{n-1}}{\Gamma(n)\Gamma(k)} \left(\frac{1}{a+S_n}\right)^{n+k} \Gamma(n+k)$$

$$\text{or} \quad f_n(S_n) = \frac{a^k}{B(n,k)} \cdot \frac{s^{n-1}-n}{(a+S_n)^{n+k}}, \ B(.,.) \text{ is Beta function.} \tag{3.3.53}$$

Finally, incorporating a dead period θ corresponding to every conception (or birth), as in (3.3.47), $\xi_n = S_n + (n-1)\theta$, then we have

$$f_n(\xi_n) = \frac{a^k}{B(n,k)} \frac{\{\xi_n - (n-1)\theta\}^{n-1}}{\{a + \xi_n - (n-1)\theta\}^{n+k}}, \; (n-1)\theta \leq \xi_n < \infty. \tag{3.3.54}$$

We observe that $\xi_n - (n-1)\theta$ conforms to a Beta distribution of Type II.

Further, the rth moment of $\xi_n - (n-1)\theta = S_n$ is

$$\mu'_r = E[\xi_n - (n-1)\theta]^r, r = 1, 2, \ldots$$

$$= \frac{a^k}{B(n,k)} \int_0^\infty \frac{\{\xi_n - (n-1)\theta\}^r \{\xi_n - (n-1)\theta\}^{n-1}}{\{a + \xi_n - (n-1)\theta\}^{n+k}} d\xi_n$$

$$\text{or} \quad \mu'_r = a^k \frac{B(r+n, k-r)}{B(n,k)}. \tag{3.3.55}$$

The relation (3.3.55) has also been alternatively derived by Biswas and Nauharia (1980) by using Dandekar's modified distribution.

3.3.7 Problems for the Development of the Model of the Interarrival Waiting Time Distribution

Let $(T_i - T_{i-1})$ be the waiting time for the ith order of conception measured form the date of $(i-1)$th order of conception $(i = 2, 3, \ldots)$; so $(T_i - T_{i-1})$ may be called 'Inter Conception' interval. If we assume that the conception rate λ varies from individual to individual even within the same interconception interval conforming to some probability distribution, say Gamma distribution, as in the earlier section; then because of weighting of λ we shall see that

(i) The renewal intervals $(T_i - T_{i-1})$ $\forall i = 2, 3, \ldots$ will cease to become i.i.d. r.v.'s. contrary to the traditional assumption.

(ii) The number of renewals even in two non-overlapping intervals will be correlated. The extent of correlation will be proportional to the variance of λ.

As a result of the correlation, because of the weighting of the process, the renewal structure is completely destroyed leading to the process to conform to infinitely divisible distributions with dependent increments.

The problem is then to obtain the interarrival distribution of such dependent processes.

A. Correlation Between T_i and T_{j-i}

Here, $\quad T_{j-i} = T_j - T_i$, for $j > i$. (3.3.56)

Let us assume the probability distribution of λ to be gamma, given by (3.3.27)

We have $\quad \text{Cov}\,(T_i, T_j) = E_\lambda(E(T_iT_j|\lambda) - E_\lambda(E(T_i|\lambda))E_\lambda(E(T_j|\lambda))$

Now $\quad E(T_iT_j|\lambda) = E(T_i(T_i + T_{j-i}|\lambda))$, from (3.3.56)

$$= E(T_i^2|\lambda) + [\text{Cov}(T_i, T_{j-1}|\lambda) + E(T_i|\lambda)E(T_{j-1}|\lambda)]$$

But Cov $(T_i, T_{j-i}|\lambda) = 0$, since T_i and T_{j-i} are two non overlapping intervals (see (3.3.63), $\because$ for given λ, $\text{Var}(\lambda) = 0$).

Further, $\quad E(T_i^2|\lambda) = \dfrac{i(i+1)}{\lambda^2}, E(T_i|\lambda) = \dfrac{i}{\lambda}, \quad \because T_i$'s are exponential.

and $\quad E(T_{j-1}|\lambda) = \frac{j-1}{\lambda}.$

Therefore $E(T_iT_j|\lambda) = i(j+1)\frac{1}{\lambda^2}$

$$E(T_iT_j) = E_\lambda[E(T_iT_j|\lambda)] = i(j+1)E\left(\frac{1}{\lambda^2}\right)$$

$$= \frac{i(j+1)a^2}{(k-1)(k-2)}, \because E\left(\frac{1}{\lambda^2}\right) = \frac{a^2}{(k-1)(k-2)}$$

$$\therefore \quad \text{Cov}\ (T_i, T_j) = \frac{a^2 i(j+k-1)}{(k-1)(k-2)}, \quad k > 2$$

$$\text{Var}\ (T_i) = \frac{a^2 i(i+k-1)}{(k-1)^2(k-2)}, \quad k > 2 \tag{3.3.57}$$

Further,
$$\begin{aligned}\text{Cov}\ (T_i, T_j) &= \text{Cov}(T_i, T_i + T_{j-i}) \\ &= E(T_i(T_i + T_{j-i})) - E(T_i)E(T_i + T_{j-i}) \\ &= E(T_i^2) + E(T_iT_{j-i}) - [E(T_i)]^2 - E(T_i)E(T_{j-i}) \\ &= E(T_i^2) - [E(T_i)]^2 + E(T_i, T_{j-i}) - E(T_i)E(T_j) \\ &= \text{Var}(T_i) + \text{Cov}(T_i, T_{j-i})\end{aligned}$$

$$\therefore \quad \begin{aligned}\text{Cov}\ (T_i, T_{j-i}) &= \text{Cov}\ (T_i, T_j) - \text{Var}\ (T_i) \\ &= \frac{a^2 i(j+k-1)}{(k-1)^2(k-2)} - \frac{a^2 i(i+k-1)}{(k-1)^2(k-2)} \\ &= \frac{a^2 i(j-i)}{(k-1)^2(k-2)}, \quad k > 2.\end{aligned} \tag{3.3.58}$$

Also
$$\begin{aligned}\text{Var}\ (T_{j-i}) &= V(T_j - T_i), \quad \text{form (3.3.56)} \\ &= \text{Var}\ (T_j) + \text{Var}\ (T_i) - 2\ \text{Cov}\ (T_i, T_j) \\ &= \frac{a^2 j(j+k-1)}{(k-1)^2(k-2)} + \frac{a^2 i(i+k-1)}{(k-1)^2(k-2)} - \frac{2a^2 i(j+k-1)}{(k-1)^2(k-2)} \\ &= \frac{a^2(j-i)(j-i+k-1)}{(k-1)^2(k-2)}\end{aligned} \tag{3.3.59}$$

$$\text{Cor}\ (T_i, T_{j-i}) = \frac{\text{Cov}\ (T_i, T_{j-i})}{\sqrt{\text{Var}\ (T_i)}\sqrt{\text{Var}\ (T_{j-i})}}. \tag{3.3.60}$$

Putting (3.3.57), (3.3.58) and (3.3.59) in (3.3.60) we get

$$\text{Cor}\ (T_i, T_{j-i}) = \frac{i(j-i)}{\sqrt{(i(i+k-1)}\sqrt{(j-i)(j-i+k-1)}}, \tag{3.3.61}$$

which is independent of a.

As a special case if $i = 1$ and $j = 2$

$$\text{Cor}\ (T_1, T_2 - T_1) = \frac{1}{k}, \tag{3.3.62}$$

the result is due to Biswas and Sehgal (1986).

B. Correlation between the number of Events in Two Non-overlapping Intervals

Let us take two non-overlapping intervals $(0, s)$ and (s, t); $t > s$.

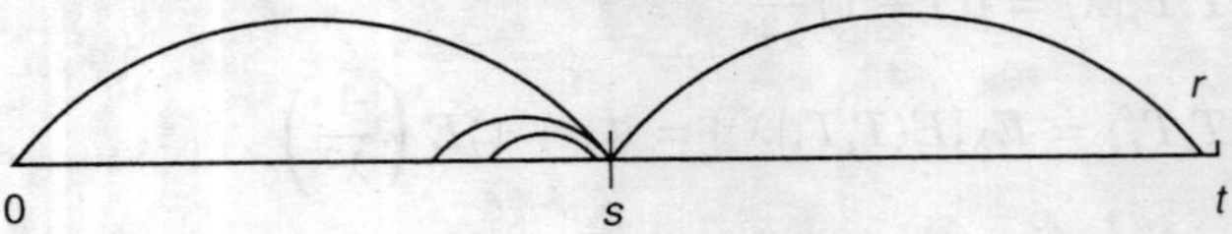

Fig. 3.1

Let $X(s)$ denote the number of conceptions in $(0, s)$, a Poisson variate with parameter $\lambda \cdot s$ and $[X(t) - X(s)]$ the number of conceptions in (s, t).

$$\begin{aligned}
\text{Then}\quad \text{Cov}\ (X(s), X(t) - X(s)) &= E_\lambda[E\{(X(s))(X(t) - X(s))\} \mid \lambda)] \\
&\quad - E_\lambda[E(X(s)) \mid \lambda]E_\lambda[E(X(t) - X(s)) \mid \lambda] \\
&= E_\lambda[\lambda \cdot s \cdot \lambda(t - s)] - E_\lambda[s \cdot \lambda]E_\lambda[\lambda(t - s)] \\
&= s(t - s)E_\lambda(\lambda^2) - s(t - s)[E_\lambda(\lambda)]^2 \\
&= s(t - s)\ \text{Var}\ (\lambda) \qquad (3.3.63)
\end{aligned}$$

The result is due to Biswas and Pachal (1981).

Remarks 3.2: The result (3.3.63) shows that

$$\text{Cov}\ [X(s), \{X(t) - X(s)\}] \to 0 \text{ as } s \to 0$$

$$\therefore \text{As} \quad s \to 0,\ \text{Cor}\ [X(s), \{X(t) - X(s)\}] \to 0.$$

By the condition of orderliness, as $s \to 0$ the infinitesimal interval cannot have more than one renewal. As $s \to 0$ the same event, under the limiting condition, at most one event may occur at the beginning of the interval s to t (see Fig. 3.1.), in which case we get the Palm Probability, that has been defined as a limiting probability of having a given number of renewals in the interval s to t given that an event (or renewal) has occurred at the beginning of the interval, as a natural sequence.

3.4 Sheps and Perrin Model of Human Reproductive Process

Here we discuss a model of Sheps and Perrin (1964) dealing with various states of the human reproductive process; thus describing the intrinsic fertility status of a woman by a Semi-Markov Process (or Markov Renewal Process). Semi-markov processes are generalizations of both Markov Chains (discrete) and Markov-Processes (continuous) with countable number of state spaces. They were independently introduced by P. Levy, W.L. Smith and L. Takac's in 1954. Roughly speaking, a Semi-Markov Process is a stochastic process under which movement occurs from one state to another (of at most a countable number of states) with the visit of successive states in the pattern of Markov Chain; while the stay in a given state is for a random length of time (unlike Markov-process), the distribution of which may depend on the present state as well as on the state to be visited next. If we denote the process by $\{X_n; t_n\}(n = 1, 2, 3, \ldots)$ with the above condition then the process is called a Markov-renewal process and then $X(t)$ is a semi-Markov process.

The formulation of Sheps and Perrin model starts with modeling a woman at $T = 0$ being in the state of S_0 which is the non-pregnant fecundable state. After a random period of time the woman may enter into the state of pregnancy denoted by S_1 from which she may enter the post-partum nonsusceptible (or non-fecundable state) to be follow by a live birth (state S_4) or a still birth (state S_3) or an early foetal death (state S_2). After a stay for a random duration of time in any of these states, it is assumed that the first phase of the sojurn is over and the woman comes back to state S_0. However, the renewal of the state S_0 of the process continues indefinitely, under the model, which is not the actual situation in reality.

A diagrammatic representation of Sheps and Perrin Model is given in Fig. 3.2.

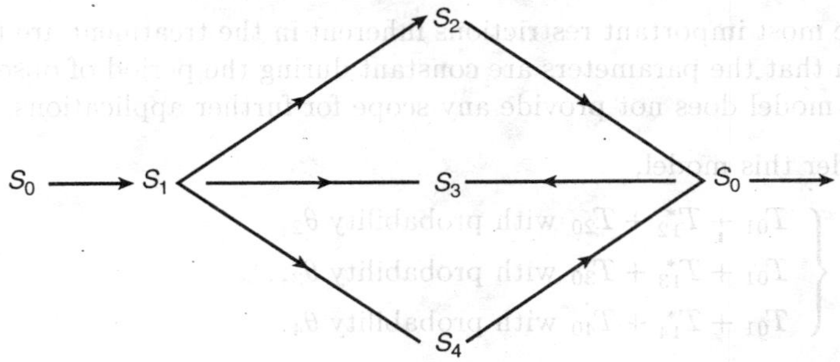

Fig. 3.2

The length of stay in each state is viewed as a random variable as in the case of Markov-Renewal Process. The details for the development of the model are presented below:

Notations:

S_0 = Non-pregnant fecundable state.

S_1 = Pregnant state.

S_2 = Post-partum infecundable State associated with abortion or early foetal loss.

S_3 = Post-partum infecundable state associated with still birth.

S_4 = Post-partum infecundable state associated with live birth.

ρ = fecundability i.e., probability of passage form S_0 to S_1).

For $i, j = 0, 1, 2, 3, 4,$

T_{ij} = the random time (in months) required for passage from S_t to S_j; these are independent *r.v.*'s.

$\mu_{ij} = E(T_{ij})$ = mean passage time from S_i to S_j

$\sigma_{ij}^2 = \text{Var}(T_{ij})$ = variance of passage time from S_i to S_j

For $i = 2, 3, 4,$

T_{ij}^* = Gestation period, $j = 2, 3, 4.$

θ_i = Probability of direct passage form S_i to S_j and

$\nu_i,\ \xi_i^2$ = the mean and variance, respectively, of the length of the stay in S_i i.e. gestation period given that S_i is reached from S_1 (3.4.1)

$\eta_i = \nu_i + \mu_{io}$ = mean of gestation period plus mean of post-partum infecundable period for outcome i (3.4.2)

$\lambda_i^2 = \xi_i^2 + \sigma_{io}^2$ = Variance of gestation period plus post partum infecundable period for outcome i. (3.4.3)

Assuming for the present that the relevant parameters remain constant during the period of observation it may be shown that the resulting Stochastic Process is a Markov renewal process.

Note that

(i) The model does not permit the probabilities of conception or foetal loss or duration of infecundability to depend on such salient factors as age, birth order or differences among individuals.

(ii) The most important restrictions inherent in the treatment are the assumption that the parameters are constant during the period of observation and the model does not provide any scope for further applications.

Now under this model,

$$T_{00} = \begin{cases} T_{01} + T_{12}^* + T_{20} \text{ with probability } \theta_2, \\ T_{01} + T_{13}^* + T_{30} \text{ with probability } \theta_3, \\ T_{01} + T_{14}^* + T_{40} \text{ with probability } \theta_4. \end{cases} \tag{3.4.4}$$

$\therefore \quad E[T_{o1}] = \mu_{01} = \dfrac{(1-\rho)}{\rho}$, (being the mean of geometric distribution with parameter ρ) (3.4.5)

and $\operatorname{Var}[T_{01}] = \sigma_{01}^2 = \dfrac{(1-\rho)}{\rho^2}$, (being the variance of geometric distribution with parameter ρ) (3.4.6)

Thus the expected value of T_{00}, the recurrence time for the fecundable nonpregnant state, is a weighted mean of the expected time spent in each of three paths leading to the recurrence,

i.e.
$$\begin{aligned} \mu_{00} = E(T_{00}) &= \sum_{i=2}^{4} \theta_i E[T_{01} + T_{1i}^* + T_{i0}] \\ &= (\theta_2 + \theta_3 + \theta_4) E(T_{01}) + \sum_{i=2}^{4} \theta_i E(T_{1i}^* + T_{i0}) \\ &= E(T_{01}) + \sum_{i=2}^{4} \theta_i (\nu_i + \mu_{i0}) \end{aligned}$$

$(\because \theta_2 + \theta_3 + \theta_4 = 1, E(T_{1i}^*) = \nu_i, E(T_{i0}) = \mu_{i0})$

$$\therefore \quad \mu_{00} = E(T_{00}) = \frac{(1-\rho)}{\rho} + \sum_{i=2}^{4} \theta_i \eta_i, \quad \text{from (3.4.5) and (3.4.2)} \tag{3.4.7}$$

$$\begin{aligned} \text{Similarly, } E(T_{00}^2) &= \sum_{i=2}^{4} \theta_i E(T_{01} + T_{1i}^* + T_{i0})^2 \\ &= \sum_{i=2}^{4} \theta_i E(T_{01}^2) + \sum_{i=2}^{4} \theta_i E[T_{1i}^* + T_{i0}]^2 + 2\sum_{i=2}^{4} \theta_i E[T_{01}(T_{1i}^* + T_{i0})] \\ &= E(T_{01}^2) + \sum_{i=2}^{4} \theta_i E(T_{1i}^* + T_{i0})^2 + 2\sum_{i=2}^{4} \theta_i E(T_{01}) E(T_{1i}^* + T_{i0}) \end{aligned}$$

$\because T_{ij}$'s are independent, (3.4.8)

$$\text{and}\quad \{E(T_{00})\}^2 = \left\{\sum_{i=2}^{4}\theta_i E(T_{01}+T_{1i}^*+T_{i0})\right\}^2$$

$$= \left\{\sum_{i=2}^{4}\theta_i E(T_{01})\right\}^2 + \left\{\sum_{i=2}^{4}\theta_i E(T_{1i}^*+T_{i0})\right\}^2 + 2E(T_{01})\sum_{i=2}^{4}\theta_i E(T_{1i}^*+T_{i0})$$

$$= \{E(T_{01})\}^2 + \sum_{i=2}^{4}\theta_i^2[E(T_{1i}^*+T_{i0})]^2 + 2\sum_{i<j}\theta_i\theta_j[E(T_{1i}^*+T_{i0})E(T_{1j}^*+T_{j0})]$$

$$+ 2E(T_{01})\sum_{i=2}^{4}\theta_i E(T_{1i}^*+T_{i0})$$

$$= [E(T_{01})]^2 + \left(\sum_{i=2}^{4}\theta_i\right)\left(\sum_{i=2}^{4}\theta_i\{E(T_{1i}^*+T_{i0})\}^2\right)$$

$$-\sum_{\substack{i=2\\ i<j}}^{4}\theta_i\theta_j\{E(T_{1i}^*+T_{i0})\}^2 - \sum_{\substack{i=2\\ i<j}}^{4}\theta_i\theta_j\{E(T_{1j}^*+T_{j0})\}^2$$

$$+2\sum_{\substack{i=2\\ i<j}}^{4}\theta_i\theta_j E(T_{1i}^*+T_{i0})E(T_{1j}^*+T_{j0}) + 2E(T_{01})\sum_{i=2}^{4}\theta_i E(T_{1i}^*+T_{i0})$$

$$= [E(T_{o1})]^2 + \sum_{i=2}^{4}\theta_i[E(T_{1i}^*+T_{i0})]^2 - \sum_{\substack{i,j=2\\ i<j}}^{4}\theta_i\theta_j[E(T_{1i}^*+T_{i0})]^2$$

$$-\sum_{\substack{i,j=2\\ i<j}}^{4}\theta_i\theta_j[E(T_{1j}^*+T_{j0})]^2 + 2\sum_{\substack{i,j=2\\ i<j}}^{4}\theta_i\theta_j E(T_{1i}^*+T_{i0})E(T_{1j}^*+T_{j0})$$

$$+2\sum_{i=2}^{4}\theta_i E(T_{o1})E(T_{1i}^*+T_{i0}). \tag{3.4.9}$$

Subtracting (3.4.9) from (3.4.8), we get

$$\text{Var}\,(T_{00}) = E(T_{00}^2) - [E(T_{00})]^2$$

$$= \{E(T_{01}^2) - [E(T_{01})]^2\} + \sum_{i=2}^{4}\theta_i E(T_{1i}^*+T_{i0})^2$$

$$+2\sum_{i=2}^{4}\theta_i E(T_{01})E(T_{1i}^*+T_{i0}) - \sum_{i=2}^{4}\theta_i[E(T_{1i}^*+T_{i0})]^2$$

$$+\sum_{\substack{i,j=2\\ i<j}}^{4}\theta_i\theta_j[E(T_{1i}^*+T_{i0})]^2 + \sum_{\substack{i,j=2\\ i<j}}^{4}\theta_i\theta_j[E(T_{1j}^*+T_{j0})]^2$$

$$-2\sum_{\substack{i,j=2\\ i<j}}^{4}\theta_i\theta_j E(T_{1i}^*+T_{i0})E(T_{1j}^*+T_{j0}) - 2\sum_{i=2}^{4}\theta_i E(T_{01})E(T_{1i}^*+T_{i0})$$

$$= \text{Var } (T_{01}) + \sum_{i=2}^{4} \theta_i \{E(T_{1i}^* + T_{i0})^2 - [E(T_{1i}^* + T_{i0})]^2\}$$

$$+ \sum_{\substack{i,j=2 \\ i<j}}^{4} \theta_i \theta_j \{E(T_{1i}^* + T_{i0}) - E(T_{1j}^* + T_{j0})\}^2$$

$$= \text{Var } (T_{01}) + \sum_{i=2}^{4} \theta_i \text{ Var } (T_{1i}^* + T_{i0}) + \sum_{\substack{i,j=2 \\ i<j}}^{4} \theta_i \theta_j \{E(T_{1i}^* + T_{i0})$$

$$- E(T_{1i}^* + T_{j0})\}^2. \quad (3.4.10)$$

Now $\text{Var } (T_{01}) = \dfrac{(1-\rho)}{\rho^2}$ from (3.4.5), $\text{Var } (T_{1i}^*) = \xi_i^2$ form (3.4.1),

$\text{Var } (T_{i0}) = \sigma_{i0}^2,\ i = 2, 3, 4$

and $\text{Cov } (T_{1i}^*, T_{i0}) = 0$ (This assumption is not strictly justified in this model). Putting $\lambda_i^2 = \xi_i^2 + \sigma_{i0}^2;\ i = 2, 3, 4$ and $\therefore\ \eta_i = \nu_i + \mu_{i0}$, form (3.4.2) where $\nu_i = E(T_{1i}^*)$ and $\mu_{i0} = E(T_{i0})$, (3.4.10) becomes

$$\text{Var } (T_{oo}) = \frac{(1-\rho)}{\rho^2} + \sum_{i=2}^{4} \theta_i \lambda_i^2 + \sum_{\substack{i,j=2 \\ i<j}}^{4} \theta_i \theta_j (\eta_i - \eta_j)^2. \quad (3.4.11)$$

A quantity of special interest is T_{44}, the waiting time between successive live births per female. This time can be represented as follows:

$$T_{44} = T_{40} + T_{01} + T_{11.\bar{4}}^{(1)} + T_{11.\bar{4}}^{(2)} + \cdots + T_{11.\bar{4}}^{(N)} + T_{14}^*, \quad (3.4.12)$$

where $T_{11.\bar{4}}^{(k)}$ represents the time spent in the kth consecutive recurrence of state S_1 accomplished without passage through State S_4. The total number of times, N, which the female cycles between pregnancies without the occurrence of a live birth is a random variable and is easily seen to have (in this model) a geometric distribution with

$$\text{Mean} = E(N) = \frac{\theta_2 + \theta_3}{\theta_4} \text{ and Variance} = \text{Var } (N) = \frac{(\theta_2 + \theta_3)}{\theta_4^2}. \quad (3.4.13)$$

$$\text{Now} \quad T_{11.\bar{4}} = \begin{cases} T_{12}^* + T_{20} + T_{01} \text{ with probability } \dfrac{\theta_2}{\theta_2 + \theta_3}, \\ T_{13}^* + T_{20} + T_{01} \text{ with probability } \dfrac{\theta_2}{\theta_2 + \theta_3}. \end{cases} \quad (3.4.14)$$

$$E(T_{11.\bar{4}}) = \frac{1}{\theta_2 + \theta_3} \sum_{i=2}^{3} \theta_i E(T_{1i}^* + T_{i0} + T_{01})$$

$$= \frac{\theta_2}{\theta_2 + \theta_3} E(T_{12}^* + T_{20} + T_{01}) + \frac{\theta_3}{\theta_2 + \theta_3} E(T_{13}^* + T_{30} + T_{01})$$

$$= \frac{\theta_2}{\theta_2 + \theta_3} E(T_{12}^* + T_{20}) + \frac{\theta_2}{\theta_2 + \theta_3} E(T_{13}^* + T_{30}) + \frac{(\theta_2 + \theta_3)}{(\theta_2 + \theta_3)} E(T_{01})$$

$$= \frac{\theta_2}{\theta_2 + \theta_3} (\nu_2 + \mu_{20}) + \frac{\theta_3}{\theta_2 + \theta_3} (\nu_3 + \mu_{30}) + E(T_{01})$$

$(\because E(T_{i0}) = \mu_{i0},\ E(T_{1i}^*) = \nu_i)$

$$= \frac{\theta_2}{\theta_2+\theta_3}\eta_2 + \frac{\theta_3}{\theta_2+\theta_3}\eta_3 + \frac{(1-\rho)}{\rho}, \text{ form (3.4.2) and (3.4.5)} \tag{3.4.15}$$

Further, $E(T^2_{11.\bar{4}}) = \frac{1}{\theta_2+\theta_3}\sum_{i=2}^{3}\theta_i E(T^*_{1i}+T_{i0}+T_{01})^2$

$$= \frac{1}{\theta_2+\theta_3}\left\{\sum_{i=2}^{3}\theta_i[E(T^*_{1i}+T_{i0})^2 + E(T_{01})^2 + 2E(T_{01})E(T^*_{1i}+T_{i0})]\right\}$$

$$= \frac{1}{\theta_2+\theta_3}\left\{\sum_{i=2}^{3}\theta_i E(T_{01})^2 + \sum_{i=2}^{3}\theta_i E(T^*_{1i}+T_{i0})^2 + 2E(T_{01})\sum_{i=2}^{3}\theta_i E(T^*_{1i}+T_{i0})\right\}$$

$$= \frac{\left(\sum_{i=2}^{3}\theta_i\right)}{\theta_2+\theta_3}E(T^2_{01}) + \frac{1}{\theta_2+\theta_3}\sum_{i=2}^{3}\theta_i E(T^*_{1i}+T_{i0})^2 + \frac{2E(T_{01})}{\theta_2+\theta_3}\sum_{i=2}^{3}\theta_i E(T^*_{1i}+T_{i0})$$

or $$E(T^2_{11.\bar{4}}) = E(T^2_{01}) + \frac{1}{\theta_2+\theta_3}\sum_{i=2}^{3}\theta_i E(T^*_{1i}+T_{i0})^2 + \frac{2E(T_{01})}{\theta_2+\theta_3}\sum_{i=2}^{3}\theta_i E(T^*_{1i}+T_{i0}), \tag{3.4.16}$$

Similarly, $[E(T_{11.\bar{4}})]^2 = \left\{\frac{1}{\theta_2+\theta_3}\sum_{i=2}^{3}\theta_i E(T^*_{1i}+T_{i0}+T_{01})\right\}^2$

$$= \frac{1}{(\theta_2+\theta_3)^2}\left\{\sum_{i=2}^{3}\theta_i E(T^*_{1i}+T_{i0}) + \sum_{i=2}^{3}\theta_i E(T_{01})\right\}^2$$

$$= \frac{1}{(\theta_2+\theta_3)^2}\left\{\left[\sum_{i=2}^{3}\theta_i E(T^*_{1i}+T_{i0})\right]^2 + (E(T_{01}))^2\left(\sum_{i=2}^{3}\theta_i\right)^2 + 2E(T_{01})\left(\sum_{i=2}^{3}\theta_i\right)\sum_{i=2}^{3}\theta_i E(T^*_{1i}+T_{i0})\right\}$$

$$= \frac{1}{(\theta_2+\theta_3)^2}\sum_{i=2}^{3}\theta_i^2[E(T^*_{1i}+T_{i0})]^2 + \frac{2}{(\theta_2+\theta_3)^2}\cdot\theta_2\theta_3 E(T^*_{12}+T_{20})(T^*_{13}+T_{30}) + [E(T_{01})]^2 + \frac{2E(T_{01})}{(\theta_2+\theta_3)}\sum_{i=2}^{3}\theta_i E(T^*_{1i}+T_{i0})$$

$$= \frac{1}{(\theta_2+\theta_3)^2}\left\{\left(\sum_{i=2}^{3}\theta_i\right)\sum_{i=2}^{3}\theta_i[E(T_{1i}^*+T_{i0})]^2 - \theta_2\theta_3 E(T_{12}^*+T_{20})^2 - \theta_2\theta_3 E(T_{13}^*+T_{30})^2\right\}$$

$$+ \frac{2\theta_2\theta_3}{(\theta_2+\theta_3)^2}E(T_{12}^*+T_{20})E(T_{13}^*+T_{30}) + [E(T_{01})]^2$$

$$+ \frac{2E(T_{01})}{(\theta_2+\theta_3)}\sum_{i=2}^{3}\theta_i E(T_{1i}^*+T_{i0}) \qquad (3.4.17)$$

$$\therefore \ \text{Var}\,(T_{11.\bar{4}}) = E(T_{11.\bar{4}}^2) - [E(T_{11.\bar{4}})]^2$$

$$= E(T_{01}^2) - [E(T_{01})]^2 + \frac{1}{\theta_2+\theta_3}\sum_{i=2}^{3}\theta_i\{E(T_{1i}^*+T_{i0})^2 - [E(T_{1i}^*+T_{i0})]^2\}$$

$$- [E(T_{1i}^*+T_{i0})]^2\} + \frac{\theta_2\theta_3}{(\theta_2+\theta_3)^2}\{E(T_{12}^*+T_{20}) - E(T_{13}^*+T_{30})\}^2,$$

from (3.4.16) and (3.4.17)

$$= \text{Var}\,(T_{01}) + \frac{1}{(\theta_2+\theta_3)}\sum_{i=2}^{3}\theta_i \ \text{Var}\,(T_{1i}^*+T_{i0})$$

$$+ \frac{\theta_2\theta_3}{(\theta_2+\theta_3)^2}[(\nu_2+\mu_{20}) - (\nu_3+\mu_{30})]^2$$

$$\therefore \ \text{Var}(T_{11.\bar{4}}) = \frac{(1-\rho)}{\rho^2} + \frac{1}{(\theta_2+\theta_3)}\sum_{i=2}^{3}\theta_i\lambda_i^2 + \frac{\theta_2\theta_3}{(\theta_2+\theta_3)^2}(\eta_2-\eta_3)^2,$$

from (3.4.3) and (3.4.2). (3.4.18)

Also from (3.4.12) we have

$$E(T_{44}) = E(T_{40}) + E(T_{01}) + E\left(T_{11.\bar{4}}^{(1)}\right) + E\left(T_{11.\bar{4}}^{(2)}\right) + \cdots$$

$$+ E\left(T_{11.\bar{4}}^{(N)}\right) + E(T_{14}^*), \text{ from (3.4.5) and (3.4.1)} \qquad (3.4.19)$$

$$= \mu_{40} + \frac{(1-\rho)}{\rho} + \underbrace{E(T_{11.\bar{4}}) + E(T_{11.\bar{4}}) + \cdots + E(T_{11.\bar{4}})}_{\text{upto some terms } N, \text{ which is a random variable}} + \nu_4 \qquad (3.4.20)$$

$$= \mu_{40} + \frac{(1-\rho)}{\rho} + E(N)E(T_{11.\bar{4}}) + \nu_4 \qquad (3.4.21)$$

$$= \mu_{04} + \frac{(1-\rho)}{\rho} + E(N)\left\{\frac{\sum_{i=2}^{3}\theta_i\eta_i}{\theta_2+\theta_3} + \left(\frac{1-\rho}{\rho}\right)\right\} + \nu_4, \text{ (from (3.4.15))}$$

Also we know that,

$$E(N) = \frac{\theta_2+\theta_3}{\theta_4}, \quad \text{(form (3.4.13)) and Var}(N) = \frac{\theta_2\theta_3}{\theta_4^2} \qquad (3.4.22)$$

$$\therefore\ E(T_{44}) = \left[\mu_{40} + \frac{(1-\rho)}{\rho}\right] + \left[\frac{\sum_{i=2}^{3}\theta_i\eta_i}{\theta_2+\theta_3} + \frac{(1-\rho)}{\rho}\right]\cdot\frac{\theta_2+\theta_3}{\theta_4} + \nu_4$$

$$= \mu_{40} + \frac{(1-\rho)}{\rho} + \frac{\sum_{i=2}^{3}\theta_i\eta_i}{\theta_4} + \frac{(1-\rho)}{\rho}\cdot\frac{(\theta_2+\theta_3)}{\theta_4} + \nu_4$$

$$= \nu_4 + \mu_{40} + \frac{(1-\rho)}{\rho}\left\{1 + \frac{(\theta_2+\theta_3)}{\theta_4}\right\} + \frac{\sum_{i=2}^{3}\theta_i\eta_i}{\theta_4}$$

$$= \eta_4 + \frac{(1-\rho)}{\rho}\left\{\frac{(\theta_2+\theta_3+\theta_4)}{\theta_4}\right\} + \frac{\sum_{i=2}^{3}\theta_i\eta_i}{\theta_4}$$

$$= \frac{\theta_4\eta_4}{\theta_4} + \frac{(1-\rho)}{\rho}\cdot\frac{1}{\theta_4} + \frac{\sum_{i=2}^{3}\theta_i\eta_i}{\theta_4}, \qquad \because\ \theta_2+\theta_3+\theta_4 = 1$$

$$\text{or } E(T_{44}) = \frac{1}{\theta_4}\left\{\frac{1-\rho}{\rho} + \sum_{i=2}^{4}\theta_i\eta_i\right\}. \tag{3.4.23}$$

Precisely in a similar way

$$E(T_{22}) = \mu_{22} = \text{Average waiting time between mis-carriages}$$

$$= \frac{1}{\theta_2}\left\{\frac{(1-\rho)}{\rho} + \sum_{i=2}^{4}\theta_i\eta_i\right\} \text{ and} \tag{3.4.24}$$

$$E(T_{33}) = \mu_{33} = \text{Average waiting time between still births}$$

$$= \frac{1}{\theta_3}\left\{\frac{(1-\rho)}{\rho} + \sum_{i=2}^{4}\theta_i\eta_i\right\} \tag{3.4.25}$$

Now to find the variance of T_{44}, the waiting time between live births,

$$\text{Var }(T_{44}) = E(T_{44}^2) - [E(T_{44})]^2,$$

first consider

$$T_{44} - E(T_{44}) = T_{44} - \left[\mu_{40} + \frac{(1-\rho)}{\rho} + E(N)E(T_{11.\bar{4}}) + \nu_4\right], \text{ from (3.4.19)}$$

$$= (T_{40} - \mu_{40}) + \left(T_{01} - \frac{1-\rho}{\rho}\right) + (T_{14}^{*} - \nu_4)$$

$$+ \left(T_{11.\bar{4}}^{(1)} - E(T_{11.4})\right) + \cdots + \left(T_{11.\bar{4}}^{(N)} - E(T_{11.\bar{4}})\right) + NE(T_{11.\bar{4}}) - E(N)E(T_{11.\bar{4}}),$$

$$\text{from (3.4.12) and (3.4.5).} \tag{3.4.26}$$

Squaring both sides and taking expectation, we get

$$\begin{aligned}\text{Var }(T_{44}) &= \sigma_{44}^2 = E(T_{44} - E(T_{44}))^2 \\ &= E(T_{40} - \mu_{40})^2 + E(T_{01} - \mu_{01})^2 + E(T_{14}^* - \nu_4)^2 \\ &\qquad + E(N)E[T_{11.\bar{4}} - E(T_{11.\bar{4}})]^2 + [E(T_{11.\bar{4}})]^2 E[N - E(N)]^2 \\ &= \lambda_4^2 + \sigma_{01}^2 + E(N)\text{Var}(T_{11.\bar{4}}) + [E(T_{11.\bar{4}})]^2 \text{Var}(N) \qquad (3.4.27)\end{aligned}$$

$$\because E(T_{40} - E(T_{40}))^2 = \sigma_{40}^2, E(T_{14}^* - \nu_4)^2 = \xi_4^2 \text{ and } \xi_4^2 + \sigma_{40}^2 = \lambda_4^2 \qquad (3.4.28)$$

$$\therefore \text{ Var }(T_{44}) = \lambda_4^2 + \frac{(1-\rho)}{\rho^2} + \frac{(\theta_2+\theta_3)}{\theta_4}\text{Var}(T_{11.\bar{4}}) + \frac{(\theta_2+\theta_3)}{\theta_4^2}\{E(T_{11.\bar{4}})\}^2$$

$$= \lambda_4^2 + \frac{(1-\rho)}{\rho^2} + \frac{(\theta_2+\theta_3)}{\theta_4}\left\{\frac{(1-\rho)}{\rho^2} + \frac{1}{(\theta_2+\theta_3)}\sum_{i=2}^{3}\theta_i\lambda_i^2 \right.$$

$$\left. + \frac{\theta_2\theta_3}{(\theta_2+\theta_3)^2}(\eta_2-\eta_3)^2\right\} + \frac{(\theta_2+\theta_3)}{\theta_4^2}\left\{\frac{\theta_2\eta_2}{\theta_2+\theta_3} + \frac{\theta_3\eta_3}{\theta_2+\theta_3} + \frac{(1-\rho)}{\rho}\right\}^2,$$

(using (3.4.15) and (3.4.18)).

$$= \lambda_4^2 + \frac{(1-\rho)}{\rho^2}\left\{1 + \frac{\theta_2+\theta_3}{\theta_4}\right\} + \frac{1}{\theta_4}\sum_{i=2}^{3}\theta_i\lambda_i^2$$

$$+ \frac{\theta_2\theta_3}{(\theta_2+\theta_3)\theta_4}\cdot(\eta_2-\eta_3)^2 + \frac{(\theta_2+\theta_3)}{\theta_4^2}\left\{\frac{\theta_2^2\eta_2^2}{(\theta_2+\theta_3)^2} + \frac{\theta_3^2\eta_3^2}{(\theta_2+\theta_3)^2}\right.$$

$$\left. + \frac{(1-\rho)^2}{\rho^2} + \frac{2\theta_2\theta_3}{(\theta_2+\theta_3)^2}\eta_2\eta_3 + \frac{2\theta_2\eta_2}{(\theta_2+\theta_3)}\frac{(1-\rho)}{\rho} + \frac{2\theta_3\eta_3}{(\theta_2+\theta_3)}\cdot\frac{(1-\rho)}{\rho}\right\}$$

$$= \frac{\theta_4\lambda_4^2}{\theta_4} + \frac{(1-\rho)}{\rho^2}\cdot\frac{1}{\theta_4} + \frac{1}{\theta_4}\sum_{i=2}^{3}\theta_i\lambda_i^2 + \frac{\theta_2\theta_3}{\theta_4(\theta_2+\theta_3)}(\eta_2-\eta_3)^2$$

$$+ \frac{\theta_2^2}{\theta_4^2(\theta_2+\theta_3)}\eta_2^2 + \frac{\theta_3^2}{\theta_4^2(\theta_2+\theta_3)}\eta_3^2 + \frac{(\theta_2+\theta_3)}{\theta_4^2}\cdot\frac{(1-\rho)^2}{\rho^2}$$

$$+ \frac{2\theta_2\theta_3}{\theta_4^2(\theta_2+\theta_3)}\eta_2\eta_3 + \frac{2\theta_2\eta_2}{\theta_4^2}\cdot\frac{(1-\rho)}{\rho} + \frac{2\theta_3\eta_3}{\theta_4^2}\cdot\frac{(1-\rho)}{\rho}$$

$$= \frac{(1-\rho)}{\theta_4\cdot\rho} + \frac{1}{\theta_4}\sum_{i=2}^{4}\theta_i\lambda_i^2 + \frac{\theta_2\theta_3}{\theta_4(\theta_2+\theta_3)}(\eta_2-\eta_3)^2 + \frac{\theta_2^2\eta_2^2}{\theta_4^2(\theta_2+\theta_3)}$$

$$+ \frac{\theta_3^2\eta_3^2}{\theta_4^2(\theta_2+\theta_3)} + \frac{(\theta_2+\theta_3)}{\theta_4^2}\frac{(1-\rho)^2}{\rho^2} + \frac{2\theta_2\theta_3}{\theta_4^2(\theta_2+\theta_3)}\eta_2\eta_3$$

$$+ \frac{2\theta_2\eta_2}{\theta_4^2}\cdot\frac{(1-\rho)}{\rho} + \frac{2\theta_3\eta_3}{\theta_4^2}\frac{(1-\rho)}{\rho}$$

$$= \frac{1}{\theta_4}\cdot\frac{(1-\rho)}{\rho} + \frac{1}{\theta_4}\sum_{i=2}^{4}\theta_i\lambda_i^2 + \frac{(\theta_2+\theta_3)}{\theta_4^2}\frac{(1-\rho)^2}{\rho^2} + \frac{2}{\theta_4}\frac{(1-\rho)}{\rho}\sum_{i=2}^{3}\theta_i\eta_i$$

$$+ \frac{\theta_2\theta_3}{\theta_4(\theta_2+\theta_3)}(\eta_2-\eta_3)^2 + \frac{1}{\theta_4^2(\theta_2+\theta_3)}\{\theta_2^2\eta_2^2 + \theta_3^2\eta_3^2 + 2\theta_2\theta_3\eta_2\eta_3\}$$

$$\therefore\ \mathrm{Var}(T_{44}) = \frac{1}{\theta_4}\cdot\frac{(1-\rho)}{\rho} + \frac{1}{\theta_4}\sum_{i=2}^{4}\theta_i\lambda_i^2 + \frac{(\theta_2+\theta_3)}{\theta_4^2}\frac{(1-\rho)^2}{\rho^2} + \frac{2}{\theta_4}\frac{(1-\rho)}{\rho}\sum_{i=2}^{3}\theta_i\eta_i$$

$$+\frac{\theta_2\theta_3}{\theta_4(\theta_2+\theta_3)}(\eta_2-\eta_3)^2 + \frac{1}{\theta_4^2(\theta_2+\theta_3)}\{\theta_2\eta_3+\theta_3\eta_3\}^2 \quad (3.4.29)$$

It is also of interest to determine the expected length and variance of the waiting time T_{04} between the onset of marriage and the first live birth. This is easily done since

$$T_{44} = T_{40} + T_{04} \quad (3.4.30)$$

and therefore $\mu_{04} = E(T_{04}) = \mu_{44} - \mu_{40}$

$$= \nu_4 + \mu_{40} + \frac{1}{\theta_4}\left(\theta_2\eta_2 + \theta_3\eta_3 + \frac{(1-\rho)}{\rho}\right) - \mu_{40}$$

$$= \nu_4 + \frac{1}{\theta_4}\left\{\theta_2\eta_2 + \theta_3\eta_3 + \frac{(1-\rho)}{\rho}\right\} \quad (3.4.31)$$

And $\mathrm{Var}\,(T_{04}) = \sigma_{04}^2 = \sigma_{44}^2 - \sigma_{40}^2$, form (3.4.30) as T_{40} and T_{04} are independent.

$$= \xi_4^2 + \frac{(1-\rho)}{\rho^2} + \frac{(\theta_2+\theta_3)}{\theta_4}\cdot\mathrm{Var}(T_{11.\bar{4}}) + \frac{(\theta_2+\theta_3)}{\theta_4^2}[E(T_{11.\bar{4}})]^2,$$

from (3.4.27), (3.4.28), (3.4.22) and (3.4.6).

$$= \xi_4^2 + \frac{(1-\rho)}{\rho^2}\cdot\frac{1}{\theta_4} + \frac{1}{\theta_4}\sum_{i=2}^{3}\theta_i(\lambda_i^2+\eta_i^2) + \frac{(\theta_2+\theta_3)}{\theta_4}\cdot\frac{(1-\rho)}{\rho^2}$$

$$+\frac{2}{\theta_4}\cdot\left(\frac{1-\rho}{\rho}\right)\sum_{i=2}^{3}\theta_i\eta_i, \text{ from (3.4.18), (3.4.17) and notations} \quad (3.4.32)$$

The analogous expression for the first moment of the waiting time between marriage and the first mis-carriage or still birth (i.e. of T_{02} and T_{03}) may similarly be derived. Application of the above results to the study of intervals to the first live birth and between consecutive births also have been described by Sheps and Perrin.

References

1. Berclay, G.W. (1958): Techniques of population analysis, John Wiley & Sons Inc. New York.
2. Brass, William (1958): The distribution of births in Human Population, Population studies; Vol. 12, pp. 51–72.
3. Biswas, S. (1973): A note on the generalization of William Brass model: Demography, August, 1973 Vol. 10, No. 3, pp. 450–452.
4. Biswas, S. (1975): On a more generalised probability model of the waiting time of conception based on censored sampling form a mixed population; Sankhya, Series B, Vol; 37 Part 3, August, 1975, pp. 343–354.
5. Biswas, S. (1980): On the extension of some results of counter models with Poisson inputs and their applications; Journal of Indian Statistical Association. Vol. 18, pp. 45–53.

6. Biswas, S. and Nauharia Indu (1980): A note on the development of some interrupted waiting time distribution; Pure and Applied Mathematical Sciences. Vol. XI

7. Biswas, S. and Pachal, T.K. (1981): On the application of Palm probability of obtaining inter-arrival time distribution in weighted Poisson Process; Calcutta Statistical Association Bulletin, Vol. 32, No. 125-126, pp. 111–123.

8. Biswas, S. and Sehgal, V.K. (1986): On the correlation between inter-arrival delays of shocks; Microelectronics and Reliability, Vol. 28, No. 8, pp. 189–192.

9. Biswas, S. and Srestha, Ganga (1984): Waiting time distribution for first conception leading to a live birth, International Report IC/85/98, International Centre of Theoretical Physics, Trieste, Italy.

10. Biswas, S. and Srestha, Ganga (1985): A probability model of the waiting time distribution between consecutive conceptions based on the data of live birth, Metron, Vol. XLIV, Nos. 1-4, page 195–206.

11. Chiang, C.L. (1971): A stochastic model on Human fertility, multiple transition probabilities; Biometrics Vol. 27, pp. 345–356.

12. Dandekar, V.M. (1955): Certain modified forms of Binomial and Poisson distributions, Sankhyā, Series B, Vol. 15, pp. 237–250.

13. Dharmadhikari, S.W. (1964): A generalization of the stochastic model considered by V.M. Dandekar; Sankhya, series A, Vol. 25, pp. 31–38.

14. Iosufescu, M. and Tautu P. (1973): Stochastic Processes and applications in Biology and Medicine II Models: Biomathematics, Vol. 4, Springer-Verlag, Berlin; Helberg, New York.

15. Keyfitz, N. (1977): Introduction to the mathematics of population-with revision, Addison Wesley & Co. London.

16. Neyman, J. (1949): On the problem of estimating the number of schools of fish, University of California, Publication in Statistics I, pp. 21–32.

17. Neyman, J. (1949): Contributions to the theory of the χ^2-test; Proceedings of Berkley Symposium in Mathematics and Statistics, University of California Press, pp. 239-273.

18. Pathak, K.B. (1966): A probability distribution for number of conceptions: Sankhya, Series B, Vol. 28 pp. 213–218.

19. Perrin, E.B. and Sheps, M.C. (1964): Human reproduction A Stochastic Processes: Biomathematics, Vol. 20, pp. 28–45.

20. Poti, S. J. and Biswas, S. (1963): A study of child Health During the First year of Life; Sankhya, Ser. B, Vol. 36, Parts 1 and 2, pp. 85–120.

21. Potter, R.G. and Parker, M.P. (1964): Predicting the time required to conceive, Population Studies, Vol. XVIII.

22. Rogers, A. (1975): Introduction to Mathematical Demography, John Wiley and Sons, New York.

23. Sheps, M.C. and Perrin, E.B. (1966): Further results from Human Fertility model with a variety of pregnancy outcomes, Human Biology, Vol. 8, pp. 180–193.

24. Sheps M.C., Menken, J.A. and Radick, A.P. (1969): Probability models for family building-An analytical review; Demography, Vol. 8, pp. 161–183.

25. Singh S.N. (1964a): On the time of first birth, Sankhyā, Series B, Parts I and II, pp. 95–102.

26. Singh, S.N. (1964a): A Probability model for couple Fertility, Sankhyā, Series B, Vol. 21, pp. 89–94.

27. Takacs, Lajos. (1960): Stochastic Processes, Problems and Solutions: Translated by P. Zador, John Wiley & Sons Inc, New York.

4

Techniques of Demographic Analysis: Population Growth Indices

4.0 Introduction

Given the indices of mortality and fertility as discussed in Chapter-2 and Chapter-3 for a natural population, a question that naturally arises is whether the tendency of the given population is to increase, decrease or to remain more or less stationary over time. Therefore, before any idea about the growth of a population is taken it is necessary to evolve proper indices of population growth. We have seen in Chapter-3 that a fairly good index of population growth is given by "Net Reproduction Rate". But the validity of the Net Reproduction Rate is again subjected to the limitation of fertility and mortality conditions maintaining 'status quo'. Besides that the Net Reproduction Rate essentially reflects the growth in female population. The parallel index for male is difficult to construct. Thus to estimate growth of a population, a good approach would be to evolve certain theoretical models (deterministic or stochastic) describing the growth of populations. Growth parameter may often be estimable from the model itself.

4.1 Measurements of Population Growth

Let us take the simplest situation in which we have a population of N individuals and we assume that population remains closed for Migration i.e. there is neither immigration nor emigration.

The modeling for natural populations have three basic characterizations as follows:

(i) The population over time may show that the average density of population is being maintained at a constant level over a long period of time, unless there is a major environmental change. In other words, population under such a setup neither die out nor explode. This, phenomenon, is called 'Balancing', by Nicholson (1957).

(ii) In the second characterization, the growth of a population need not necessarily remain at a constant level but the same may fluctuate around a constant mean value randomly.

(iii) Finally, a third characterization is given by superimposition of a random cycle of oscillation on the type of random variation, already considered in the second type of characterization, Thus the third characterization is a generalization of the second.

With this setup we propose to develop a simple deterministic model. Here, we ignore such factors as environmental conditions etc. while developing a deterministic model. Further, in a deterministic model the probabilistic considerations relating to the variation of N (which is an integer) is ignored and one can reasonably assume N to be a continuous variable (vide Moran (1961)).

Let $N(t)$ be the population at time t, $B(t)$ and $D(t)$ be the birth rate and death rate, respectively, at time t. Then the relative growth of the population at time t is

$$\frac{1}{N(t)}\frac{dN(t)}{dt} = B(t) - D(t) \tag{4.1.1}$$

$$= r(t), \quad \text{(say)} \tag{4.1.2}$$

Note that $B(t)$ and $D(t)$ are defined w.r.t. the whole population, since we are ignoring the distinction between male and female population.

If $r(t) = r, a$ constant independent of t, then from (4.1.2)

$$\frac{1}{N(t)}\frac{dN(t)}{dt} = r, \tag{4.1.3}$$

$$\Rightarrow \quad \frac{d\log N(t)}{dt} = r \quad \Rightarrow \quad \log N(t) = r\cdot t + a \text{ (an integration constant)}$$

$$\Rightarrow \quad N(t) = Ae^{rt}, \quad \text{where} \quad A = e^{a} \tag{4.1.4}$$

This model (4.1.4) was given by Malthus (1798). This is called Malthusian geometric law of population growth.

Here as $t \to \infty$, $N(t) \to \infty$, which is unrealistic, due to the resources and geographic conditions of the region.

4.2 A Population Density Dependent Growth Model

We may note that the assumptions in (4.1.1) that $B(t)$ and $D(t)$ are independent of N, are not realistic i.e. the reason of a population being at a given level must be explained by the density of the population. In view of this we write (4.1.1) as

$$\left.\begin{aligned}\frac{1}{N(t)}\frac{dN(t)}{dt} &= B(N,t) - D(N,t)\\ &= r(N,t)\end{aligned}\right\} \tag{4.2.1}$$

4.2.1 The Logistic Model for Population Growth

Verhulst (1838) took into account the limitation of Malthus model (4.1.4) and postulated that the relative growth rate $r(t)$ is a decreasing function of $N(t)$.

If M is the maximum population that a region can support, then according to Verhulst $r(t)$ may be taken as

$$r(N,t) = r\left[1 - \frac{N(t)}{M}\right], \tag{4.2.2}$$

where $r(> 0)$ and M are constants.

So, (4.2.1) becomes

$$\frac{1}{N}\frac{dN}{dt} = r[1 - a\cdot N], \quad \text{where} \quad N = N(t) \quad \text{and} \quad a = 1/M. \tag{4.2.3}$$

$$\Rightarrow \quad \frac{dN}{N(1 - a\cdot N)} = r\ dt \quad \Rightarrow \quad \left[\frac{1}{N} + \frac{a}{1 - a\cdot N}\right] dN = r\cdot dt$$

Integrating both side we get

$$\log N - \log(1 - a \cdot N) = r \cdot t + c \text{ (an integration constant).}$$

$$\Rightarrow \quad \log \frac{N}{1 - a \cdot N} = r \cdot t + c \Rightarrow \frac{N}{1 - a \cdot N} = A \cdot e^{r \cdot t}, \quad \text{where} \quad A = e^c$$

$$\Rightarrow \quad N = \frac{1/a}{1 + \frac{1}{a \cdot A} e^{-r \cdot t}}$$

$$\text{or} \quad N(t) = \frac{k}{1 + b.e^{-r \cdot t}} \quad \text{where} \quad k = 1/a \quad \text{and} \quad b = 1/(a \cdot A). \tag{4.2.4}$$

This is a solution (4.2.3).

The equation (4.2.4) is known as Logistic law of Growth; k, b and r are three parameters of the distribution. This is discussed in the following sections.

The initial value at $t = 0$, i.e. $N(0)$ is given by

$$N(0) = \frac{k}{1 + b} \tag{4.2.5}$$

The limiting value of $N(t)$ as $t \to \infty$, converges as

$$\lim_{t \to \infty} N(t) = N(\infty) = k, \text{ from (4.2.4).} \tag{4.2.6}$$

Now, from (4.2.5) and (4.2.6)

$$N(0) < N(\infty) \;\; \Rightarrow \;\; 1 + b > 1 \Rightarrow b > 0$$

4.2.2 Properties of Logistic Curve

Let us denote $N(t) = y(t) = y$, then the equation of logistic curve (4.2.4) becomes

$$y(t) = \frac{k}{1 + b.e^{-r \cdot t}} \tag{4.2.7}$$

$$\text{Then} \quad \frac{dy}{dt} = \frac{k \cdot b \cdot r \cdot e^{-r \cdot t}}{(1 + b \cdot e^{-r \cdot t})^2} = \frac{b.e^{-r \cdot t} r \cdot k}{(1 + b \cdot e^{-r \cdot t})(1 + b \cdot e^{-r \cdot t})}.$$

Now, from (4.2.7)

$$1 + b.e^{-r \cdot t} = \frac{k}{y} \Rightarrow b.e^{-r \cdot t} = \frac{k}{y} - 1$$

$$\Rightarrow \quad \frac{b.e^{-r \cdot t}}{1 + b.e^{-r \cdot t}} = \left(\frac{k}{y} - 1\right) \Big/ \left(\frac{k}{y}\right)$$

$$\therefore \quad \frac{dy}{dt} = r \cdot y \left(\frac{k}{y} - 1\right) \Big/ \left(\frac{k}{y}\right) = r \cdot y \left(1 - \frac{y}{k}\right) \tag{4.2.8}$$

$$\text{Hence,} \quad \frac{d^2 y}{dt^2} = r \cdot \frac{dy}{dt}\left(1 - 2\frac{y}{k}\right) = r^2.y\left(1 - \frac{y}{k}\right)\left(1 - 2\frac{y}{k}\right) \tag{4.2.9}$$

From(4.2.8), $\frac{dy}{dt} = 0$ at $y = 0$ and $y = k$ i.e. at $t = -\infty$ and $t = \infty$, from (4.2.7).

Thus $y = 0$ and $y = k$ are two asymptotes of the logistic curve (4.2.7).

Further, from (4.2.8)

$$0 < y < k \quad \Rightarrow \quad \frac{dy}{dt} > 0$$

i.e. logistic curve $y(t)$ is a monotonically increasing curve.

Again, from (4.2.9),

$$\frac{d^2y}{dt^2} = 0 \quad \Rightarrow \quad 2\frac{y}{k} = 1 \quad \Rightarrow \quad y = k/2, \textbf{ is the point of inflection.}$$

The logistic curve $y(t)$ given by (4.2.7) **may be represented** as given below; for $y < k$, it is a S-shaped curve.

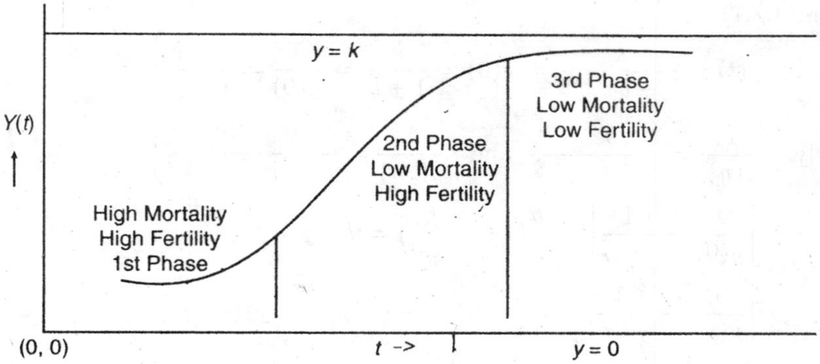

Fig. 4.1

4.2.3 A Critical Review of the Logistic Model

Verhulst (1838) evolved the Logistic curve while experimenting with growth of the insects under controlled environmental condition. Also Logistic curve derived by other similar assumption has, in the past, given good fit to the population data of several countries as well as to other kinds of data. As a matter of fact, the Logistic model with three parameters often gives a closed fit to the population data taken over periods of time. But as cautioned by Fellar (1940) closed fit of the data with the logistic does not necessarily give any evidence that the underlying process of growth follows a logistic pattern. As such the population projection by fitting of logistic curve remains often a very questionable issue. A result due to Leslie (1948) in this respect may appear to be quite interesting. Leslie has shown that if the age specific fertility and mortality rates are also density dependent i.e. they depend on the size of the population then the population growth rate will acquire a logistic pattern provided the initial age distribution from which we start is stable or stationary. The implication of this result is that if the initial age distribution is not stable then the growth rate will not acquire a logistic from. But Indian population, which is far from having a stable age distribution, has given quite good fit to the logistic distribution. This is quite misleading Demographically. It is, therefore, advisable to examine critically the theoretical premises in support of a model relating to a population growth curve, (especially logistic) before the same is employed for the measurement of the population growth or projection purpose.

Further, it is true that a logistic model for predicting the growth of a population has all the defects and limitations, which a single equation model has while predicting the course of a population.

4.3 Methods of Fitting Logistic Curves

In order to fit the Logistic curve (4.2.7) we are to estimate the parameters k, b and r. There are many methods of fitting a Logistic curve, we discuss here a few more common ones.

4.3.1 Method of Three Points

The simplest method of fitting a Logistic curve is to choose three equidistant points $t = 0,\ t = n$ and $t = 2n$ with corresponding populations $y(0),\ y(n)$ and $y(2n)$, then estimate the parameters while making the curve pass through $y(0),\ y(n)$ and $y(2n)$. The procedure is as follows:

From (4.2.7)

$$\left.\begin{array}{l} y(t) = \dfrac{k}{1+be^{-r\cdot t}} \Rightarrow y(0) = \dfrac{k}{1+b} \Rightarrow \dfrac{1}{y(0)} = \dfrac{1+b}{k} \\ \text{Similarly,}\quad \dfrac{1}{y(n)} = \dfrac{1+be^{-r\cdot n}}{k} \quad \text{and} \quad \dfrac{1}{y(2n)} = \dfrac{1+be^{-r\cdot 2n}}{k} \end{array}\right\} \tag{4.3.1}$$

$$\text{Then,}\quad \left[\frac{1}{y(0)} - \frac{1}{y(n)}\right] = \frac{b}{k}(1-e^{-r\cdot n}) = d_1,\ \text{(say)} \tag{4.3.2}$$

$$\left[\frac{1}{y(n)} - \frac{1}{y(2n)}\right] = \frac{b}{k}e^{-r\cdot n}(1-e^{-r\cdot n}) = d_2,\ \text{(say)} \tag{4.3.3}$$

$$\therefore \quad \frac{d_1}{d_2} = e^{r\cdot n} \Rightarrow \hat{r} = \frac{1}{n}\log\left(\frac{d_1}{d_2}\right) \tag{4.3.4}$$

$$\text{Again,}\quad \frac{d_1}{d_2} = e^{r\cdot n} \Rightarrow \frac{d_1-d_2}{d_1} = 1-e^{r\cdot n} \Rightarrow \frac{d_1-d_2}{d_1} = \frac{kd_1}{b},\ \text{from (4.3.2)}$$

$$\Rightarrow \frac{d_1-d_2}{d_1^2} = \frac{k}{b} \Rightarrow \frac{d_1^2}{d_1-d_2} = \frac{b}{k} = \left(\frac{1}{y(0)} - \frac{1}{k}\right),\ \text{from(4.3.1)} \tag{4.3.5}$$

$$\Rightarrow \frac{1}{k} = \frac{1}{y(0)} - \frac{d_1^2}{d_1-d_2}$$

$$\Rightarrow \hat{k} = 1/\left[\frac{1}{y(0)} - \frac{d_1^2}{d_1-d_2}\right] \tag{4.3.6}$$

Further, from (4.3.5)

$$\hat{b} = \left(\frac{k}{y(0)} - 1\right) \tag{4.3.7}$$

Thus, the estimates of $b,\ r$ and k viz. $\hat{b},\ \hat{r}$ and $\hat{k}$ can be obtained by using (4.3.7), (4.3.4) and (4.3.6), respectively.

Then, the fitted curve is given by

$$y(t) = \frac{\hat{k}}{1+\hat{b}e^{-\hat{r}\cdot t}} \tag{4.3.8}$$

Remark 4.1: This method has its obvious short comings. The three points taken here can be chosen arbitrarily which will give different set of estimates of (k, b, r). Of course, this method has an advantage of quick and easy estimation. An improvement of this method is the following method of 'Sums of Reciprocals'.

4.3.2 Method of Sums of Reciprocals

This method utilizes the entire set of data provided the number of observations N is a multiple of 3.

Let, $N = 3m$ and $t = 0, 1, 2, \ldots, (N-1)$, are N points of observations.

Let, $$S_1 = \sum_{t=0}^{m-1} \frac{1}{y(t)}; S_2 = \sum_{t=m}^{2m-1} \frac{1}{y(t)} \quad \text{and} \quad S_3 = \sum_{t=2m}^{N-1} \frac{1}{y(t)} \tag{4.3.9}$$

For the Logistic curve (4.3.1), $\dfrac{1}{y(t)} = \dfrac{1+be^{-r\cdot t}}{k}$

Hence, $$S_1 = \sum_{t=0}^{m-1} \frac{1}{y(t)} = \sum_{t=0}^{m-1} \frac{1+be^{-r\cdot t}}{k}$$

$$= \frac{m}{k} + \frac{b}{k}\left[1 + e^{-r} + e^{-2\cdot r} + \cdots + e^{-(m-1)}\right]$$

$$= \frac{m}{k} + \frac{b}{k}\frac{(1-e^{m\cdot r})}{(1-e^{-r})} = \frac{m}{k} + \frac{b}{k}c \text{ (say)}, \tag{4.3.10}$$

where, $$c = \frac{(1-e^{m\cdot r})}{(1-e^{-r})}. \tag{4.3.11}$$

Similarly, $$S_2 = \frac{m}{k} + \frac{b}{k}c\, e^{-m\cdot r} \tag{4.3.12}$$

and $$S_3 = \frac{m}{k} + \frac{b}{k}c\, e^{-2m\cdot r} \tag{4.3.13}$$

Then, $$S_1 - S_2 = \frac{b\cdot c}{k}(1 - c^{-m\cdot r}) = D_1, \text{ (say)} \tag{4.3.14}$$

and $$S_2 - S_3 = \frac{b\cdot c}{k}e^{-m\cdot r}(1 - e^{-m\cdot r}) = D_2, \text{ (say)} \tag{4.3.15}$$

$$\Rightarrow \quad e^{-m\cdot r} = \frac{D_1}{D_2} \Rightarrow \hat{r} = \frac{1}{m}\log\frac{D_1}{D_2} \tag{4.3.16}$$

Further, $$\frac{D_2}{D_1} = e^{-m\cdot r} \quad \text{so that} \quad \frac{D_1 - D_2}{D_1} = 1 - e^{-m\cdot r} = \frac{k\cdot D_1}{bc}, \quad \text{from (4.3.14)}$$

$$\therefore \quad \frac{D_1^2}{D_1 - D_2} = \frac{bc}{k} = S_1 - \frac{m}{k} = d, \text{ (say), from (4.3.10)} \tag{4.3.17}$$

$$\hat{k} = \frac{m}{S_1 - d}. \tag{4.3.18}$$

Again, $$c = \frac{(1-e^{m\cdot r})}{(1-e^{-r})}, \quad \text{from (4.3.11)}$$

$$= \frac{kD_1}{b\cdot c(1-e^{-r})}, \quad \text{from (4.3.14)}$$

$$= \frac{D_1}{d\cdot(1-e^{-r})}, \quad \text{from (4.3.17).} \tag{4.3.19}$$

Having estimated $\hat{r}$ and $\hat{k}$, from (4.3.16) and (4.3.18), we get an estimate of b from the relation

$$\hat{b} = \frac{kD_1^2}{c(D_1 - D_2)} = \frac{kS_1 - m}{c}, \text{ where } c \text{ is given by (4.3.19)} \tag{4.3.20}$$

Remark 4.2: Method of three points and sums of reciprocals are basically the same in procedure. In the former we choose 3 arbitrary equidistant points where as in the latter we consider 3 sets of figures of equal length.

4.3.3 Rhodes Method of Fitting a Logistic Curve

Rhodes (1940) used the method of regression of the reciprocal of an observation, say $y(t+h)$, on the reciprocal of another nearer observation, say $y(t)$.

Let the population figures be given for N equidistant points of time, say $t = 0, 1, 2, \ldots, (N-1)$. If the observed population figures were given exactly by the Logistic equation (4.2.7), then we have,

$$\frac{1}{y(t)} = \frac{1+be^{-r\cdot t}}{k} \tag{4.3.21}$$

and $$\frac{1}{y(t+h)} = \frac{1+be^{-r\cdot(t+h)}}{k} = \frac{1+be^{-r\cdot h}}{k} + e^{-r\cdot h}\left(\frac{1+be^{-r\cdot t}}{k}\right)$$

$$\frac{1}{y(t+h)} = \frac{1+be^{-r\cdot h}}{k} + e^{-r\cdot h}\frac{1}{y(t)}, \text{ from (4.3.21)} \tag{4.3.22}$$

or $$Z(t) = A + B\,X(t) \tag{4.3.23}$$

where $$Z(t) = \frac{1}{y(t+h)},\ X(t) = \frac{1}{y(t)},\ A = \frac{1+be^{-r\cdot h}}{k} \text{ and } b = e^{-r\cdot h} \tag{4.3.24}$$

Rhodes observed that the parameters A and B of the above regression of $\frac{1}{y(t+h)}$ on $\frac{1}{y(t)}$ as given in (4.3.23) should not be estimated by ordinary least squares method, because both the variables $\frac{1}{y(t+h)}$ and $\frac{1}{y(t)}$ are subject to errors.

Rhodes (1940) obtained the estimates of B and A as

$$\hat{B} = \sqrt{\frac{\sum_{i=1}^{N-1}(Z_i - \overline{Z})^2}{\sum_{i=1}^{N-1}(X_i - \overline{X})^2}} \quad \text{and} \quad \hat{A} = \overline{Z} - \hat{B}\overline{X} \tag{4.3.25}$$

where $\overline{X} = \frac{1}{N}\sum_{i=0}^{N-1} X_i$ and $\overline{Z} = \frac{1}{N}\sum_{i=0}^{N-1} Z_i$.

The constants k and r of the Logistic equation are estimated given the estimates of A and B from (4.3.25), as

$$\hat{B} = e^{-\hat{r}\cdot h}, \text{ from (4.3.24)}$$

$$\Rightarrow \quad \hat{r} = \frac{1}{h}[-\log \hat{B}]. \tag{4.3.26}$$

and $$\hat{k} = \frac{1}{\hat{A}}(1 - e^{-\hat{r}\cdot h}), \text{ from (4.3.24)} \tag{4.3.27}$$

To obtain $\hat{b}$ we proceed as follows: From (4.3.21)

$$\frac{k}{y(t)} = 1 + be^{-r\cdot t} \Rightarrow \log\left(\frac{k}{y(t)} - 1\right) = \log b - r\cdot t$$

$$\Rightarrow \quad \frac{1}{N}\sum_{t=0}^{N-1}\log\left(\frac{k}{y(t)}-1\right) = \log b - \frac{r}{N}\sum_{t=0}^{N-1} t = \log b - \frac{rN(N-1)}{2N}$$

$$\Rightarrow \quad \frac{1}{N}\sum_{t=0}^{N-1}\log\left(\frac{\hat{k}}{y(t)}-1\right) + \frac{\hat{r}(N-1)}{2} = \log\hat{b}. \qquad (4.3.28)$$

On substituting the estimates of r and k from (4.3.26) and (4.3.27) in (4.3.28), the estimate of b is obtained and substituting these values in (4.2.7) we get the fitted Logistic curve.

4.3.4 Fisher and Hotelling Methods of Fitting Logistic Curves

Here, instead of regressing $\dfrac{1}{y(t+h)}$ on $\dfrac{1}{y(t)}$, regression of $\dfrac{d\log y(t)}{dt} = Z(t)$, say, on $y(t)$ was considered. These two methods differ in expressing Z in terms of observations. Fisher uses Sterling's Central Difference formula neglecting the 3rd and higher differences whereas Hotelling did not use any such approximation.

In both the methods we consider

$$y(t) = \frac{k}{1+be^{-r\cdot t}} \quad \Rightarrow \quad \frac{dy(t)}{dt} = \frac{rkbe^{-r\cdot t}}{(1+be^{-r\cdot t})^2} = ry(t)\left[1-\frac{y(t)}{k}\right] \qquad (4.3.29)$$

$$\Rightarrow \quad \frac{1}{y(t)}\frac{dy(t)}{dt} = Z(t),\ \text{(say)} = r\left[1-\frac{y(t)}{k}\right]. \qquad (4.3.30)$$

If we are able to estimate Z then the estimated values of Z can be regressed on $y(t)$; which provides estimates of r and k. Up to this point Fisher's and Hotelling's method remain the same. However, the method of estimation of r and k by the two methods are different.

Estimation of Z

(i) By Fisher's technique

From (4.3.30),

$$Z_1(t) = Z_i = \left(\frac{1}{y}\frac{dy}{dt}\right)_{t=i} = \left(\frac{d\log y}{dt}\right)_{t=i} = r\left(1-\frac{y_i}{k}\right), \text{ where } y \equiv y(t). \qquad (4.3.31)$$

Fisher expanded $\left(\dfrac{d\log y}{dt}\right)_{t=i}$ by Stirling's central difference formula, under the assumption that the third and higher order differences can be neglected, as

$$\left(\frac{d\log y}{dt}\right)_{t=i} = \frac{1}{2}\{\Delta\log y_i + \Delta\log y_{i-1}\} + O(\Delta^3) \qquad (4.3.32)$$

$$\cong \frac{1}{2}\{\log y_{i+1} - \log y_i + \log y_i - \log y_{i-1}\}$$

$$= \frac{1}{2}\log\frac{y_{i+1}}{y_{i-1}};\ i = 1, 2, \ldots, N-2.$$

Hence, the estimate of Z_1 is given by

$$\hat{Z}_i = \frac{1}{2}\log\frac{y_{i+1}}{y_{i-1}}; \qquad (4.3.33)$$

(ii) By Hotelling's technique

Hotelling considered 3 consecutive values of population and calculated $Z(t) = \frac{1}{y(t)}, \frac{dy(t)}{dt}$, corresponding to the observation in the middle as follows:

For 3 consecutive values of population, say, y_{i-1}, y_i, y_{i+1}, we have from (4.3.29)

$$\frac{1}{y_{i-1}} = \frac{1}{k} + \frac{be^{-r(i-1)}}{k}$$

$$\frac{1}{y_i} = \frac{1}{k} + \frac{be^{-r \cdot i}}{k} \Rightarrow \frac{1}{y_i} - \frac{1}{k} = \frac{be^{-r \cdot i}}{k} \tag{4.3.34}$$

and $$\frac{1}{y_{i+1}} = \frac{1}{k} + \frac{be^{-r(i+1)}}{k} \Rightarrow \frac{1}{y_{i+1}} - \frac{1}{k} = \frac{be^{-r \cdot i}}{k} e^{-r}$$

$$= \left(\frac{1}{y_i} - \frac{1}{k}\right) e^{-r}, \text{ from (4.3.34)} \tag{4.3.35}$$

Similarly, $$\left(\frac{1}{y_i} - \frac{1}{k}\right) = \left(\frac{1}{y_{i-1}} - \frac{1}{k}\right) e^{-r} \tag{4.3.36}$$

or $$\frac{1}{y_i} = \frac{1 - e^{-r}}{k} + \frac{e^{-r}}{y_{i-1}} \tag{4.3.37}$$

From (4.3.35), $$\frac{1}{y_{i+1}} = \frac{1 - e^{-r}}{k} + \frac{e^{-r}}{y_i} \tag{4.3.38}$$

So, from (4.3.37) and (4.3.38) we get

$$e^{-r} = \frac{\left(\frac{1}{y_i} - \frac{1}{y_{i+1}}\right)}{\left(\frac{1}{y_{i-1}} - \frac{1}{y_i}\right)} = \frac{y_{i-1}}{y_{i+1}} \cdot \frac{y_{i+1} - y_i}{y_i - y_{i-1}} = q \text{ (say)}. \tag{4.3.39}$$

$$\Rightarrow \qquad \hat{r} = -\log q \tag{4.3.40}$$

Also, from (4.3.8)

$$\frac{y_i}{y_{i+1}} = \frac{y_i}{k}(1 - e^{-r}) + e^{-r}$$

$$\Rightarrow \qquad \frac{y_i}{k} = \left(\frac{y_i}{y_{i+1}} - e^{-r}\right) \Big/ (1 - e^{-r}) \tag{4.3.41}$$

Now, from (4.3.30)

$$Z_i = r\left(1 - \frac{y_i}{k}\right) = -\log q \left(1 - \frac{(y_i/y_{i+1}) - q}{1 - q}\right), \text{ from (4.3.41).} \tag{4.3.42}$$

Hence, and estimate of Z_i is given by

$$\hat{Z}_i = \frac{\log q}{q - 1} \frac{(y_{i+1} - y_i)}{y_{i+1}} \tag{4.3.43}$$

Thus, once the estimate of Z_i is obtained, either from (4.3.33) or from (4.3.43), we can estimate r and k from the following regression equation by OLS method,

$$Z_i = r\left(1 - \frac{y_i}{k}\right). \tag{4.3.44}$$

It may be noted that the methods of Fisher of Hotelling do not consider the estimation of the parameter b, separately. However, once the estimates of r and k are obtained, by either of the methods, one can estimate b as in Rhodes method from (4.3.28).

4.3.5 Yule's Method

Here, two consecutive values of the population, say $y(i)$ and $y(i+1)$, are used to regress $\frac{y(i+1)-y(i)}{y(i+1)}$ on $y(i)$ instead of $\frac{d\log y(t)}{dt}$ on $y(t)$ as in the last section.

$$\text{From (4.3.22),}\quad \frac{1}{y(i+1)} = \frac{1+be^{-r}}{k} + e^{-r}\frac{1}{y(i)}, \tag{4.3.45}$$

Changing the sign and adding $\frac{1}{y(i)}$ on both sides we get

$$\left(\frac{1}{y_i} - \frac{1}{y_{i+1}}\right) = -\frac{1-e^{-r}}{k} + \frac{1}{y_i}(1-e^{-r})$$

$$\Rightarrow \quad \frac{y_{i+1}-y_i}{y_i y_{i+1}} = (1-e^{-r})\left(\frac{1}{y_i} - \frac{1}{k}\right) = \frac{1-e^{-r}}{y_i}\left(1-\frac{y_i}{k}\right) \tag{4.3.46}$$

$$\Rightarrow \quad \frac{y_{i+1}-y_i}{y_{i+1}} = (1-e^{-r})\left(1-\frac{y_i}{k}\right) = 1-e^{-r} - \frac{1-e^{-r}}{k}y_i \tag{4.3.47}$$

$$\text{or} \quad Z_i = A + By_i \tag{4.3.48}$$

$$\text{where}\quad Z_i = \frac{y_{i+1}-y_i}{y_{i+1}},\ A = 1-e^{-r} \quad \text{and} \quad B = \frac{1-e^{-r}}{k} \tag{4.3.49}$$

A and B can be estimated as usual by OLS, whence we get estimates of r and k. Estimate of b is obtained by Rhodes method from (4.3.28).

4.3.6 Nair's Method

This is a variation of Yule's method. For two consecutive values y_i and y_{i+1} of the population, we have

$$\frac{1}{y_i} = \frac{1}{k} + \frac{be^{-r\cdot i}}{k} \quad \text{and} \quad \frac{1}{y_{i+1}} = \frac{1}{k} + \frac{be^{-r(i+1)}}{k}, \text{ see (4.3.34)}$$

$$\Rightarrow \quad \frac{1}{y_i} + \frac{1}{y_{i+1}} = \frac{2}{k} + \frac{b}{k}e^{-r\cdot i}(1+e^{-r})$$

$$\Rightarrow \quad \frac{2}{k} - \left(\frac{1}{y_i} + \frac{1}{y_{i+1}}\right) = -\frac{b}{k}e^{-r\cdot i}(1+e^{-r}) \tag{4.3.50}$$

$$\text{Similarly,}\quad \frac{1}{y_i} - \frac{1}{y_{i+1}} = \frac{b}{k}e^{-r\cdot i}(1+e^{-r}) \tag{4.3.51}$$

Substituting from (4.3.50) in (4.3.51) we get

$$\frac{1}{y_i} - \frac{1}{y_{i+1}} = \left(\frac{e^{-r}-1}{e^{-r}+1}\right)\left\{\frac{2}{k} - \left(\frac{1}{y_i} + \frac{1}{y_{i+1}}\right)\right\}. \tag{4.3.52}$$

Nair actually extended Yule's method by regressing $Z_i = \frac{1}{y_i} - \frac{1}{y_{i+1}}$ on $X_i = \frac{1}{y_i} + \frac{1}{y_{i+1}}$ instead of $\frac{1}{y_i}$. Then we can write (4.3.52) as $Z_i = A + B\ X_i$

where, $A = \left(\dfrac{e^{-r}-1}{e^{-r}+1}\right)\left\{\dfrac{2}{k}\right\}$ and $B = -\left(\dfrac{e^{-r}-1}{e^{-r}+1}\right) = \dfrac{1-e^{-r}}{1+e^{-r}}$ (4.3.53)

The estimates of A and B provide the estimates of r and k that of b can be obtained from (4.3.28).

4.3.7 Comparison of Different Methods of Fitting Logistic Curves

The first two methods use three equidistant points and can profitably be employed if the number of observations is a multiple of 3. The other methods employ regression technique. Rhodes method employs regression of reciprocal of an observation on reciprocal of a nearer observation by taking into account the fact that both the variables are subjected to errors. This method appears to be statistically more sound than other methods; it provides consistent estimates of the parameters in the case of failure of OLS in regression in the presence of errors in both the variables. Fisher and Hotelling both utilized regression of $\dfrac{d\log y(t)}{dt}$ on $y(t)$; Hotelling's method, which combines 3 point method with regression, seems to be practically more feasible and frequently used. Yule's method, which regresses $\dfrac{y(i+1)-y(i)}{y(i+1)}$ on $y(i)$, is more exact than regressing the approximation of $\dfrac{d\log y(t)}{dt}$ on $y(t)$. Nair's method is an extension of Yule's method.

4.4 Other Population Growth Models

Hitherto we restricted to Logistic model and its generalization to describe the dynamics of Population Growth. Now we intend to take up the alternative variants based on

(i) Classical law of Population growth by Malthus, which assumes geometrical compound interest rate of conversion of Population (both annual and instantaneous).

(ii) Models of Population growth based on Stable growth rate (i.e. time independent birth, death and growth rates) applicable for all populations in the third phase of fig. 4.1. (i.e. low fertility and low mortality rates without any secular trend) by Lotka and Dublin.

(iii) Models of Population growth based on Quasi stable population growth rate (i.e. one of the components of growth is time independent) by Coale Ansley.

Let us define a function R_k as

$$R_k = \int_0^\infty x^k p(x) f(x)dx, \quad k = 0, 1, 2\ldots \tag{4.4.1}$$

where $p(x)$ = probability of surviving upto age x.

$f(x)$ = Probability of giving a birth in the age group (x to $x+1$).

$f(x)dx$ = probability of giving birth in the age group (x to $x+dx$), in the female cohort.

and w = upper bound of the age of child bearing and can reasonably be replaced by ∞.

By putting $k = 0$ in (4.4.1) we get the,

Net Reproduction rate (NRR) per woman $= Ro$, see (3.1.7) (4.4.2)

Here $p(x)f(x) = \psi(x)$, (say) (4.4.3)

is called the 'Net Maternity Function'.

Now $\psi(x)$ is not a proper probability density function as $\int_0^\infty p(x)f(x)dx \neq 1$.

But if we define

$$\phi(x) = \frac{p(x)f(x)}{\int_0^\infty p(x)f(x)dx}, \tag{4.4.4}$$

then $\phi(x)$ is the proper density function of the r.v. X, representing the age of giving birth.

The mean age of giving birth (or child bearing) is

$$\mu_1' = E(X) = \frac{\int_0^\infty xp(x)f(x)dx}{\int_0^\infty p(x)f(x)dx}, \text{ from (4.4.2)}$$

$$= \frac{R_i}{R_0}, \text{ from (4.4.1).} \tag{4.4.5}$$

Remark 4.3: Since X represents the random gap between a mother of age X and her new born female child (age '0'), so $E(X)$ is also known as the mean length of generation. The mean length of generation and mean age of childbearing are approximately same in a stable population.

The kth raw moment of X is given by

$$\mu_k' = \int_0^\infty x^k\psi(x)dx / \int_0^\infty \psi(x)dx \tag{4.4.6}$$

We can easily see that variance of X is

$$\mu_2 = \sigma_x^2 = \frac{R_2}{R_0} - \left(\frac{R_1}{R_0}\right)^2 \tag{4.4.7}$$

With the above discussions and notations we discuss the following Population Growth Model.

4.4.1 Compound Interest Law of Population Growth with Constant Annual Growth Rate

Here population means female population. The population growth may be given by the compound interest formula viz.

$$P(M) = P(0)(1+r)^M \tag{4.4.8}$$

where $M =$ mean length of generation, in general it will be a function of r.

$P(M) =$ Population after the length of one generation measured from the origin of the earlier generation.

$P(0) =$ Initial size of population $= k$

$r =$ annual growth rate per individual (or female) per year.

Now, we have

$$P(M) = k \cdot R_0, \tag{4.4.9}$$

where $R_0 = \int_0^\infty p(x)f(x)dx = \text{NRR per woman, from (4.4.1)}$ (4.4.10)

Thus $P(M)$ is the NRR per k women and is used as *replacement index of the population* (see Chapter-3).

Then, from (4.4.9) and (4.4.8)

$$k \cdot R_0 = k \cdot (1+r)^M \Rightarrow R_0 = (1+r)^M$$

$$\Rightarrow \quad (R_0)^{1/M} = 1 + r \Rightarrow \hat{r} = (R_0)^{1/M} - 1$$

where $\hat{r}$ is the estimate of r.

Now $M = \dfrac{R_1}{R_0}$, from (4.4.5)

$$\hat{r} = \left[R_0^{R_0/R_1} - 1\right] \tag{4.4.11}$$

4.4.2 Population Growth with Instantaneous Growth Rate

We assume a law where the population is compounded instantaneously i.e. rate of change of population is proportional to the population:

$$\frac{dP(t)}{dt} = rP(t), \text{ where } r \text{ is a proportionality constant}$$

$$\Rightarrow \quad \frac{d}{dt}\log P(t) = r \quad \Rightarrow \quad P(t) = Ce^{rt}, \text{ where } C \text{ is an integration constant}$$

$$\Rightarrow \quad P(t) = P(0)e^{rt} \tag{4.4.12}$$

$$\therefore \quad P(M) = P(0)e^{rM} \tag{4.4.13}$$

Now, From (4.4.11)

$$P(M) = P(0) \cdot R_0$$

$$\therefore \quad P(0)e^{rM} = P(0) \cdot R_0 \quad \Rightarrow \quad \log R_0 = rM$$

$$\text{or} \quad \hat{r} = \frac{1}{M}\log R_0 \tag{4.4.14}$$

4.4.3 Lotka and Dublin's Model: Stable Population Analysis

A population is stable if its growth is independent of time (i.e. both fertility and mortality rates are independent of time) and the population is closed to migratory movements.

Lotka and Dublin's stable population theory consists in assuming that in the 3rd phase of population growth of Fig. 4.1. (i.e. low fertility and low mortality) the structural form of the population will be characterized by the following:

(i) Birth rate is independent of t.

(ii) Date rate is independent of t.

(iii) The age distribution between ages (x to $x + dx$) is independent of t.

(iv) The population is closed to migration.

Let $C(x,t)dx$ = The proportion of population in the age group (x to $x+dx$) at time t.

Note that x is a r.v. representing age, but t is calender time, it is not a r.v.

r = Growth rate.

$B(t)$ = Total no. of births by time t.

$P(t)$ = Population at time t.

$b(t) = \dfrac{B(t)}{P(t)}$ = Birth rate per individual at time t

$p(x)$ = Probability of surviving up to age x (independent of t)

$f(x)dx$ = Probability of giving birth between the age x to $x+dx$ (independent of t).

$d(x)dx$ = Probability of dying in (x to $x+dx$).

We have the basic identity

$$P(t)C(x,t) = B(t-x)p(x), \tag{4.4.15}$$

Since L.H.S. = Total population at time $t \times$ Proportion of population of age x at time t

= Population of age x at time t.

and R.H.S. = Number of births at time $(t-x) \times$ Probability of surviving up to age x.

= Population of age x at time t,

Multiplying both sides by $f(x)$ and integrating in $(0,\infty)$ we get

$$\int_0^\infty P(t)C(x,t)f(x)dx = \int_0^\infty B(t-x)p(x)f(x)dx \tag{4.4.16}$$

Note that, the R.H.S. represents the birth at time t. Hence, we can write

$$B(t) = \int_0^\infty B(t-x)p(x)f(x)dx = \int_0^\infty B(t-x)\psi(x)dx. \tag{4.4.17}$$

where $\psi(x) = p(x)f(x)$, known as the Net maternity function.

This is an integral equation.

Lotka and Dublin assumed a trial solution of the form.

$$B(t) = \sum_{n=0}^{\infty} Q_n \exp(r_n t), \tag{4.4.18}$$

where Q_n is the population at the beginning of nth year (treated here as constant) and r_n is the growth rate in the nth year, $n = 0, 1, \ldots$

Substituting (4.4.18) in (4.4.17)

$$\sum_{n=0}^{\infty} Q_n \exp(r_n t) = \int_0^\infty \sum_{n=0}^{\infty} Q_n\{r_n(t-x)\}\psi(x)dx, \text{ see (4.4.5)}$$

$$\Rightarrow \quad \sum_{n=0}^{\infty} Q_0 \exp(r_n t) = \sum_{n=0}^{\infty} Q_n \exp(r_n t) \int_0^\infty \exp(-r_n x)\psi(x)dx. \tag{4.4.19}$$

It can be shown that $r_0, r_1, \ldots$ corresponds to the roots of the integral equation

$$\int_0^\infty \exp(-rx)\psi(x)dx = 1 \tag{4.4.20}$$

This is known as Lotka's integral equation. It has been shown by Lotka that of the infinite number of roots $r_0, r_1, \ldots$ of (4.4.20), only one is real and rest all are complex.

We show below that the real part of any complex root will be less than the only real root (say) r_0.

If $r_k = \alpha_k + i\beta_k, k = 1, 2, \ldots$ is a complex root of (4.4.20), then

$$\int_0^\infty \exp\{-(\alpha_k + i\beta_k)x\}\psi(x)dx = 1, \text{ where } i = \sqrt{-1}$$

$$\Rightarrow \quad \int_0^\infty \exp\{(-\alpha_k x)\{\cos\beta_k x - i\sin\beta_k x\}\psi(x)dx = 1.$$

Equating real and imaginary parts on both sides, we get

$$\int_0^\infty \exp\{(-\alpha_k x)\{\cos\beta_k x\}\psi(x)dx = 1 \tag{4.4.21}$$

when $r = r_0$ (real) then (4.4.20) gives

$$\int_0^\infty \exp(-r_0 x)\psi(x)dx = 1 \tag{4.4.22}$$

Since $\cos\beta_k x \le 1$, comparing (4.4.21) and (4.4.22) we get

$$\exp(-r_0 x) \le \exp(-\alpha_k x)$$

Hence, $r_0 \ge \alpha_k$ (4.4.23)

Now (4.4.18) can be written as

$$B(t) = Q_0 \exp(r_0 t) + \sum_{n=1}^{\infty} Q_n \exp(r_n t).$$

Since $r_0 \ge \alpha_k, k = 1, 2, \ldots$ it follows that for $k = 1, 2, \ldots$

$$|\exp(r_k t)| = |\exp(\alpha_k t)\{\cos\beta_k t + i\sin\beta_k t\}|$$
$$= |\exp(\alpha_k t)| \le |\exp(r_0 t)| \tag{4.4.24}$$

and as $t \to \infty, \exp(\alpha_k t)$ will be negligible in comparison to $\exp(r_0 t)$, for $k = 1, 2, \ldots$

Note that $t \to \infty$ means the process has started since a very long time.

Hence, as $t \to \infty$

$$B(t) \cong Q_0 \exp(r_0 t) \tag{4.4.25}$$

$$\therefore \quad B(t - x) \cong Q_0 \exp\{r_0(t - x)\} \Rightarrow B(t) = B(t - x)\exp(r_0 x)$$

Dropping the suffix we have

$$B(t) = B(t - x)\exp(rx) \tag{4.4.26}$$

where r stands for the real root of Lotka integral equation (4.4.20).

Now to obtain the real root of (4.4.20), let

$$y = \int_0^\infty \exp(-rx)\psi(x)dx \tag{4.4.27}$$

$$\Rightarrow \quad \frac{dy}{dr} = -\int_0^\infty x\exp(-rx)\psi(x)dx$$

$$= -\left\{\frac{\int_0^\infty x\exp(-rx)\psi(x)dx}{\int_0^\infty \exp(-rx)\psi(x)dx}\right\}\int_0^\infty \exp(-rx)\psi(x)dx$$

$= -yA(r)$ (say), where $A(r)$ is the quantity in { } above. (4.4.28)

Thus we have the differential equation

$$\frac{dy}{dr} = -yA(r) \Rightarrow \frac{dy}{y} = -A(r)dr \tag{4.4.29}$$

$\Rightarrow \quad \log y = -\int A(r)dr + \log y_0$ (an integration constant).

$$\Rightarrow \quad y = y_0 \exp\left\{-\int A(r)dr\right\} \tag{4.4.30}$$

For $y = 1, y_0 = \exp\left\{\int A(r)dr\right\}$, from (4.4.30) (4.4.31)

Now, when $r = 0, dr = 0$

$\therefore \quad y = y_0$, from (4.4.30).

$\therefore \quad y_0 = \int_0^\infty \psi(x)dx$, from (4.4.27)

$= R_0 = NRR$ per woman, from (4.4.2) (4.4.32)

$\therefore \quad \log R_0 = \log y_0 = \int A(r)dr$, from (4.4.31). (4.4.33)

Further, from (4.4.28)

$$A(r) = \int_0^\infty x\exp(-rx)\psi(x)dx \Big/ \int_0^\infty \exp(-rx)\psi(x)dx$$

$$= \int_0^\infty x\psi(x)\left\{\sum_{j=0}^\infty (-1)^j\frac{(rx)^j}{j!}\right\}dx \Big/ \int_0^\infty \psi(x)\sum_{j=0}^\infty (-1)^j\frac{(rx)^j}{j!}dx$$

$$= \sum_{j=0}^\infty \frac{(-1)^j r^j}{j!}\int_0^\infty x^{j+1}\psi(x)dx \Big/ \sum_{j=0}^\infty \frac{(-1)^j r^J}{j!}\int_0^\infty x^j\psi(x)dx$$

$\therefore \quad A(r) = \sum_0^\infty (-1)^j \frac{r^j}{j!}R_{j+1} \Big/ \sum_0^\infty (-1)^j\frac{r^J}{j!}R_j$, from (4.4.1) and (4.4.3) (4.4.34)

Let $A(r) = \alpha + \beta r + \gamma r^2 + \cdots$ (say) (4.4.35)

Now putting $r = 0$ in the R.H.S. of (4.4.34) and (4.4.35) and then comparing them, we get

$$\alpha = \frac{R_1}{R_0} \tag{4.4.36}$$

and then differentiating them w.r.t. r and putting $r = 0$, we get

$$\beta = \left(\frac{R_1}{R_0}\right)^2 - \frac{R_2}{R_0}. \tag{4.4.37}$$

Again differentiating and putting $r = 0$, we have

$$\gamma = \frac{1}{2}\left[\frac{R_3}{R_0} - \frac{3R_1R_2}{R_0^2} + 2\left(\frac{R_1}{R_0}\right)^3\right], \text{ etc.} \tag{4.4.38}$$

Further, from (4.4.35), neglecting higher powers of r, we get

$$A(r) \cong \alpha + \beta\, r. \tag{4.4.39}$$

Now, $\log R_0 = \int A(r)dr$, from (4.4.33)

$$= \int (\alpha + \beta\, r)dr = \alpha r + \beta\frac{r^2}{2}$$

or $\beta r^2 + 2\alpha r - 2\log R_0 = 0$, which gives

$$r = \frac{-2\alpha \pm \sqrt{4\alpha^2 + 8\beta \log R_0}}{2\beta}$$

and substituting the values of α, β from (4.4.36), (4.4.37)

$$r = \frac{-2\dfrac{R_1}{R_0} \pm \sqrt{4\left(\dfrac{R_1}{R_0}\right)^2 + 8\left[\left(\dfrac{R_1}{R_0}\right)^2 - \dfrac{R_2}{R_0}\right]\log R_0}}{2\left[\left(\dfrac{R_1}{R_0}\right)^2 - \dfrac{R_2}{R_0}\right]}, \tag{4.4.40}$$

where, r is the growth rate (or parameter) of the stable population.

Now, R_0, R_1 and R_2 are estimated as

$\hat{R}_0 = \sum P(x)f(x) = \text{NRR}$, from (4.4.2)

$\hat{R}_1 = \text{Mean age of childbearing} \times \hat{R}_0$, from (4.4.5)

and $\hat{R}_2 = \mu_2' \times \hat{R}_0$, from (4.4.7).

Then substituting these estimates of R_0, R_1 and R_2 in (4.4.40), two real values of r will be obtained, one positive and one negative.

4.4.4 Certain Important Deductions for Stable Population Analysis

(i) To obtain $b(t)$ and $C(x;t)$ for stable population

We have $B(t) = B(t-x)\exp(rx)$, see (4.4.26)

Also $P(t)C(x,t)dx = B(t-x)p(x)dx$, see (4.4.15)

$$\Rightarrow \quad C(x,t) = \frac{B(t-x)}{P(x)} \tag{4.4.41}$$

$\therefore \quad C(x,t) = B(t)\exp(-rx)p(x)/P(t)$, using (4.4.24)

or $\quad C(x,t) = b(t)\exp(-rx)p(x) \qquad (4.4.42)$

$\because \quad b(t) = B(t)/P(t)$

Now since $\int_0^\infty C(x,t)dx = 1$, from definition of $C(x,t)$.

Hence, from (4.4.42) we get

$$b(t)\int_0^\infty \exp(-rx)p(x)dx = 1$$

$$\therefore \quad b(t) = 1/\left[\int_0^\infty \exp(-rx)p(x)dx\right] \qquad (4.4.43)$$

$\therefore$ From (4.4.42)

$$C(x,t) = \exp(-rx)p(x)/\left[\int_0^\infty \exp(-rx)p(x)dx\right] \qquad (4.4.44)$$

Relations (4.4.43) and (4.4.44) shows that birth rate and the age distribution are independent of t. Further, since

$$\text{Death rate } = d(t) = r - b(t) \qquad (4.4.45)$$

and since, by the assumption, in a stable population growth rate r is independent of time, hence the death rate $d(t)$ is also independent of time.

Thus we conclude that, in a stable population

(1) Birth and death rates are independent of time,

(2) $C(x,t) = C(x)$ i.e. age distribution is also independent of time

Birth and death rates as well as age distribution may undergo changes but these changes are of random nature.

(ii) Death rate in a stationary population

A population is stationary when $r = 0$, in which case the stationary age distribution is given by

$$C(x) = p(x)/\int_0^\infty p(x)dx, \text{ from (4.4.44).} \qquad (4.4.46)$$

The life table population is a stationary population. For a life table

$$p(x) = l(x)/l_0, \text{ see (2.5.6).} \qquad (4.4.47)$$

$$\therefore \quad C(x) = l(x)/\int_0^\infty l(x)dx, \qquad (4.4.48)$$

which represents the life table age distribution.

Again for a stationary population birth rate (b) = death rate (d). Hence, substituting from (4.4.47) in (4.4.43) for $r = 0$, we get

$$b = d = l_0/\int_0^\infty l(x)dx. \qquad (4.4.49)$$

Now from (2.7.7),

$$T_0 = \int_0^\infty l(x)dx \tag{4.4.50}$$

$\therefore \quad d = l_0/T_0 = 1/e_0^0$, from (2.7.10)

where e_0^0 is the complete expectation of life. Thus the life table death rate is usually denoted by $1/e_0^0$.

(iii) Determination of Q_0 under stable population analysis.

Consider a general formulation

$$B(t) = G(t) + \int_0^t B(t_x)\psi(x)dx, \tag{4.4.51}$$

where $G(t)$ is the number of births to women already born at time $t = 0$.

The term $G(t) = 0$ for $t \geq \beta$ i.e. no birth can occur to women of age $\geq \beta$.

or $$G(t) = B(t) - \int_0^t B(t-x)p(x)\cdot i(x)dx. \tag{4.4.52}$$

Taking Laplace Transform of (4.4.52) on both sides

$$\Rightarrow \quad L(G(t)) = \int_0^\infty \exp(-r_0 t)G(t)dt = \int_0^\beta \exp(-r_0 t)G(t)dt$$

$$(\because G(t) = 0 \,\forall\, t \geq \beta)$$

$$= \int_0^\beta \exp(-r_0 t)\left[B(t) - \int_0^t B(t-x)\psi(x)dx\right], \text{ using (4.4.51)}$$

$$= \int_0^\beta \exp(-r_0 t)\left[Q_0 \exp(r_0 t) + \sum_{n=1}^\infty Q_n \exp(r_n t) - \int_0^t \left\{Q_0 \exp[r_0(t-x)] + \sum_{n=1}^\infty Q_n \exp[r_n(t-x)]\right\}\psi(x)dx\right] dt$$

using (4.4.18)

$$= \int_0^\beta \exp(-r_0 t)\left[Q_0 \exp(r_0 t) + \sum_{n=1}^\infty Q_n \exp(r_n t) - \int_0^t Q_n \exp[r_0(t-x)]\psi(x)dx + \int_0^t \sum_{n=1}^\infty Q_n \exp[r_n(t-x)]\psi(x)dx\right] dt$$

$$= \int_0^\beta \exp(-r_0 t)\left[Q_0 \exp(r_0 t) - \int_0^t Q_0 \exp\{r_0(t-x)\}\psi(x)dx\right] dt + C_0' \tag{4.4.53}$$

where $$C_0' = \int_0^\beta \exp(-r_0 t)\left[\sum_{n=1}^\infty Q_n \exp(r_n t) - \int_0^t \left\{\sum_{n=1}^\infty Q_n \exp[r_n(t-x)]\psi(x)dx\right\}\right]. \tag{4.4.54}$$

Therefore, $L(G(t)) = \int_0^\beta \left\{ Q_0 - \int_0^t Q_0 \exp(-r_0 x)\psi(x)dx \right\} dt + C_0'$

$$= Q_0 \int_0^\beta \left\{ 1 - \int_0^t \exp(-r_0 x)\psi(x)dx \right\} dt + C_0' \tag{4.4.55}$$

Since r_0 is the real root of the equation

$$\int_\alpha^\beta e^{-rx} p(x) f(x) dx = \int_0^\infty e^{-rx} p(x) f(x) dx = 1$$

or

$$\int_0^\beta e^{-rx}\psi(x)dx = 1,$$

we may write

$$L(G(t)) = Q_0 \int_0^\beta \left\{ \int_0^\beta \exp(-r_0 x)\psi(x)dx - \int_0^t \exp(-r_0 x)\psi(x)dx \right\} dt + C_0'$$

$$= Q_0 \int_0^\beta \int_t^\beta \exp(-r_0 x)\psi(x)dx\, dt + C_0' \tag{4.4.56}$$

Next we shall change the order of integration in the above double integral.

Note that in the case given above sums of the type of vertical strips given in the Fig. 4.2 were taken. Now we shall take sums of the horizontal type of strips. The relevant limits are thus

$$0 \le x \le \beta \text{ and } 0 \le t \le x$$

Hence (4.4.56) becomes

$$L(G(t)) = Q_0 \int^\beta \int_0^x \exp(-r_0 x)\psi(x)dt\; dx + C_0'$$

$$= Q_0 \int_0^\beta \left\{ \exp(-r_0 x)\psi(x) \int_0^x dt \right\} dx + C_0' \tag{4.4.57}$$

$$= Q_0 \int_0^\beta \exp(-r_0 x\psi(x) \cdot x dx + C_0' \tag{4.4.58}$$

$$\Rightarrow \quad Q_0 = \frac{L(G(t))}{\int_0^\beta x \exp(-r_0 x)\psi(x)dx}$$

$$= \frac{\int_0^\beta \exp(-r_0 t) G(t) dt}{\int_0^\beta x \exp(-r_0 x)\psi(x)dx}, \tag{4.4.59}$$

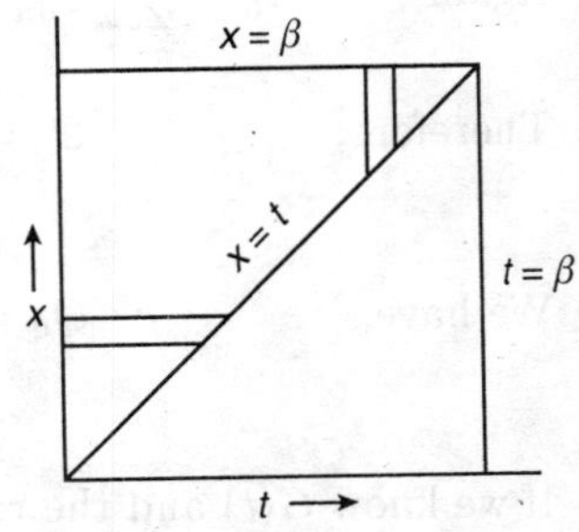

Fig. 4.2

provided $C_0' = 0$.

Now to show that $C_0' = 0$, we see that C_0' is made up of sums of such terms as:

$$C_{0n}' = Q_n \int_0^\beta \exp(-r_0 t) \left[\exp r_n t - \int_0^t \exp(r_n(t-x)]\psi(x)dx \right] dt, \; n \ne 0$$

$$= Q_n \int_0^\beta \left[\exp[r_n - r_0)t] - \int_0^t \exp[(r_n - r_0)t] \exp(-r_n x)\psi(x)dx\right] dt$$

$$= Q_n \int_0^\beta \exp[(r_n - r_0)t] \left\{1 - \int_0^t \exp(-r_n x)\psi(x)dx\right\} dt.$$

We put $\int_0^\beta \exp(-r_n x)\psi(x)dx = 1.$

Again since r_n is also a root of the Lotka's integral equation, we get

$$C'_{0n} = Q_n \int_0^\beta \exp[(r_n - r_0)t] \left\{\int_0^\beta \exp(-r_n x)\psi(x)dx - \int_0^t \exp(-r_n x)\psi(x)dx\right\} dt$$

$$= Q_n \int_0^\beta \exp[(r_n - r_0)t] \left\{\int_t^\beta \exp(-r_n x)\psi(x)dx\right\} dt \tag{4.4.60}$$

Again changing the order of integration

$$C'_{0n} = Q_n \int_0^\beta \left\{\exp(-r_n x)\psi(x) \int_0^x \exp[(r_n - r_0)t]dt\right\} dx$$

$$= \frac{Q_n}{r_n - r_0} \int_0^\beta \exp(-r_n x)\psi(x)[\exp\{r_n - r_0)\}x - 1]dx$$

$$= \frac{Q_n}{r_n - r_0} \left\{\int_0^\beta \exp(-r_0 x)\psi(x)dx - \int_0^\beta \exp(-r_n x)\psi(x)dx\right\}$$

$$= 0, \tag{4.4.61}$$

because r_u and r_0 are both roots of Lotka's integral equation (4.4.20)

i.e. $$\int_0^\beta \exp(-r_0 x)\psi(x)dx = \int_0^\beta \exp(-r_n x)\psi(x)dx = 1$$

Again $C'_0 = \sum_{n=1}^{\infty} C'_{0n}$, and each $C'_{0n} = 0$, from (4.4.61)

Therefore, $$C'_0 = 0. \tag{4.4.62}$$

We have, $$Q_0 = \frac{\int_0^\beta \exp(-r_0 t)G(t)dt}{\int_0^\beta x\exp(-r_0 x)\psi(x)dx}, \text{ from (4.4.59)} \tag{4.4.63}$$

If we know $G(t)$ and the rate of survivorship f and fertility $p(x)[\psi(x) = p(x) \cdot f(x)]$ then one can calculate Q_0 from r_0.

A special case

Let us take the daughters of B_0 births occurred at $t = 0$.

Then $G(t) = B_0(t)p(t)f(t) = B_0\psi(t)$

In this case, $$Q_0 = \frac{B_0 \int_0^\beta \exp(-r_0 t)\psi(t)dt}{\int_0^\beta x\exp(-r_0 x)\psi(x)dx} = \frac{B_0}{\int_0^\beta x\exp(-r_0 x)\psi(x)dx} \quad (4.4.64)$$

Theorem 4.4.1: If $M(r)$ (see(4.4.8)) is the mean length of generation and M_0 is the mean age of childbearing in a stable population, then

(a) $r = \dfrac{\log R_0}{M(r)}$, (4.4.65)

(b) $M_0 \cong M(r) + \dfrac{r\sigma^2}{2}$, (4.4.66)

where R_0 is the NRR, r is the growth rate and σ^2 is the variance of the age distribution.

Proof of (a): From (4.4.29)

$$\int \frac{dy}{y} = -\int A(r)dr + c, \text{ where } c \text{ is an integration constant.} \quad (4.4.67)$$

Now, by M.V. Theorem (A.1.13) (taking $f = A(r)$ and $g = 1$)

$$\int A(r)dr = M(r)\cdot r \quad (4.4.68)$$

So (4.4.67) gives

$$\log y = -M(r)\cdot r + c \quad (4.4.69)$$

Also, $\log y]_{r=0} = \log R_0$, from (4.4.32)

$\Rightarrow \quad c = \log R_0$

Then (4.4.69) becomes

$$\log y = -M(r)\cdot r + \log R_0$$

By (4.4.20), $y = 1$

$$\therefore \quad \log y = \log 1 = 0 \;\Rightarrow\; \log R_0 = M(r)\cdot r \;\Rightarrow\; r = \frac{\log R_0}{M(r)}$$

This is another proof of (4.4.14).

Proof of (b): From (4.4.68) and (4.4.35)

$$M(r)\cdot r = \alpha r + \beta\frac{r^2}{2} + \gamma\frac{r^3}{3} + \delta\frac{r^4}{4} + \cdots$$

$$\Rightarrow \quad M(r) = \alpha + \beta\frac{r}{2} + \gamma\frac{r^2}{3} + \delta\frac{r^3}{4} + \cdots, \quad (4.4.70)$$

where α, β, γ are given by (4.4.36) to (4.4.38).

Since the coefficients of γ, δ etc. are small ($\because$ r is much smaller than 1),

$$\therefore \quad M(r) \cong \alpha + \beta\frac{r}{2} \quad \text{or} \quad M(r) \cong \frac{R_1}{R_0} - \frac{r}{2}\left[\frac{R_2}{R_0} - \left(\frac{R_1}{R_0}\right)^2\right]$$

or $\quad M(r) \cong M_0 - \dfrac{r}{2}\sigma^2$, from (4.4.5) and (4.4.7).

Remark 4.4: $M(r) = M_0$ holds exactly for a stationary population (i.e. for $r = 0$).

4.4.5 Growth Rate of a Sectionally Stable Population

We assume that within a given age group, the population at age x, say $P(x)$, is

$$P(x) = e^{-rx} l(x), \tag{4.4.71}$$

where r may be different in different age sectors i.e. r will depend on age sector.

Theorem 4.4.2: Show that

$${}_nr_x = \frac{1}{2n} \log \left(\frac{{}_nP_{x-n}/{}_nL_{x-n}}{{}_nP_{x+n}/{}_nL_{x+n}} \right). \tag{4.4.72}$$

where ${}_nr_x$ = Growth rate r in the age sector (x to $x+n$)

${}_nP_x$ = Population size in the age sector (x to $x+n$)

${}_nL_x$ = Person years lived by the cohort $l(x)$ in (x to $x+n$)

$$= \int_0^n l(x+t)dt \tag{4.4.73}$$

All the quantities here are for the calendar year Z.

Proof: From (4.4.71)

$${}_nP_x = \int_0^n e^{-r(x+t)} l(x+t)dt, \text{ where } r = {}_nr_x.$$

$$\therefore \quad {}_nP_{x-n} = \int_0^n e^{-r(x-n+t)} l(x-n+t)dt$$

Then using Generalised Mean Value theorem (A.1.13), and choosing an A such that

$$A\,e^{-\frac{rn}{2}} \in \left[e^{-r\cdot 0}, e^{-r\cdot n}\right], \text{ we get}$$

$${}_nP_{x-n} = A\,e^{-r\left(x-n+\frac{n}{2}\right)} \int_0^n l(x+n+t)dt$$

$$= A\,e^{-r\left(x-\frac{n}{2}\right)} {}_nL_{x-n}, \text{ from (4.4.73).}$$

Similarly, $\quad {}_nP_{x+n} = Ae^{-r\left(x+\frac{3n}{2}\right)} {}_nL_{x+n}$

$$\therefore \quad \left(\frac{{}_nP_{x-n}/{}_nL_{x-n}}{{}_nP_{x+n}/{}_nL_{x+n}} \right) = e^{2r\cdot n}.$$

$$\Rightarrow \quad \log \left(\frac{{}_nP_{x-n}/{}_nL_{x-n}}{{}_nP_{x+n}/{}_nL_{x+n}} \right) = 2r \cdot n$$

$$\Rightarrow \quad \frac{1}{2n} \log \left(\frac{{}_nP_{x-n}/{}_nL_{x-n}}{{}_nP_{x+n}/{}_nL_{x+n}} \right) = {}_nr_x. \quad //$$

4.4.6 Construction of Abridged Life Table by Successive Iteration for a Stable Population

It is possible to construct a more accurate Life Table, by successive iteration (based on age specific mortality rates obtained from (say) registration data or other less reliable sources), given that the population is stable. The methodology is due to Keyfitz (1966).

The Life table central rate of mortality in the age group (x to $x+n$) is given by

$${}_nm_x = {}_nd_x/{}_nL_x. \tag{4.4.74}$$

The age-specific mortality rare for a stable population (Life Table population is a stationary population) is defined as

$$ {}_nM_x = {}_nD_x / {}_nP_x, \tag{4.4.75}$$

where ${}_nD_x$ = Number of deaths in the age sector (x to $x+n$) in the stable population.

Let ${}_nM'_x$ = Age-specific death rate in (x to $x+n$), applicable to the (stable) population with growth rate r (as calculated from (4.4.72)) and ${}_nM_x$ = Observed age-specific death rate in (x to $x+n$).

We start with a defective draft of a Life table and suppose the annual rate of mortality in the defective table is ${}_nq_x$. Then the revised estimate of ${}_nq_x$, say ${}_nq^*_x$, is obtained by

$$ {}_nq^*_x = {}_nq_x \frac{{}_nM_x}{{}_nM'_x} \tag{4.4.76}$$

Again if the stable population, between ages (x to $x+dx$), is given by

$$P(x)dx = e^{-rx} l(x)dx, \tag{4.4.77}$$

then we have

$$ {}_nP'_x = \int_0^n P(x+t)dt = \int_0^n \exp\{-r(x+t)\} l(x+t)dt \tag{4.4.78}$$

Now, ${}_nd_x = \int_0^n l(x+t)\mu(x+t)dt$, from (2.5.2), (2.6.11) and (3) of Sec. 2.7.1

Hence, the corresponding quantity for the stable population (4.4.77) is given by

$$\begin{aligned} {}_nD'_x &= \int_0^n e^{-r(x+t)} l(x+t)\mu(x+t)dt \\ &= \int_0^n e^{-r(x+t)} l(x+t)\left[-\frac{1}{l(x+t)}\frac{dl(x+t)}{dt}\right]dt \\ &= -\int_0^n e^{-r(x+t)} dl(x+t). \end{aligned}$$

Integrating by parts, we get

$$\begin{aligned} {}_nD'_x &= -\left[e^{-r(x+t)} l(x+t)|_0^n + \int_0^n re^{-r(x+t)} l(x+t)dt\right] \\ &= e^{-r\cdot x} l(x) - e^{-r(x+n)} l(x+n) - r\int_0^n e^{-r(x+t)} l(x+t)dt \\ &= e^{-r\cdot x} l(x) - e^{-r(x+n)} l(x+n) - r\,{}_nP'_x, \text{ from (4.4.78)} \end{aligned}$$

$$ {}_nM'_x = \frac{{}_nD'_x}{{}_nP'_x} = \frac{e^{-r\cdot x} l(x) - e^{-r(x+n)} l(x+n)}{{}_nP'_x} - r$$

$$ {}_nM'_x = \frac{e^{-r\cdot x} l(x) - e^{-r(x+n)} l(x+n)}{\int_0^n e^{-r(x+t)} l(x+t)dt} - r, \text{ from (4.4.78)} \tag{4.4.79}$$

but this is not in explicit form.

An explicit form of ${}_nM'_x$ (not containing the integral) may be obtained by approximating it by a cubic through the points $P(x-t)$, $P(x)$, $P(x+5)$ and $P(x+10)$ etc.

We use the approximation

$$\int_0^n l(x+t)dt = {}_nL_x, \text{ from (4.4.73)}$$
$$= \frac{n}{2}[l(x)+l(x+n)] + \frac{n}{24}[{}_nd_{x+n} - {}_nd_{x-n}]. \tag{4.4.80}$$

This is obtained by approximating $l(x+t)$ by a cubic curve [see Wolfenden (1954)]

$$= \frac{13n}{24}[l(x)+l(x+n)] - \frac{n}{24}[l(x-n)+l(x+2n)$$

$\because \quad {}_nd_x = l(x) - l(x+n).$

Hence,
$$\int_0^n e^{-r(x+t)}l(x+t)dt = \frac{13n}{24}\left[e^{-rx}l(x) + e^{-r(x+n)}l(x+n)\right]$$
$$-\frac{n}{24}\left[e^{-r(x-n)}l(x-n) + e^{-r(x+2n)}l(x+2n)\right]$$
$$= K(x,n,r); \text{ (say)}. \tag{4.4.81}$$

Then, from (4.4.79) we get the first approximation as

$${}_nM_x^{(1)} = \frac{e^{-r\cdot x}l(x) - e^{-r(x+n)}l(x+n)}{K(x,n,r)} - r \tag{4.4.82}$$

Thus, the objective has been to make each draft of the Life table produce an ${}_nM'_x$ which is directly comparable with the ${}_nM_x$ of the data (i.e. observed). This permits a correction of the ${}_nq_x$ column of the Life table. Denoting the improved estimate of ${}_nq_x$ by ${}_nq_x^{(1)}$ in the first iteration, we have

$${}_nq_x^{(1)} = {}_nq_x^{(0)}\frac{{}_nM_x}{{}_nM_x^{(1)}}, \tag{4.4.83}$$

where ${}_nq_x^{(0)}$ is the entry in the original defective Life table.

Using ${}_nq_x^{(1)}\ \forall\, x$ and n we construct improved Life table, from which we again make an improved estimate of ${}_nM'_x$, say ${}_nM_x^{(2)}$, and get

$${}_nq_x^{(2)} = {}_nq_x^{(1)}\frac{{}_nM_x}{{}_nM_x^{(2)}}. \tag{4.4.84}$$

We go on continuing with this iteration till $\left|{}_nM_x - {}_nM_x^{(k)}\right|$ is as small as we please, where ${}_nM_x^{(k)}$ is the kth iterated value of ${}_nM'_x$.

4.5 Population Projection Techniques

Projection of population implies forecasting of the population in the years to come, on the basis of suitable assumptions on the determinants of population growth such as future patterns in (i) Fertility (ii) Mortality and (iii) Migration. The success of projection lies not only on the technique of population projection but also on how far realistic are the assumptions relating to demographic factors affecting the size and growth of population in the future years to come. Needless is to mention that projection process plays a vital role in the National planning, as no planning of resources is complete without a knowledge of the size of the population growing over time.

The technique of population projection can be divided into the following classes:

(1) *Projection by Using a single Mathematical Model*

Projection by Logistic, Gompertz curves or any other deterministic model which have been believed by demographers to describe the natural growth of population, is projection by using a single mathematical model. However, experience shows that a single equation model, however realistic to the premises of the same in building the model, is hardly successful in forecasting the long term trend of the population over time. Although, we have given due importance to Mathematical models, for making population projection, their validities are always to be taken with a degree of reservation. However, for short term projections it has been observed that their performance is more or less fair.

(2) *Components method of population Projection*

Projection by taking into consideration the components of the population growth, is known as the component method of population projection. This technique is understandable and empirically valid since this takes into account each component of growth individually: but again success of the projection depends on the reality of the assumptions for the technique, in future years to come.

(3) *Population Projection Employing the Techniques of Stable or Quasi-stable Populations*

Quasi-stable population shows departure from stability because of departure of assumptions in respect to either fertility or mortality, not both (see Section 4.6).

Of the above three methods, the most realistic and therefore most applicable method is (2) viz. Component Method.

Below, we out line the technique of component method which is employed in all National Population Projections. This method is more widely known as Thomas Frejka's Component Method of Population Projection.

4.5.1 Frejka's Component Method

Factors affecting the changes in the future population are governed by the following quantities:

(A) Survivors of the present population in the year of the projection, say at every Tth year, $T = 0, t, 2t, \ldots$

(B) the number of births which will occur from the current period $T = 0$ (the base year of projection) to a period $T = t$ and their survivors at $T = t$ and so on.

(C) Adjustment of migratory disturbances i.e. estimating the emigrants over the immigrants or immigrants over emigrants.

Techniques of the these studies are described below:

(A) Technique for Estimating the Survivors of the Present Population After a time of t years for a t yearly projection

Let ${}_nP_x^Z$ be the population in the age sector (x to $x+n$) in the calendar year Z. Let the survivor of this population after t years be ${}_nP_{x+t}^{Z+t}$. Here we can assume that

$$\frac{{}_nP_{x+t}^{Z+t}}{{}_nP_x^Z} = \frac{{}_nL_{x+t}}{{}_nL_x}, \tag{4.5.1}$$

where ${}_nL_x$ and ${}_nL_{x+t}$ pertain to a life table valid for the period Z to $Z+t$.

$$\therefore \quad {}_nP_{x+t}^{Z+t} = {}_nP_x^Z \frac{{}_nL_{x+t}}{{}_nL_x} = {}_nP_x^Z \; {}_n\pi_x^t, \text{ (say)}, \tag{4.5.2}$$

where ${}_n\pi_x^t \left(= \dfrac{{}_nL_{x+t}}{{}_nL_x} \right)$ is known as the survival factor.

Applying (4.5.2) for all x and n one can project the first component viz. the survivors of the present population (i.e. for $T = 0$) at $T = t$.

(B) Estimation of new births in $(0, t)$ and their survivors at $T = t$.

Let t be a simple multiple of n (the length of the age group) i.e. $t = kn$.

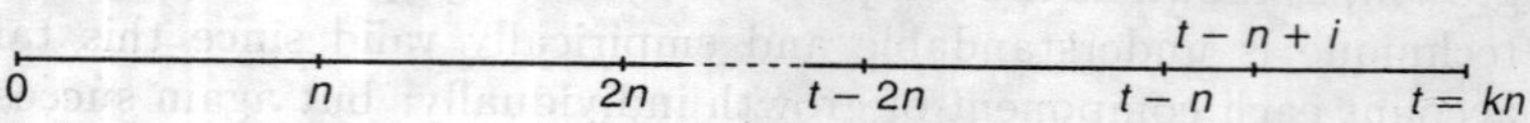

Fig. 4.3

Consider the $(t-n+i)$th year (see Fig. 4.3) and let $B^{(t-n+i)}, i = 0, 1, 2, \ldots$ be the number of births in this year. The babies born in the $(t-n+i)$th year would be exposed to the risk for a period of $\left[t - (t-n+i) + \frac{1}{2}\right] = n - i + \frac{1}{2}$ years on the average.

The expected number of survivors of $B^{(t-n+i)}$ births at $T = t$ is

$$B^{(t-n+i)} {}_{n-1+\frac{1}{2}}p_0 = B^{(t-n+i)} \left(\frac{l_{n-i+1/2}}{l_0} \right) = B^{(t-n+i)} \left(\frac{L_{n-i}}{l_0} \right), \text{from (2.5.8)} \tag{4.5.3}$$

Thus pulling the births in the interval $(t-n, t)$ the expected number of survivors of those at $T = t$ is

$$= \sum_{i=0}^{n} B^{(t-n+i)} \left(\frac{L_{n-i}}{l_0} \right)$$

Similarly, the expected number of survivors at time $T = t$ of all the births in the interval $(t-2n, t-n)$ is

$$= \sum_{i=0}^{n} B^{(t-2n+i)} \left(\frac{L_{2n-i}}{l_0} \right).$$

Expected number of survivors of the births in $(t-3n, t-2n)$ is

$$= \sum_{i=0}^{n} B^{(t-3n+i)} \left(\frac{L_{3n-i}}{l_0} \right)$$

and so on. Proceeding in this way the expected number of survivors at $T = t$ of all births in $(0, t)$ is

$$= \sum_{i=0}^{n} B^{(t-n+i)} \left(\frac{L_{n-i}}{l_0} \right) + \sum_{i=0}^{n} B^{(t-2n+i)} \left(\frac{L_{2n-i}}{l_0} \right) + \cdots + \sum_{i=0}^{n} B^{(i)} \left(\frac{L_{t-i}}{l_0} \right) \tag{4.5.4}$$

For simplicity if we assume that $B^{(i)} = B$ for $j = 0, 1, \ldots, t : t = kn$, then the expected number of survivors of all berths in $(0, t)$, recorded at time $T = t$ is

$$= B \sum_{i=0}^{n} \sum_{x=1}^{k} \left(\frac{L_{xn-i}}{l_0} \right) \tag{4.5.5}$$

Now, to estimate B we require the age specific fertility rates.

The number of births $B^{(j)}$ in the year j is obtained as

$$B^{(j)} = \sum_{x=w_1}^{w_2} f_x^{(j)} N_x^{(j)}, \tag{4.5.6}$$

where, $N_x^{(j)}$ = number of women in the age group (x to $x+1$) in year j,

$f_x^{(j)}$ = Probability of giving birth in the year j, for a woman in the age group (x to $x+1$),

and w_1 and w_2 are the lower and upper bounds of the child bearing period.

Putting the value of $B^{(j)}$ from (4.5.6) in (4.5.4) or $B^{(j)} = B$ for $\forall\ j = 0, 1, \ldots, t$ in (4.5.5), we get the estimate of the survivors of all births in $(0, t)$ recorded at $T = t$.

(C) Migration

There are a large number of arbitrary elements creeping in while performing adjustment of migration in the projection process. Because factors leading to in or out migration are influenced to a considerable extent by social and economical conditions pertaining to the country of origin as well as that of destination. Further, the migratory pattern is also influenced by Governmental policies. However, on the basis of recent migration trends regarded as also the future patterns to continue, one can estimate the extent of projected annual migration (both emigration and immigration). The distribution of net migration (emigration over immigration, say) by age and sex is usually assumed to remain unchanged over future years in making the adjustment due to migration in Projection process.

To simplify the computation of net migration during a period of t calendar years, it may be assumed to be concentrated on the last day of the period.

In this way, births and deaths among the migrants is not taken in to account during the 't' year period of projection. But since the error in the estimation of migrants usually remains small, this crudity in the assumption, while making adjustment due to migration, does not vitiate the Projected population much.

4.5.2 Representation of Component Method by the Use of Leslie Matrix

Here we present the component method of population projection by the use of Leslie matrix (Leslie (1945), (1948))

It is assumed here that the population is stable i.e. the fertility and mortality parameters are independent of time.

The factors given in 4.5.1 (A) and (B) affecting the change in the future population leads to a set of linear equations assuming that migratory disturbances are nil.

For projection purpose the following assumptions are made:

(1) The projection is made every fifth year i.e. $t = 5$

(2) For projection purpose, only female population is considered.

(3) child bearing age is taken from 15 to 45 years.

We denote,

(i) ${}_nP_x^{(t)}$ = Female population in the age group (x to $x+n$), at time t.

(ii) nF_x = age specific fertility rate in the age group (x to $x+n$), independent of t.

To write the linear equations we note that the female population in the age group (0,5) at time $(t+5)$ is the female population born and survive during the time t to $t+5$. Since the child bearing age is taken from 15 to 45 at an interval of 5 years, so the average number of female in the age group (15 to 20) during the time t to $t+5$ is $\frac{1}{2}\left[{}_5P_{15}^{(t)} + {}_5P_{15}^{(t+5)}\right]$. So, the number of females born to these females is obtained by multiplying it by the corresponding fertility rate ${}_5F_{15}$ and out of these the number which survive is obtained by multiplying it with the corresponding average number of survivals of infants during a span of 5 years viz. $\frac{{}_5L_0}{l_0}$, which is fixed and does not depend on the age of mothers.

We may recall that ${}_nL_x$ is the person years lived by the life table population in the age group (x to $x+n$), (see (2.7.5)).

So, the number of females born and survive to average number of females in the age group (15 to 20) in time t to $t+5$ is

$$= \frac{1}{2}\left[{}_5P_{15}^{(t)} + {}_5P_{15}^{(t+5)}\right] {}_5F_{15}\left(\frac{{}_5L_0}{l_0}\right) \tag{4.5.7}$$

Similarly, the number of females born and survive to the females of age group (20 to 25) in time t to $t+5$

$$= \frac{1}{2}\left[{}_5P_{20}^{(t)} + {}_5P_{20}^{(t+5)}\right] {}_5F_{20}\left(\frac{{}_5L_0}{l_0}\right) \tag{4.5.8}$$

etc.

Lastly, the number of females borne and survive to the females of age group (40 to 45) in time t to $t+5$

$$= \frac{1}{2}\left[{}_5P_{40}^{(t)} + {}_5P_{40}^{(t+5)}\right] {}_5F_{40}\left(\frac{{}_5L_0}{l_0}\right) \tag{4.5.9}$$

So the number of females in the age group (0 to 5) at time $(t+5)$ is

$$\begin{aligned}{}_5P_0^{(t+5)} &= \frac{1}{2}\left[{}_5P_{15}^{(t)} + {}_5P_{15}^{(t+5)}\right] {}_5F_{15}\left(\frac{{}_5L_0}{l_0}\right) + \frac{1}{2}\left[{}_5P_{20}^{(t)} + {}_5P_{20}^{(t+5)}\right] {}_5F_{20}\left(\frac{{}_5L_0}{l_0}\right) \\ &\quad + \cdots + \frac{1}{2}\left[{}_5P_{40}^{(t)} + {}_5P_{40}^{(t+5)}\right] {}_5F_{40}\left(\frac{{}_5L_0}{l_0}\right)\end{aligned} \tag{4.5.10}$$

Now the number of females in the age group (5 to 10) at time $(t+5)$ is only those who were in the age group (0 to 5) at time t and survived, so

$${}_5P_5^{(t+5)} = {}_5P_0^{(t)}\left(\frac{{}_5L_5}{{}_5L_0}\right), \tag{4.5.11}$$

Similarly $${}_5P_{10}^{(t+5)} = {}_5P_5^{(t)}\left(\frac{{}_5L_{10}}{{}_5L_5}\right) \tag{4.5.12}$$

$${}_5P_{15}^{(t+5)} = {}_5P_{10}^{(t)}\left(\frac{{}_5L_{15}}{{}_5L_{10}}\right) \tag{4.5.13}$$

and so on. Lastly

$${}_5P_{40}^{(t+5)} = {}_5P_{35}^{(t)}\left(\frac{{}_5L_{40}}{{}_5L_{35}}\right) \tag{4.5.14}$$

$${}_5P_{45}^{(t+5)} = {}_5P_{40}^{(t)}\left(\frac{{}_5L_{45}}{{}_5L_{40}}\right) \tag{4.5.15}$$

Now substituting on l.h.s. of (4.5.10) the population values in the year $t+5$ in terms of that in the year t from the relations (4.5.13) on words, we get.

$$\frac{1}{2}\left[{}_5P_{15}^{(t)} + {}_5P_{10}^{(t)}\,\frac{{}_5L_{15}}{{}_5L_{10}}\right]{}_5F_{15}\left(\frac{{}_5L_0}{l_0}\right) + \frac{1}{2}\left[{}_5P_{20}^{(t)} + {}_5P_{15}^{(t)}\,\frac{{}_5L_{20}}{{}_5L_{15}}\right]{}_5F_{20}\left(\frac{{}_5L_0}{l_0}\right)$$
$$+\cdots+\frac{1}{2}\left[{}_5P_{40}^{(t)} + {}_5P_{35}^{(t)}\,\frac{{}_5L_{40}}{{}_5L_{35}}\right]{}_5F_{40}\left(\frac{{}_5L_0}{l_0}\right) = {}_5P_0^{(t+5)} \tag{4.5.16}$$

Rearranging the terms in the increasing order of ${}_5P_{10}^{(t)}, {}_5P_{15}^{(t)}, {}_5P_{20}^{(t)}, \ldots, {}_5P_{40}^{(t)}$, we get,

$$\frac{{}_5L_0}{2l_0}\left[\frac{{}_5L_{15}}{{}_5L_{10}}{}_5F_{15}\right]{}_5P_{10}^{(t)} + \frac{{}_5L_0}{2l_0}\left[{}_5F_{15} + \frac{{}_5L_{20}}{{}_5L_{15}}{}_5F_{20}\right]{}_5P_{15}^{(t)} + \frac{{}_5L_0}{2l_0}\left[{}_5F_{20} + \frac{{}_5L_{25}}{{}_5L_{20}}{}_5F_{25}\right]$$
$${}_5P_{20}^{(t)} + \cdots + \frac{{}_5L_0}{2l_0}\left[{}_5F_{35} + \frac{{}_5L_{40}}{{}_5L_{35}}{}_5F_{40}\right]{}_5P_{35}^{(t)} + \frac{{}_5L_0}{2l_0}({}_5F_{15}){}_5P_{40}^{(t)} = {}_5P_0^{(t+5)} \tag{4.5.17}$$

The equations (4.5.17), (4.5.11) to (4.5.15) can be written in the matrix form as

$$\begin{bmatrix} {}_5P_0^{(t+5)} \\ {}_5P_5^{(t+5)} \\ {}_5P_{10}^{(t+5)} \\ \cdots \\ {}_5P_{45}^{(t+5)} \end{bmatrix} = \begin{bmatrix} 0 & 0 & \frac{{}_5L_0}{2l_0}\left(\frac{{}_5L_{15}}{{}_5L_{10}}{}_5F_5\right) & \frac{{}_5L_0}{2l_0}\left({}_5F_{15}+\frac{{}_5L_{20}}{{}_5L_{15}}{}_5F_{20}\right) & \cdots & \frac{{}_5L_0}{2l_0}({}_5F_{40}) \\ \frac{{}_5L_5}{{}_5L_0} & 0 & 0 & 0 & \cdots & 0 \\ 0 & \frac{{}_5L_{10}}{{}_5L_5} & 0 & 0 & \cdots & 0 \\ \cdots & \cdots & \cdots & \cdots & \cdots & \cdots \\ 0 & 0 & 0 & 0 & \cdots & \frac{{}_5L_{45}}{{}_5L_{40}} \end{bmatrix} \begin{bmatrix} {}_5P_0^{(t)} \\ {}_5P_5^{(t)} \\ {}_5P_{10}^{(t)} \\ \cdots \\ {}_5P_{40}^{(t)} \end{bmatrix}$$

or $$P^{(t+5)} = LP^{(t)}, \tag{4.5.19}$$

where $P^{(t)}$ and $P^{(t+5)}$ represent the population age vector in the year t and $(t+5)$, respectively and L is the Leslie matrix consisting of elements which are functions of fertility rate and mortality rate which are independent of time.

Thus with time independent Leslie matrix, we have

$$\left.\begin{aligned} P^{(t+5)} &= LP^{(t)} \\ P^{(t+10)} &= LP^{(t+5)} = L^2P^{(t)} \\ P^{(t+15)} &= LP^{(t+10)} = L^3P^{(t)}, \text{ etc.} \\ \text{Thus, for any integer } k,\ P^{(t+5.k)} &= L^kP^{(t)}, \end{aligned}\right\} \tag{4.5.20}$$

which shows that the sequence $P^{(t)}, P^{(t+5)}, \ldots$, constitutes a simple Markov chain.

4.5.3 Properties of Time Independent Leslie Matrix

We describe the projection process for a closed population with time independent fertility and mortality parameters and projection is made every year. Let,

(i) n_{xi} = Number of persons in the age sector (x to $x+1$) at time $t=i$, $i=0, 1, 2, \ldots, m$, where '0' represents the base year (i.e. the starting year of projection) and 'm' is the upper bound of the age of survival.

(ii) f_x = Probability of giving birth in the age sector (x to $x+1$)

(iii) p_x = Probability or surviving between the age (x to $x+1$)

Both f_x and p_x are (calender) time independent parameters and are related to a stable female population only.

$$\text{Then, } n_{01} = \sum_{x=0}^{m} f_x n_{x0} = \sum_{x=15}^{45} f_x n_{x0},$$

since $f_x = 0$ for $x < 15$ and $x > 45$, as per assumptions in the last section.

$$\begin{aligned} n_{11} &= p_0 n_{00} \\ n_{21} &= p_1 n_{10} \\ \cdots &\quad \cdots \\ n_{m1} &= p_{m-1} n_{(m-1)0} \end{aligned} \tag{4.5.21}$$

Using matrix notations we have

$$\begin{pmatrix} n_{01} \\ n_{11} \\ n_{21} \\ \cdot \\ \cdot \\ \cdot \\ n_{m1} \end{pmatrix}_{(m+1)\times 1} = \begin{pmatrix} f_0 & f_1 & \cdots & f_{m-1} & f_m \\ p_0 & 0 & \cdots & 0 & 0 \\ 0 & p_1 & \cdots & 0 & 0 \\ \cdot & \cdot & & \cdot & \cdot \\ \cdot & \cdot & \cdots & \cdot & \cdot \\ \cdot & \cdot & & \cdot & \cdot \\ 0 & 0 & \cdots & p_{m-1} & 0 \end{pmatrix}_{(m+1)\times(m+1)} \begin{pmatrix} n_{00} \\ n_{10} \\ n_{20} \\ \cdot \\ \cdot \\ \cdot \\ n_{(m-1)0} \\ n_{m0} \end{pmatrix}_{(m+1)\times 1} \tag{4.5.22}$$

$$\text{or} \qquad N_1 = LN_0 \tag{4.5.23}$$

where N_1 denotes the column vector on the l.h.s. and L denotes the matrix and N_0 denotes the column vector on the $r \cdot h \cdot s$ of (4.5.22). L is the time independent Leslie matrix of order $(m+1) \times (m+1)$.

Precisely in the same way

$$\left.\begin{aligned} N_2 &= LN_1 = L^2 N_0 \\ N_3 &= L^3 N_0 \\ \cdots &\quad \cdots \quad \cdots \quad \cdots \\ N_t &= L^t N_0 \end{aligned}\right\} \tag{4.5.24}$$

$$\text{where} \quad N_t = (n_{0t}, n_{1t}, \ldots, n_{mt})' \tag{4.5.24}$$

$$= \text{Projected population after } t \text{ years} \tag{4.5.25}$$

N_0 = Population at the base year.

Next let us assume that k is the upper bound of the child bearing age interval (since the discussion is for any biological population, not necessarily human population only, hence no lower age limit is considered) and we concentrate our attention only on $n_{0t}, n_{1t}, \ldots, n_{kt}$, for $t = 0, 1, 2, \ldots, m$. Then, the corresponding Leslie matrix will be of order $(k+1) \times (k+1)$, which is a condensed form of $(L)_{(m+1)\times(m+1)}$ in (2.5.23). Let us denote this condensed matrix also by L; we note that it is non-singular and none of its characteristic roots are zero.

Theorem 4.5.1: Leslie matrix L has only one positive characteristic root and the rest of the roots are either negative or complex. Further, if λ_1 is the positive root, then $|\lambda_i| \leq \lambda_1$ where $\lambda_i (i = 2, 3, \ldots)$ are, either negative or complex roots. In other words λ_1 is the most dominant.

Proof : From (4.5.24) We have

$$N_t = (L)^t_{(k+1)\times(k+1)} N_0 \tag{4.5.26}$$

If λ is the characteristic root of L, then $|L - \lambda I| = 0$ is the characterize equation, which can be written as

$$\lambda^{k+1} - f_0\lambda^k - f_1p_0\lambda^{k-1} - f_2p_0p_1\lambda^{k-2} - \cdots - f_kp_0p_1\ldots p_{k-1} = 0. \tag{4.5.27}$$

where f_x are p_x are defined in (ii) and (iii).

or $\quad \Phi(\lambda) = f_0/\lambda + f_1p_0/\lambda^2 + f_2p_0p_1/\lambda^3 + \cdots + f_kp_0p_1\ldots p_{k-1}/\lambda^{k+1} = 1.$ (4.5.28)

It can be seen that the number of change of sign in (4.5.27) in only one and hence by Descarte's rule of sign, in Theory of Equation, there can almost be one positive value of λ; the remaining values are either negative or complex. Again $\Phi(\lambda)$ in (4.5.28) is a continuous function of λ in $(0 \leq \lambda \leq \infty)$ and $\Phi(0) = \infty$, $\Phi(\infty) = 0$, so there exists at least one positive root of (4.5.28). Let us denote the positive root by λ_1 and other kind of roots by$\lambda_i, i = 2, 3, \ldots, (k+1)$, then we can write

$$\lambda_i^{-1} = e^{\alpha}(\cos\beta + i\sin\beta) \tag{4.5.29}$$

$$\Rightarrow \quad \lambda_i^{-r} e^{r\cdot\alpha}(\cos r\beta + i\sin i\beta), i = 2, 3, \ldots (k+1).$$

With this notation, from (4.5.28) we get

$$\Phi(\lambda) = f_0e^{\alpha}(\cos\beta + i\sin\beta) + f_1p_0e^{2\alpha}(\cos 2\beta + i\sin 2\beta) + \cdots + f_kp_0p_1\ldots p_{k-1}e^{(k+1)\alpha}\{\cos(k+1)\beta + i\sin(k+1)\beta\} = 1.$$

Equating the real and imaginary parts on both sides, we get

$$f_0e^{\alpha}\cos\beta + f_1p_0e^{2\alpha}\cos 2\beta + \cdots + f_kp_0p_1\ldots p_{k-1}e^{(k+1)\alpha}\cos(k+1)\beta = 1$$

Now $\because |\cos r\beta| \leq 1, \forall\, r = 1, 2, \ldots$, it follows that

$$f_0e^{\alpha} + f_1p_0e^{2\alpha} + \cdots + f_kp_0p_1\ldots p_{k-1}e^{(k+1)\alpha} \geq 1 \tag{4.5.30}$$

Further, from (4.5.29)

$$\left|\lambda_i^{-1}\right|^r = e^{r\alpha}, i = 2, 3, \ldots, (k+1),$$

then (4.5.30) becomes

$$f_0\left|\lambda_i^{-1}\right| + f_1p_0\left|\lambda_i^{-2}\right| + \cdots + f_kp_0p_1\ldots p_{k-1}\left|\lambda^{-(k+1)}\right| \geq 1, i = 2, 3, \ldots \tag{4.5.31}$$

Comparing (4.5.28) and (4.5.31) we see that

$$\frac{1}{|\lambda_i|} \geq \frac{1}{\lambda_1} \Rightarrow \lambda_1 \geq |\lambda_i|, i = 2, 3, \ldots, (k+1). \tag{4.5.32}$$

Hence, λ_1 is most dominant. //

Theorem 4.5.2: With the above notations we can show that

$N_t = \lambda_1 N_{t-1}$ holds asymptotically i.e. for large t. (4.5.33)

Proof: Since $(L)_{(k+1)\times(k+1)}$ is non-singular, $\exists$ a non-singular matrix P such that

$$P^{-1}LP = \text{ diag }(\lambda_1, \lambda_2, \ldots, \lambda_{k+1}) = \Lambda \text{ (say)}$$

$$\Rightarrow \quad L = P\Lambda P^{-1} \;\Rightarrow\; L^2 = P\Lambda^2 P^{-1}.$$

In general, $L^t = P\Lambda^t P^{-1}$ (4.5.34)

Now, $N_t = (L)^t N_0$, from (4.5.26)

or $N_t = P\Lambda^t P^{-1} N_0$, using (4.5.34)

$$\Rightarrow \quad \frac{N_t}{\lambda_1^t} = P\frac{\Lambda^t}{\lambda_1^t}P^{-1}N_0 \tag{4.5.35}$$

Now, $(P^{-1})_{(k+1)\times(k+1)}(N_0)_{(k+1)\times 1}$ is a $(k+1)$ rowed column vector.

$$\text{So let,} \quad P^{-1}N_0 = \begin{pmatrix} C \\ * \\ . \\ . \\ * \end{pmatrix},$$

i.e. the first element of $P^{-1}N_0$ is C whereas other elements are arbitrary and denoted by *. Since, the values of * does not affect the proof, hence these are kept as *.

So, (4.5.35) becomes

$$\frac{N_t}{\lambda_1^t} = P \begin{pmatrix} \frac{\lambda_1^t}{\lambda_1^t} & 0 & \cdots & 0 \\ 0 & \frac{\lambda_2^t}{\lambda_1^t} & \cdots & 0 \\ \cdots & \cdots & \cdots & \cdots \\ 0 & 0 & \cdots & \frac{\lambda_{k+1}^t}{\lambda_1^t} \end{pmatrix} \begin{pmatrix} C \\ * \\ \cdots \\ * \end{pmatrix}$$

$$\therefore \quad \lim_{t\to\infty} \frac{N_t}{\lambda_1^t} = P \text{ diag }(1, 0, 0, \ldots, 0) \begin{pmatrix} C \\ * \\ \cdots \\ * \end{pmatrix}$$

$$\because \quad \lambda_1 \geq |\lambda_i| \quad \therefore \quad \lim_{t\to\infty} \frac{\lambda_i^t}{\lambda_1^t} \to 0, i = 2, 3, \ldots, (k+1).$$

$$\therefore \quad \frac{N_t}{\lambda_1^t} = P \begin{pmatrix} C \\ * \\ \cdots \\ * \end{pmatrix} \text{ holds as } t \to \infty \tag{4.5.36}$$

Similarly, $$\frac{N_{t-1}}{\lambda_1^{t-1}} = P \begin{pmatrix} C \\ * \\ \cdots \\ * \end{pmatrix} \text{ as } t \to \infty \tag{4.5.37}$$

So, from (4.5.36) and (4.5.37)

$$\frac{N_t}{\lambda_1^t} = \frac{N_{t-1}}{\lambda_1^{t-1}} \tag{4.5.38}$$

$$\Rightarrow \qquad N_t = \lambda_1 N_{t-1}, \text{ for large } t. \qquad //$$

Further, from (4.5.33)

$$\begin{pmatrix} n_{0t} \\ n_{1t} \\ \cdots \\ \cdots \\ n_{kt} \end{pmatrix} = \lambda_1 \begin{pmatrix} n_{0t-1} \\ n_{1t-1} \\ \cdots \\ \cdots \\ n_{kt-1} \end{pmatrix}, \text{ holds for large } t. \tag{4.5.39}$$

It shows that for the age distribution of a stable population

$$\frac{n_{xi}}{\sum_{x=0}^{k} n_{xi}} = \frac{\lambda_1 n_{xi-1}}{\lambda_1 \sum_{x=0}^{k} n_{xi-1}} = \frac{n_{xi-1}}{\sum_{x=0}^{k} n_{xi-1}} \text{ holds for large } t, \forall x = 0, 1, 2, \ldots, \tag{4.5.40}$$

i.e. Age distribution remains independent of t.

Theorem 4.5.3: If X_1 is the characteristic vector corresponding to characteristic root λ_1 of L, then the non-zero vector solution of

$$LX_1 = \lambda_1 X_1 \tag{4.5.41}$$

is given by

$$X_1 = \begin{pmatrix} 1 \\ \frac{p_0}{\lambda_1} \\ \frac{p_0 p_1}{\lambda_1^2} \\ \cdots \\ \frac{p_0 p_1 \cdots p_{k-1}}{\lambda_1^k} \end{pmatrix}_{(k+1)\times 1} \tag{4.5.42}$$

where λ_1 is the positive characteristic root of L.

Proof: From (4.5.41) we have $[L - \lambda_1 I]X_1 = O \Rightarrow LX_1 = \lambda_1 X_1$

$$\Rightarrow \begin{pmatrix} f_0 & f_1 & \cdots & f_{k-1} & f_k \\ p_0 & 0 & \cdots & 0 & 0 \\ 0 & p_1 & \cdots & 0 & 0 \\ \cdot & \cdot & & \cdot & \cdot \\ \cdot & \cdot & \cdots & \cdot & \cdot \\ \cdot & \cdot & & \cdot & \cdot \\ 0 & 0 & \cdots & p_{k-1} & 0 \end{pmatrix}_{(k+1)\times(k+1)} \begin{pmatrix} x_0 \\ x_1 \\ x_2 \\ \cdot \\ \cdot \\ \cdot \\ x_k \end{pmatrix}_{(k+1)\times 1} = \lambda_1 \begin{pmatrix} x_0 \\ x_1 \\ x_2 \\ \cdot \\ \cdot \\ \cdot \\ x_k \end{pmatrix}_{(k+1)\times 1} \tag{4.5.43}$$

From (4.5.43) we get

$$f_0x_0 + f_1x_1 + \cdots + f_kx_k = \lambda_1 x_0 \tag{4.5.44}$$

$$\left.\begin{aligned} &\text{and} && p_0x_0 = \lambda_1 x_1 \Rightarrow x_1 = p_0x_0/\lambda_1, \\ & && p_1x_1 = \lambda_1 x_2 \Rightarrow x_2 = p_1x_1/\lambda_1 = p_0p_1x_0/\lambda_1^2. \\ &\text{Similarly,} && x_3 = p_0p_1p_2x_0/\lambda_1^3, \ldots, x_{k-1} = p_0p_1\ldots p_{k-2}x_0/\lambda_1^{k-1}, \\ & && x_k = p_0p_1\ldots p_{k-1}x_0/\lambda_1^k, \end{aligned}\right\} \tag{4.5.45}$$

where x_0 can be obtained from (4.5.44).

$$\text{Hence,} \quad X_1 = \begin{pmatrix} x_0 \\ p_0x_0/\lambda_1 \\ p_0p_1x_0/\lambda_1^2 \\ \ldots \\ p_0p_1\cdots p_{k-1}x_0/\lambda_1^k \end{pmatrix} \tag{4.5.46}$$

$$\text{Also,} \quad (L-\lambda_1 I)_{(k+1)\times(k+1)}(X_1)_{(k+1)\times 1} = (O)_{(k+1)\times 1}$$

has one linearly independent solution.

Now, Number of columns of $(L-\lambda_1 I)$ − Rank of $(L-\lambda_1 I)$
= Number of linearly independent solution vector.

or $(k+1)$ − Rank of $(L-\lambda_1 I) = 1 \Rightarrow$ Rank of $(L-\lambda_1 I) = k$.

The number of linearly independent vector solution being only one, we can write the vector solution of $LX_1 = \lambda_1 X_1$ as (4.5.46) where x_0 is given by (4.5.44). //

We give below two theorems on the Leslie matrix L, without proof.

Theorem 4.5.4: A sufficient condition that λ_1 is the only positive root of L is that λ_1 is strictly dominant.

This result holds for any matrix with non-negative elements; L is such a matrix.

Theorem 4.5.5: If two consecutive entries f_j and f_{j+1} in the first row of L are not zero, then the positive characteristic root of L is strictly dominant.

4.5.4 Population Density Dependent Leslie Matrix

Theorem 4.5.6: suppose the mortality and the fertility parameters viz. $1-p_x$ and f_x in the Leslie matrix L are also population density dependent (i.e. these also depend on the size of the population N at any time in addition to being age dependent). Then the population growth rate is Logistic (Leslie (1948)).

Proof: Let

$$L' = y^{-1}L = \begin{pmatrix} y^{-1}f_0 & y^{-1}f_1 & \cdots & y^{-1}f_{k-1} & y^{-1}f_k \\ y^{-1}p_0 & 0 & \cdots & 0 & 0 \\ 0 & y^{-1}p_1 & \cdots & 0 & 0 \\ . & . & & . & . \\ . & . & \cdots & . & . \\ . & . & & . & . \\ 0 & 0 & \cdots & y^{-1}p_{k-1} & 0 \end{pmatrix} \tag{4.5.47}$$

where $y = a + bN$, (4.5.48)

a and b are constants and N is the size of the population and L is given by (4.5.22). Here, the birth parameters (f_x) and the death parameters $q_x(p_x = 1 - q_x)$ are multiplied by y^{-1} to make them population density dependent.

If $X(t+1)$ be the population age vector at time $(t+1)$, then

$$X(T+1) = L'X(t) = y^{-1}LX(t) = y^{-1}\lambda_1 X(t), \tag{4.5.49}$$

since λ_1 is the most dominant characteristic root of L (Theorem 4.5.1)

Adding (4.5.49) over all ages we get,

$$N(t+1) = y^{-1}\lambda_1 N(t), \tag{4.5.50}$$

Now, as $t \to \infty$,

$$N(t+1) = N(t) = N \text{ (say)} \Rightarrow \text{growth rate is zero.} \tag{4.5.51}$$

So, if $N \to \infty$ (i.e. when $t \to \infty$) the growth rate essentially becomes zero. Then, from (4.5.51)

$$N(t+1) - N(t) = 0 \Rightarrow \frac{N(t+1) - N(t)}{N(t)} = 0$$

From (4.5.50)

$$\Rightarrow \quad \frac{\lambda_1 y^{-1}N(t) - N(t)}{N(t)} = \frac{(\lambda_1 y^{-1} - 1)N(t)}{N(t)} = 0$$

$$\Rightarrow \quad (\lambda_1 y^{-1} - 1) = 0 \Rightarrow \lambda_1 y^{-1} = 1 \Rightarrow y = \lambda_1. \tag{4.5.52}$$

So, as $n \to \infty$, the most dominant characteristic root of L',

$$\text{say} \quad \lambda_1' = \lambda_1 y^{-1} = 1, \text{ from (4.5.52),} \tag{4.5.53}$$

Next, if $N \to 0$ in $y = a + bN$, without loss of generality, we may take $a = 1$.

So, as $N \to 0, y = a + bN \to y = 1 + bN$ (4.5.54)

On the other hand if N is very large, say $N = N_0$, as $t \to \infty$

$$\text{i.e.} \quad \lim_{t\to\infty} N = N_0 \tag{4.5.55}$$

So, from (4.5.52), (4.5.54) and (4.5.55)

$$\lambda_1 = 1 + bN_0 \Rightarrow b = \frac{(\lambda_1 - 1)}{N_0}. \tag{4.5.56}$$

Substituting (4.5.56) in (4.5.54) we get

$$y = 1 + \frac{(\lambda_1 - 1)}{N_0}N \ . \tag{4.5.57}$$

Now, substituting the value of y in (4.5.50) we get

$$N(t+1) = \frac{\lambda_1 N(t)}{\left[1 + \left(\frac{\lambda_1 - 1}{N_0}\right) n(t)\right]}. \tag{4.5.58}$$

(4.5.58) is a difference equation in $N(t)$ and it can be verified that its solution is

$$N(t) = \frac{K}{1 + \rho e^{-rt}}, \text{ where } K,\ \rho \text{ and } r \text{ are appropriate constants,} \tag{4.5.59}$$

which shows that the growth rate is logistic. //

Remarks 4.6.4: However, if the initial age distribution is not in stable form, then $N(t)$ will not increase in the Logistic form, as shown. Although this un-justifies the fitting of Logistic curve, in such a situation, often it has been found that the Logistic curve produces a good fit; this is indeed misleading [Feller (1949), p. 52]. Similar is the case with several empirically good- fitting of population data by Logistic.

4.6 Quasi Stable Population Analysis

In Quasi stable population out of two factors fertility and mortality one is time dependent and the other is independent of time. There are basically two variants of Quasi stable Population. It may happen that one of the two components, Fertility and Mortality, is independent of time; whereas the other component has a time trend. For example, in India, many Demographers believe that Fertility rate is time independent, not showing much movement over time; whereas the Mortality has a time dependent declining trend. A second variant of the Quasi Stable Population would correspond to a situation when the mortality component remains stable whereas the Fertility shows a time trend. Recently there is development of models of different variants of Quasi Stable Population by Coale in (1976). In view of the importance of the Quasi Stable Models providing Demographic basis for many developing countries, Coale's analysis acquires formidable importance in the analysis and Projection of different categories of Quasi Stable Population. Below we present a short outline of the Quasi stable analysis. For details the readers are referred to Coale (1976).

4.6.1 Quasi Stable Population Analysis with Changing Mortality (Coale's model)

Coale's assumptions for the development of the model are as follows:

(i) If $\mu(x, t_1)$ and $\mu(x, t_2)$ are the mortality rates at the age x and times t_1 and t_2, respectively,

then $$\mu(x, t_1) - \mu(x, t_2) = k\Delta\mu(x) \tag{4.6.1}$$

where k is a constant and $\Delta\mu(x)$ is non changing overtime but changes with age x.

(ii) The mortality is approximated by a curve which declines sharply from 0 to 5 and remains at almost the same level from 5 to 45 and the section of the curve corresponding to the age 45 and above rises linearly, Fig. 4.4.

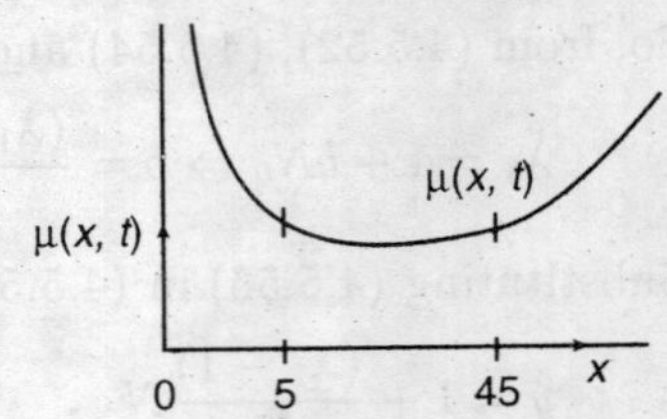

Fig. 4.4

(iii) The mortality change is linear with respect to time at any age x i.e.,

$$\mu(x, 0) - \mu(x, t) = t\Delta\mu(x) \tag{4.6.2}$$

The characteristics of $\mu(x)$ in three assumptions given above is fairly accurate; though the changes in mortality from 5 to 45 are not strictly constant, nor is the rising portion after 45 is precisely linear. Finally, one may also note that it is not necessary that rising starts only at $x = 45$. In order to show the effect of changing mortality depends on age distribution, a function $\gamma(x)$ has been defined as

$$\gamma(x) = \Delta\mu(x) - [\Delta\mu(x)]_{x=(5 \text{ to } 45)} \tag{4.6.3}$$

where $[\Delta\mu(x)]_{x=5 \text{ to } 45}$ is the value of $\Delta\mu(x)$ in the level portion from 5 to 45.

This is called min $\Delta\mu$. It is easy to see how $\gamma(x)$ behaves; $\gamma(x)$ falls from a maximum at age 0 to 0 at age 5 and remains steadily at zero level from 5 to 45 and then rise linearly form zero after age 45.

The annual change of the form $\gamma(x)$ has the same effect on the age distribution as $\Delta\mu(x)$ itself. We need only consider the two non zero portions of $\gamma(x)$. The horizontal portion of $\Delta\mu(x)$ (which is the portion when $\gamma(x)$ is zero) has no effect on the age distribution. This can be easily seen by the fact that a constant $\Delta\mu(x)$ implies that the change is of the form of equal increment (positive or negative) in $\mu(x)$. Consider now the proportion ${}_d\pi_x$ of populations surviving between ages x to $x+d$ which is

$${}_d\pi_x = \exp\left(-\int_x^{x+d} \mu(\tau)d\tau\right), {}_d\pi_x \text{ is a survival function.} \tag{4.6.4}$$

If the change in mortality is such that the new mortality rate

$$\mu_2(\tau) = \mu_1(\tau) - S \text{ at all ages } x, \text{ i.e. } S \text{ is a constant,} \tag{4.6.5}$$

where $\mu_1(\tau)$ corresponds to the rate at the old mortality schedule (or rate).

Now, if $\mu(\tau) = \mu_1(\tau) = \mu_2(\tau) + S$, (4.6.6)

then $${}_d\pi_x = \exp\left(-\int_x^{x+d} [\mu_2(\tau) + S]d\tau\right)$$

or, $${}_d\pi_x(e^{Sd}) = \exp\left(-\int_x^{x+d} \mu_2(\tau)d\tau\right), \tag{4.6.7}$$

in the age group x to $(x+d)$ under the new schedule (or rate) of Mortality.

The implication of the result is that every year there will be e^S more survivors (or more people) at every age i.e. the total number of people will increase but the age distribution remains the same. Again the decreased mortality would cause additional births proportional to e^S, since in every age group there are e^S more persons. Thus the age distribution in the age group (5 to 45) corresponding to the horizontal portion of the curve may be ignored while considering new age distributions as a result of declining mortality.

4.6.2 Effect of Declining Mortality when $\gamma(x)$ is Falling Steeply to Zero

Let, $\pi(5,t)$ = Proportion surviving from birth to the age 5 according to a cross section life table at time t

and $p(5,t)$ = Proportion surviving to age 5 in the cohort reaching 5 at time t, (4.6.8)

Reaching age at time $t \Rightarrow$ present time is $(t-5)$ if the age of the cohort is 0.

Reaching x at time $t \Rightarrow$ present time is $(t-5+x)$, $0 \le x \le 5$.

i.e. $$\pi(5,t) = \exp\left(-\int_0^5 \mu(x;t)dx\right), \quad 0 \le t \le 5 \tag{4.6.9}$$

and $$p(5,t) = \exp\left(\int_0^5 \mu(x, t-5+x)\right) dx \tag{4.6.10}$$

Let us consider two populations. In Population 1, we assume that the mortality is constant but fertility is increasing.

$\because \mu(x,t)$ is a constant w.r.t. t,

$$\therefore \quad \exp\left(-\int_0^5 \mu_1(x;t)dx\right) = \exp\left(-\int_0^5 \mu_1(x,t-5+x)dx\right)$$

$$= \exp\left(-\int_0^5 \mu_1(x)dx\right)$$

i.e. $\pi_1(5,t) = p_1(5,t)$ (4.6.11)

But in Population 2 we assume that mortality is constantly declining,

$$\pi_2(5,t) = \exp\left(-\int_0^5 \mu_2(x:t)dx\right),$$

since replacing t by $t-5+x$ does not alter its value

$$p_2(5,t) = \exp\left(-\int_0^5 \mu_2(x,t-5+x)dx\right)$$

$$\pi_2(5,t) > p_2(5,t), \quad \because \mu_2\ (.;.) \text{ is declining and } t \geq t-5+x; 0 \leq x \leq 5 \quad (4.6.12)$$

Next suppose at each age under five

$$\mu_2(x,t+1) = \mu_2(x,t) - \gamma(x)$$

$$(\because \gamma(x) = \mu_2(x,t) - \mu_2(x,t+1)) \text{ (see (4.6.2) and (4.6.8))} \quad (4.6.13)$$

$$\Rightarrow \quad \mu_2(x,t+n) = \mu_2(x,t) - n\gamma(x). \quad (4.6.14)$$

Further, $\pi_2(5,t+n) = \exp\left(-\int_0^5 \mu_2(x,t+n)dx\right)$

$$= \exp\left(-\int_0^5 [\mu_2(x;t) - n\gamma(x)]dx\right) \text{ by using (4.6.14)}$$

$$= \pi_2(5,t)\exp\left(n\int_0^5 \gamma(x)dx\right)$$

and $p_2(5,t) = \exp\left(-\int_0^5 \mu_2(x,t-5+x)dx\right)$

$$= \exp\left(-\int_0^5 \mu_2(x,t+(x-5))dx\right)$$

$$= \exp\left(-\int_0^5 [\mu_2(x,t) - (x-5)\gamma(x)]dx\right), \text{ by using (4.6.14)}$$

$$= \exp\left(-\int_0^5 \mu_2(x;t)dx\right)\exp\left(\int_0^5 (x-5)\gamma(x)dx\right)$$

$$= \exp\left(-\int_0^5 \mu_2(x;t)dx\right)\exp[(\xi_r - 5)]\exp\left(\int_0^5 \gamma(x)dx\right)$$

for some $\xi_r 0 \leq \xi_r \leq 5$ by M.V. theorem (A.1.13).

$$= \pi_2(5,t)\exp[(\xi_r - 5)]\exp\left(\int_0^5 \gamma(x)dx\right) \quad (4.6.15)$$

Next let $g_1(t)$ and $g_2(t)$ be the Gross reproduction rates in the two Populations. Recall that GRR is the number of female babies from $K(=1000)$ women.

From (4.6.7) there exists one constant k such that,

$$\text{for} \quad x = nK,\ g_1(t+x) = g_1(t)e^{nk} \tag{4.6.16}$$

($\because$ every year there will be e^k survivors in the first population).

$g_2(t)$ is a constant since there is no change in the fertility rate.

Let us define the age distribution of women belonging to the fertile age groups as

$$f(x) = \left[m(x,t) \Big/ \int_0^\beta m(x,t)dx \right], \tag{4.6.17}$$

where $m(x,t)$ is the number of potential mothers of age x at time t. We assume that the distribution is same for two populations, i.e.

$$f_1(x) = f_2(x) = f(x),\ \text{(say)},\ \forall x, 0 \le x \le \beta, \tag{4.6.18}$$

β being the upper bound of fertility span. Since the population may be any biological population (not necessarily human) hence lower limit may be zero. For materialization of the same, we require

$$B_1(t-5)p_1(5;t) = B_2(t-5)p_2(5;t) \tag{4.6.19}$$

where $B(t) =$ Number of births at any time t.

But for any population

$$B(t-5) = g(t-5) \int_\alpha^\beta N(x, t-5)f(x)dx, \tag{4.6.20}$$

where $N(x,t)$ is the population of age x at time t and $g(t)$ is the GRR at time t.

Since $f(x)$ and $N(x,t)$ for $x > 5$ are same for both the populations we need only $g(t-5)$ part of (4.6.20) for the number arriving at age five.

Hence for (4.6.18), substituting (4.6.20) in (4.6.19) we get

$$g_1(t-5)p_1(5,t) = g_2(t-5)p_2(5,t) \tag{4.6.21}$$

It should be noted that we started with two populations with the same number of people at each age above five ($\because$ same age distribution)

i.e. $\quad \pi_1(x,t) = \pi_2(x,t) = \pi(x,t)$, (say)

The equation (4.6.21) may be written as

$$g_1(t)e^{5k}\pi(5,t) = g_2(t)\pi(5,t)\exp\left(-(5-\xi_r)\int_0^5 \gamma(x)dx\right), \tag{4.6.22}$$

$0 \le \xi_r \le 5$ and $\gamma(x)$ is given by (4.6.13).

L.H.S. is obtained using (4.6.16) and (4.6.11), the R.H.S. by (4.6.15).

This will be satisfied if

$$\int_0^5 \gamma(x)dx = k \tag{4.6.23}$$

$$\text{and} \quad g_2(t)\exp(\xi_r k) = g_1(t) \tag{4.6.24}$$

$$\Rightarrow \quad g_2(t) < g_1(t)\xi_r,\ k > 0$$

Hence, we do have a choice of $\gamma(x)$ and k such that the two populations will have the same number arriving at age 5 annually. The condition is that the cumulative value of $\gamma(x)$ to age 5 should be k and as specified in (4.6.23) the current fertility (measured by G.R.R), in 2nd population, is lower than that of 1st populations.

In fact, rather that the annual rate of increase in the proportion surviving to age five being e^k, the increase in the proportion in addition to the increase that would occur if the annual change in mortality at all ages under five were the same as the plateau of mortality change (i.e., between the age 5 to 45) should be e^k. An approximation to the relevant annual increase in survival to age five is

$$\left[\frac{l_5'/l_5}{{}_5p_{25}'/{}_5p_{25}}\right], \tag{4.6.25}$$

where the prime indicates a value corresponding to a year later.

The relation of the age distribution of the population with declining mortality to the stable population is so clearly linked to the relation of the age distribution of the population with rising fertility to the stable population that a slight modification of the second relation (which has already been dealt with) will yield the first relation. It should be noted that the effects of the changes above the minimum level in age specific mortality rates that occur at older ages is neglected.

Though the two populations are alike above age five the population with changing fertility has a higher current fertility by a factor $\exp(k\xi_r)$. This has two implications:

(1) Above age five both the populations bear the same relation to the stable population but in case of a population with changing mortality (p_2) the relation is based on a fertility higher than the current by a factor $\exp(k\xi_r)$. Hence to find the ratio of the Population at age a to the stable population, the ratio of p_1 (Population with changing fertility) to a stable Population with a higher fertility level than the current one must be determined. This means that the relation p_1 in respect of the stable Population must be multiplied by $\exp(-\Delta r_x)$ at age x where

$$\Delta r = k\xi_r/T,$$

where ($r = r_x$) represent the intrinsic rates of growth in the age sector x of the population; T is the mean length of the generation and ξ_r is given by

$$\exp\left[\int_0^5 (x-5)\gamma(x)dx\right] = \exp[(\xi_r - 5)]\exp\left[\int_0^5 \gamma(x)dx\right]\ ;\ 0 \le \xi_r \le 5,$$

by the Mean value theorem (A.1.13); where $r(x)$ is given by (4.6.13).

(2) The proportion at age zero in case of population with changing mortality will be $\exp(-k\xi_r)$ times the proportion in case of population with changing fertility (p_1) and under age five the proportion will be $\exp(-k\xi_r/2)$ times that of p_1.

If allowance is made for these two adjustments the equations developed in the previous chapter can be adapted for approximate calculation of the effect of continuously declining mortality.

4.6.3 Adjustment for Obtaining the Age Distribution of a Quasi Stable Population with a History of Declining Mortality Relative to Stable Age Distribution

To obtain the approximate age distribution of a Population with a history of continuous declining mortality during the preceding t years the following kind of data are to be utilized.

(i) Current fertility and mortality schedule (which identify the current stable Population).

(ii) The number of years that mortality has been declining (t).

(iii) The estimated value of k viz the annual proportionate increase in fertility to which the annual increase in survival is equivalent.

(iv) The estimated value of β (the slope of the line approximating the age specific mortality rate increase above the age 50).

Estimation of k:

Let $\Delta_1 = \log\left(\frac{l_5}{l_0}\frac{l_{25}}{l_{30}}\right)\Big|_t^{t+x}$ and $\Delta_2 = \log_e\left(\frac{l_5}{l_0}\frac{l_{35}}{l_{25}}\right)\Big|_{t-x}^{t}$,

then $\hat{k} = \dfrac{\Delta_1 - \Delta_2}{x}$

Δ_1 and Δ_2 are obtainable from two Life tables.

Estimation of β:

$$\hat{\beta} = \frac{({}_5m_{65} - {}_5m_{50})|_{t-x} - ({}_5m_{65} - {}_5m_{50})|_t}{15x},$$

where ${}_5m_x$ refers to the central rate of mortality in the age sector $[x$ to $x+5]$.

Again if r is the intrinsic rate of growth and T is given by $e^{rT} = R_0$ = Net reproduction rate.

Then
$$B(t-x) = B(t)\exp\left(-r(t)x - k\left(1 - \frac{t}{T}\right)x\right), \text{ if } x < T$$
$$= B(t)\exp\left(-r(t)x - k\left(1 - \frac{t}{T}\right)\right), \text{ if } x > T\,\forall\, t < \frac{3T}{4}.$$

Again, if $t > \frac{3}{4}T$, then

$$B(t-x)\begin{cases} = B(t)\exp\left(-r(t)x - \left(\frac{k}{2}\right)x + \left(\frac{k}{2}T\right)x^2\right), & \text{if } x < m \\ = B(t)\exp\left\{-r(t)x - .09375kT + \left(-\frac{k}{2}\right)t - \left(-\frac{k}{2}\right)t^2 - k\left(1 - \frac{t}{T}\right)x\right\}, & \text{if } m < x < t \\ = B(t)\exp\left(\left\{-r(t)x - .09375kT - \left(\frac{k}{2}\right)t - \left(\frac{2}{k}\right)Tt^2 + \frac{ktx}{T}\right\}\right), & \text{if } x > t \end{cases}$$

where $m = t - \dfrac{3T}{4}$.

Denoting by $R(x,t)$ the rates of the change of the proportion at age x in a Population with a history of changing fertility at an annual rate k for t years to the Population at age x in a stable Population, and b_k and b_s being the birth rates of a population with a history of declining fertility at the rate of k for t years; and the stable population birth rate, respectively,

$$R(x;t) = \frac{b_k}{b_s}\exp\left(\left\{-k\left(1-\frac{t}{T}\right)x\right\}\right), \text{ if } m < x < t$$

$$= \frac{b_k}{b_s}\exp\left(\left\{-k\left(1-\frac{x}{T}\right)t\right\}\right), \text{ if } x > t.$$

If $t > \frac{3}{4}T$ and $m = t - \frac{3}{4}T$, then

$$R(x;t)\begin{cases} = \left(\frac{b_k}{b_s}\right)\exp\left(\left\{-.09375T + \frac{k}{2}t - \frac{k}{2}t^2 - k\left(1-\frac{t}{T}\right)x\right\}\right), \text{ if } m < x < t \\ = \left(\frac{b_k}{b_s}\right)\exp\left(\left\{-.09375kT + \frac{k}{2}t + \frac{k}{2}t^2 + \frac{ktx}{T}\right\}\right), \text{ if } x > t \\ = \left(\frac{b_k}{b_s}\right)\exp\left(\left\{-\left(\frac{k}{2}\right)x + \left(\frac{k}{2}T\right)x^2\right\}\right), \text{ if } x < m. \end{cases}$$

For appropriate $t \geq \frac{3T}{4}$ and x, the adjustment factors $W(x;t)$ are obtained as:

$$W(x;t) = R(x;t)\exp[-\{(1.27)kx/T\}]$$

and next $W_1(x;t)$ for the five yearly age groups as follows:

(i) $W_1(2.5;t) = W(2.5;t)e^{-0.63k}$

(ii) $W_1(x,t) = W(x;t)$, where $x = 7.5, 12.5, \ldots 42.5$

Again, for $y = x - 45$

$$s(y;t)\begin{cases} = l(y;t)\exp\left(\left\{-\beta\left(\frac{y^2}{2} - \frac{y^3}{3}\right)\right\}\right), \text{ for } y < t \\ = l(y;t)\exp\left\{-\beta\left[yt\left(\frac{y}{2} - \frac{t}{2}\right) + \frac{t^3}{6}\right]\right\}, \text{ for } y \geq t, \end{cases}$$

where s and l stand, respectively, for the age distribution for a stable and a population with a history of declining fertility.

$W_1(x;t)$ are thus the set of multipliers that may be applied to the stable age distribution to give the number of persons in each group in a Population with a history of declining mortality.

The proportionate age distribution by five year age intervals of this Population is

$$\frac{W_1(x;t)c(x;t)}{\sum_x W_1(x;t)c(x;t)}.$$

4.6.4 Quasi Stable Population Analysis under the Setup of Fertility Changing and Mortality Remaining Stable

We have seen from stable population analysis that the number of births at time t viz.

$$B(t) = \int_\alpha^\beta B(t-x)\ p(x)\ f(x)dx, \text{ see (4.4.17)} \tag{4.6.26}$$

If $p(x)$ and $f(x)$ (see (4.4.1)) depends on time i.e. $p(x) = p(x;t)$ and $f(x) = f(x;t)$

$$\Rightarrow \quad B(t) = \int_\alpha^\beta B(t-x)\ p(x;t)\ f(x;t)dx$$

$$= \int_\alpha^\beta p(x;t)\ f(x;t)dx \ \frac{\int_\alpha^\beta B(t-x)\ p(x;t)\ f(x;t)dx}{\int_\alpha^\beta p(x;t)\ f(x;t)dx}$$

$$= R(t)\int_\alpha^\beta B(t-x)\ \phi(x;t)dx \tag{4.6.27}$$

where $R(t) = \int_\alpha^\beta p(x;t)\ f(x;t)dx$ is the NRR per woman $= R_0$, see (4.4.32) (4.6.28)

$$\text{and}\quad \phi(x;t) = \frac{p(x;t)\ f(x;t)}{\int_\alpha^\beta p(x;t)\ f(x;t)dx},\ \alpha \le x \le \beta, \text{ from (4.4.4)} \tag{4.6.29}$$

is a proper probability density function.

Since mortality component remains stable we write

$$p(x;t) = p(x) \quad \text{and} \quad f(x,t) = f(x)e^{kt}, \tag{4.6.30}$$

since we assume that the fertility changes at a constant annual rate k

i.e. $\dfrac{df(x;t)}{dt} = kf(x;t)$

Also $R(t) \int_\alpha^\beta p(x)\ f(x)e^{kt}dx$, from (4.6.28)

$$= e^{kt}\int_\alpha^\beta p(x)\ f(x)dx \tag{4.6.31}$$

$= e^{kt}R(0)$, from (4.4.32)

and $\phi(x;t) = \dfrac{p(x)\ f(x)e^{kt}}{\left[e^{kt}\int_\alpha^\beta p(x)\ f(x)dx\right]}$, from (4.6.30)

$$= \frac{p(x)\ f(x)}{\int_\alpha^\beta p(x)\ f(x)dx} = \phi(x) \tag{4.6.32}$$

Hence (4.6.27) is reduced to

$$B(t) = R(0)e^{kt}\int_\alpha^\beta B(t-x)\ \phi(x)dx \tag{4.6.33}$$

Now by the mean value theorem (A.1.13) of Integral Calculus

$$\int_\alpha^\beta B(t-x)\ \phi(x)dx = B(t-T(t)), \tag{4.6.34}$$

where $\alpha \le T(t) \le \beta$ and $\because \int_\alpha^\beta \phi(x)dx = 1.$

$$\therefore \quad (4.6.33) \Rightarrow B(t) = R(0)e^{kt}B(t - T(t)). \tag{4.6.35}$$

To simplify the subsequent analysis the time origin is set at a moment when the NRR is unity i.e. $R(0) = 1.0$. There are two approximate solutions to (4.6.33). The first approximate solution is a fairly good approximation but based on an assumption which perhaps may be contrary to the fact. The details are given below:

(A) Method of obtaining the First approximate solution of (4.6.35)

$T(t)$ is taken to be a fixed number viz the mean length of generation T_0 as the NRR approaches unity.

Now the mean length of generation T_0 is given by

$$T_0 = \int_\alpha^\beta x\phi(x)dx = \frac{\int_\alpha^\beta xp(x)\ f(x)dx}{\int_\alpha^\beta p(x)\ f(x)dx}, \text{ from (4.6.32)} \tag{4.6.36}$$

The equation (4.6.35) is now reduced by the above assumption to

$$B(t) = e^{kt}B(t - T_0)$$

$$\Rightarrow \quad \log B(t) = kt + \log B(t - T_0) \tag{4.6.37}$$

Putting $\log B(t) = Y(t)$, (4.6.38)

we have $Y(t) - Y(t - T_0) = kt.$ (4.6.39)

(4.6.39) is a simple difference equation which may be solved by assuming that

$$Y(t) = b_0 + b_1 t + b_2 t^2 \tag{4.6.40}$$

(since the difference is linear so the function will be a quadratic)

$$\Rightarrow \quad Y(t - T_0) = b_0 + b_1(t - T_0) + b_2(t - T_0)^2$$

$$\Rightarrow \quad Y(t) - Y(t - T_0) = b_1 T_0 + b_2 T_0(2t - T_0)$$

$$\Rightarrow \quad b_1 T_0 + b_2 T_0(2t - T_0) = kt, \text{ from (4.6.39).} \tag{4.6.41}$$

Taking $2t = T_0$, for simplification, we get

$$b_1 T_0 = kt \Rightarrow b_1 = \frac{kt}{T_0} = \frac{k}{2} \tag{4.6.42}$$

Again putting $t = 0$ in $(4.180')$ $\Rightarrow b_1 T_0 - b_2 T_0^2 = 0$

$$\Rightarrow \quad b_2 = \frac{b_1}{T_0} = \frac{k}{2T_0} \text{ from (4.6.42)}$$

Again by the initial condition $Y(0) = \log B(0) = b_0$, from (4.6.40). Putting the values of b_0, b_1 and b_2 in (4.6.40) we get

$$Y(t) = \log B(0) + \frac{k}{2}t + \frac{k}{2T_0}t^2 \tag{4.6.43}$$

Now, from (4.6.38)

$$B(t) = e^{Y(t)} \Rightarrow B(t) = B(0)\exp\left(\frac{k}{2}t + \frac{k}{2T_0}t^2\right), \text{ using (4.6.43).} \tag{4.6.44}$$

The number of persons at age x in a Population with this birth sequence would be

$$B(x,t) = B(t-x)p(x), \text{ from definition,} \tag{4.6.45}$$

$$= B(0)\exp\left(\frac{k}{2}(t-x) + \frac{k}{2T_0}(t-x)^2\right)p(x), \text{ using (4.6.44)} \tag{4.6.46}$$

$$= B(0)\exp\left(\frac{k}{2}t + \frac{k}{2T_0}t^2\right) \times \exp\left(-\frac{k}{2}x - \frac{kt}{T_0}x + \frac{k}{2T_0}x^2\right)p(x) \tag{4.6.47}$$

$$\Rightarrow\ B(x,t) = B(t)\exp\left(-\frac{k}{2}x + \frac{k}{2T_0}x^2\right)e^{-r(t)x}p(x), \text{ see (4.6.49).} \tag{4.6.48}$$

where $r(t)$ is the intrinsic growth rate at time t. However, equation (4.6.48) is approximate, because

$$r(t) = \frac{\log R(t)}{T(t)}, \text{ see (4.4.63),}$$

holds for stable population.

$$\text{Now}\quad R(t) = \int_\alpha^\beta p(x;t)\, f(x;t)dx, \text{ from (4.6.28)}$$

$$= \int_\alpha^\beta p(x)\, f(x)\, e^{kt}dx, \text{ from (4.6.31)}$$

$$= e^{kt}\int_\alpha^\beta p(x)\, f(x)dx, \text{ from (4.4.5)}$$

$$= e^{kt}R_0 = e^{kt}, \quad \because R_0 = 1, \text{ by the assumption.}$$

$$\Rightarrow\quad \log R(t) = kt$$

$$\Rightarrow\quad \frac{\log R(t)}{T(t)} = \left[\frac{kt}{T(t)}\right] = r(t), \text{ as above} \tag{4.6.49}$$

Thus $\dfrac{t}{T_0}$ differs from $r(t)$ to the extent $T(t)$ differs from T_0.

The proportion of population at age x at time t is given by (see Notations is Sec. 4.4.3)

$$C(x;t) = \frac{B(x;t)}{N(t)}, \text{ by definition}$$

$$= b(t)\exp\left(-\left(\frac{k}{2}\right)x + \left(\frac{k}{2T_0}\right)x^2\right)e^{-r(t)x}p(x), \text{ from (4.6.48)} \tag{4.6.50}$$

where $b(t) = \dfrac{B(t)}{N(t)}$, (by definition Sec. (4.4.3))

But according to the stable Population analysis we have

$$C_S(x;t) = b_S(t)e^{-r(t)x}p(x), \text{ (see (4.4.40)), } S \text{ stands for stable,} \tag{4.6.51}$$

where $C_S(x;t)$ corresponds to the stable age distribution at age x

$$\frac{C(x,t)}{C_S(x;t)} = \frac{b(t)}{b_S(t)}\exp\left(-\left(\frac{k}{2}\right)x + \left(\frac{k}{2T_0}\right)x^2\right), \text{ from (4.6.50)}$$

$$\Rightarrow \quad C(x;t) = C_S(x;t)\,\frac{b(t)}{b_S(t)}\exp\left(-\left(\frac{k}{2}\right)x + \left(\frac{k}{2T_0}\right)x^2\right) \tag{4.6.52}$$

The result has been tested empirically and has been found to give fairly accurate fit. It may further be noted that though the approximation has been found to be quite useful the basic reason of the failure of the approximation to fit the actual birth sequence perfectly is that the value of $T(t)$ in equation (4.6.35) was taken to be fixed at T_0.

(B) A Second Approximate Solution of (4.6.35)

The second approximation is a modification of the first in the sense that it makes further adjustments for the 'contrary to the fact assumption' that $T(t)$ is fixed as in the case of first approximation. Here we shall not assume $T(t)$ to be fixed at T_0; but allow it to vary with the level of fertility. The adjustment is actually done in two stages; one is to use the mean length of generation (T) in stable population defined by fixed mortality and fertility schedules at time t for $T(t)$. The second adjustment is, thereafter, made by assuming that the approximate solution of $B(t)$ in

$$B(t) = B(0)\exp\left(\frac{k}{2}t + \frac{k}{2T_0}t^2\right), \text{ vide (4.6.44)},$$

introduces a systematic error between the weighted average of births between α to β years earlier and that of births T years ago

$$\text{i.e. between} \qquad \int_\alpha^\beta B(t-x)\,\phi(x)\,dx \text{ and } B(t-T)$$

by creating a systematic difference between the above. The relation (4.6.44) consider second order exponential birth sequence which is different from the birth pattern in stable population.

In stable population $B(t) = e^{kt}B(t-T)$

or $\quad B(t) = \exp([\log R/T_0]T)\;B(t-T)$, see (4.6.49), $\qquad$ (4.6.53)

T is the mean length of generation in a stable population.

In the next place, let us rewrite the equation (4.6.35)

$$B(t) = R(0)e^{kt}B(t-T(t)) \tag{4.6.54}$$

$$\text{as} \quad B(t) = R(0)e^{kt}B(t-T_0)\frac{B^*(t-T(t))}{B^*(t-T_0)} \tag{4.6.55}$$

where B^* indicates the values taken from the first approximate solution given by equation (4.6.49)

Now had B^* values been exact, the equation (4..6.55) would have reduced to (4.6.54). Since $T(t)$ is only slightly different from T_0, we may, write

$$\begin{aligned}\log\left(\frac{B^*(t-T(t))}{B^*(t-T_0)}\right) &= [\log B^*(t-T(t)) - \log B^*(t-T_0)]\\ &= (t-T(t)-t+T_0)\;\left.\frac{d\log B^*(t)}{dt}\right|_{T=t-T_0}\\ &\qquad + 0(|T(t)-T_0|), \text{ by Taylor's expansion (A.1.9)}\\ &= (T_0-T(t))\;\left.\frac{d\log B^*(t)}{dt}\right|_{T=t-T_0}\end{aligned} \tag{4.6.56}$$

Again we have from stable population analysis,

$$A(r) = \alpha + \beta r + \gamma r^2 + \cdots, \quad \text{from (4.4.35)}, \tag{4.6.57}$$

where $\alpha = \frac{R_1}{R_0} = \mu_1', \beta = \left(\frac{R_1}{R_0}\right)^2 - \frac{R_2}{R_0} = (\mu_1')^2 - \mu_2'$

and $\gamma = \left[\frac{R_3}{R_0} - 3\frac{R_1}{R_0}\frac{R_2}{R_0} + 2\left(\frac{R_1}{R_0}\right)^3\right]\frac{1}{2!}$, from (4.4.36)

$$= \left(\mu_3' - 3\mu_1'\mu_2' + 2\mu_1'^3\right)\frac{1}{2!}, \text{ from (4.4.6) with (4.4.1)}$$

and $$T = \frac{\int A(r)dr}{r}, \tag{4.6.58}$$

because $\log R_0 = \int A(r)dr$, see (4.4.33) and $r = \frac{\log R_0}{T}$, (see 4.4.65)

Then from (4.6.57) $\int A(r)dr = \alpha r + \frac{\beta r^2}{2} + \cdots$

$$\therefore \quad T = \frac{\int A(r)dr}{r} = \alpha + \frac{\beta r}{2}.$$

$$T \cong \mu_1' + \left(\mu_1'^2 - \mu_2'\right)\frac{r}{2} = T_0 - \left(\mu_2' - \mu_1'^2\right)\frac{r}{2}, \text{ from (4.6.36) and (4.4.6)}$$

$$\Rightarrow \quad T = T_0 - \frac{r}{2}\sigma^2, \text{ where } \sigma^2 \text{ is the variance of age distribution}, \tag{4.6.59}$$

Again approximating r by $\frac{kt}{T_0}$ (see (4.6.49)) we have

$$T = T_0 - \frac{kt}{2T_0}\sigma^2 \Rightarrow (T_0 - T) = \frac{kt}{2T_0}\sigma^2 \tag{4.6.60}$$

Now, $$\left.\frac{d\log B^*(t)}{dt}\right|_{t=t-T_0} = \frac{d}{dt}\left.\left(\frac{k}{2}t + \left(\frac{k}{2T_0}\right)t^2\right)\right|_{t=t-T_0}, \text{ see (4.6.44)}$$

$$= \frac{k}{2} + \left.\frac{2kt}{2T_0}\right|_{t=t-T_0} = -\frac{k}{2} + \frac{kt}{T_0}. \tag{4.6.61}$$

Substituting (4.6.60) and (4.6.61) in (4.6.56)

$$\Rightarrow \quad \log\left(\frac{B^*(t - T(t))}{B^*(t - T_0)}\right) = \frac{kt\sigma^2}{2T_0}\left(-\frac{k}{2} + \frac{kt}{T_0}\right)$$

Putting $\frac{kt}{T_0} \simeq r$, see (4.6.49) $\Rightarrow \log\frac{B^*(t-T(t))}{B^*(t-T_0)} \simeq \frac{\sigma^2 r}{2}\left(\frac{kt}{T_0} - \frac{k}{2}\right)$

$$\Rightarrow \quad \frac{B^*(t - T(t))}{B^*(t - T_0)} \simeq \exp\left(\frac{\sigma^2 r}{2}\left(\frac{kt}{T_0} - \frac{k}{2}\right)\right). \tag{4.6.62}$$

As stated earlier further adjustment is required to be done because of the fact that births, corresponding to any number of years earlier in a stable population depends on the level of fertility but we are dealing with a birth sequence of type

$B^*(t)$ and not the stable population birth sequence. We note that when $B^*(t)$ is substituted in the integral equation

$$B(t) = \int_\alpha^\beta B(t-x)\; p(x;t)\; f(x;t)\; dx, \text{ see (4.4.17),} \tag{4.6.63}$$

it fails to satisfy the same exactly.

$$B^*(t) = \int_\alpha^\beta B^*(t-x)\; p(x)\; f(x;t)\; dx \tag{4.6.64}$$

$$= B(0)\int_\alpha^\beta \exp\left(\frac{k}{2}(t-x) + \frac{k}{2T_0}(t-x)^2\right) f(x;t)\; p(x)\; dx, \text{ see (4.6.44)}$$

$$= B(0)\exp\left(\frac{k}{2}t + \frac{k}{2T_0}t^2\right)\int_\alpha^\beta \left[\exp\left(-\frac{k}{2}x + \frac{k}{2T_0}x^2 - \frac{kt}{T_0}x\right)\right] f(x;t)\; p(x)\; dx$$

$$B^*(t) = B^*(t)\int_\alpha^\beta \left(-\frac{k}{2}x + \frac{k}{2T_0}x^2 - \frac{kt}{T_0}\cdot x\right) f(x;t)\, p(x)\, dx, \text{ from (4.6.44)}$$

$$\Rightarrow \quad 1 = \int_\alpha^\beta \exp\left(-\frac{k}{2}x + \frac{k}{2T_0}x^2 - \frac{kt}{T_0}x\right) f(x;t)\; p(x)\; dx. \tag{4.6.65}$$

The modification that is intended will replace T_0 by $T(t)$. In such a case the integral equation will be satisfied, provided

$$\int_\alpha^\beta \exp\left(-\frac{k}{2}x + \frac{k}{2T(t)}x^2 - \frac{k}{T(t)}tx\right) f(x;t)\; p(x)\; dx = 1. \tag{4.6.66}$$

Again since $r(t) = \dfrac{\log R(t)}{T(t)} = \dfrac{kt}{T(t)}$, see (4.6.49)

then (4.6.66) will be reduced to

$$\int_\alpha^\beta \exp\left(-\frac{k}{2}x + \frac{k}{2T(t)}x^2 - r(t)x\right) f(x;t)\; p(x)\; dx = 1. \tag{4.6.67}$$

But according to Lotka's integral equation

$$\int_\alpha^\beta e^{-rx} p(x)\; f(x)\; dx = 1. \tag{4.6.68}$$

Thus the integral equation (4.6.68) is not satisfied unless $k = 0$.

Let us see the effect of a non zero k in the equation (4.6.67). Defining

$$Z(t,k) = \int_\alpha^\beta \exp\left(-\left(\frac{k}{2}\right)x + \left(\frac{k}{2T(t)}\right)x^2\right) e^{-r(t)x} p(x)\; f(x;t)\; dx. \tag{4.6.69}$$

We have
$$\frac{dZ(t,k)}{dk} = \int_\alpha^\beta \left[\exp\left(-\left(\frac{k}{2}\right)x + \frac{k}{2T(t)}x^2\right) e^{-r(t)x} p(x)\; f(x;t) \left(-\frac{x}{2} + \frac{x^2}{2T(t)}\right)\right] dx \tag{4.6.70}$$

$$\Rightarrow \quad \left.\frac{dZ(t,k)}{dk}\right|_{k=0} = \int_\alpha^\beta \left(-\frac{x}{2} + \frac{x^2}{2T(t)}\right) e^{-r(t)x} p(x)\; f(x;t)\; dx$$

$$= \left\{ -\frac{1}{2}\int_{\alpha}^{\beta} xe^{-r(t)x} p(x)\ f(x;t)\ dx + \frac{1}{2T(t)}\int_{\alpha}^{\beta} x^2 e^{-r(t)x} p(x)\ f(x;t)\ dx \right\}$$

$$= -\frac{\overline{A}}{2} + \frac{\sigma_x^2 + (\overline{A})^2}{2T(t)}, \tag{4.6.71}$$

where $\overline{A}$ = mean age of fertility (i.e. child bearing) in stable population

σ_x^2 = variance of fertility (child bearing) in the stable population.

If we assume $T(t) \cong \overline{A}$ i.e. the mean length of generation being approximately equal to the mean age of child bearing we get

$$\left.\frac{dZ(t)}{dk}\right|_{k=0} = -\frac{T(t)}{2} + \frac{\sigma_x^2 + (T(t))^2}{2T(t)} = \frac{-[T(t)]^2 + \sigma_x^2 + (T(t))^2}{2T(t)}$$

$$= \frac{\sigma_x^2}{2T(t)} \cong \frac{\sigma_x^2}{2T_0}, \text{ when } T(t) \text{ is approximated by } T_0. \tag{4.6.72}$$

Hence, for small values of k the value of $Z(t,k)$ given by (4.6.69) is

$$\left. Z(t,k)\right|_{k=0} = \int_{\alpha}^{\beta} e^{-r(t)x} p(x)\ f(x;t)\ dx = 1 \tag{4.6.73}$$

$$\left. Z(t,k)\right|_{k=0} = \left.\frac{\sigma_x^2 \cdot k}{2T_0}\right|_{k=0} + C = 1,$$

integrating (4.6.72) and equating with (4.6.73), where C is a constant of integration

$$\Rightarrow \quad 0 + C = 1 \ \Rightarrow \ C = 1 \tag{4.6.74}$$

Therefore, for small values of k

$$Z(t,k) = \left(1 + \frac{k\sigma_x^2}{2T_0}\right) \cong \exp\left(\frac{k\sigma_x^2}{2T_0}\right). \tag{4.6.75}$$

Hence, we may conclude that in case of a stable population where $B(t)$ is a first order exponential (i.e. when $\exp(\cdot)$ contains t only),

$$B(t) = \int_{\alpha}^{\beta} B(t-x)\ \phi(x;t)\ dx, \text{ see (4.6.26)}$$

$$= \int_{\alpha}^{\beta} B(t-x)\ \phi(x)\ dx = B(t - T(t)), \text{ see (4.6.34)}$$

exactly, where $f(x)$ is given by (4.6.32).

However, when $B(t)$ is a second order exponential (i.e. when $\exp(\cdot)$ contains t and t^2 both) of the from

$$B(t) = B(0)\exp\left(\frac{k}{2}t + \left(\frac{k}{2T}\right)t^2\right), \text{ see (4.6.44)}, \tag{4.6.76}$$

$B(t-x)$ is concave upwards or downwards depending on the sign of k.

Now, $\displaystyle\int_{\alpha}^{\beta} B(t-x)\ \phi(x)\ dx < B(t - T(t))$, if $k < 0$

and $\geq B(t - T(t))$ if $k > 0$.

The needed adjustment for

$$\int_{\alpha}^{\beta} B(t-x)\ \phi(x)\ dx = B(t-T(t))$$

is, therefore

$$B^*(t-T(t)) \exp\left(\frac{k}{2}\right) \frac{\sigma^2}{T} = \int_{\alpha}^{\beta} B(t-x)\ \phi(x)\ dx = B(t-T(t)).$$

We have from (4.6.55)

$$B(t) = R(0)e^{kt} B(t-T_0) \left[\frac{B^*(t-T(t))}{B^*(t-T_0)}\right] \exp\left(\frac{k}{2}\right) \frac{\sigma^2}{T_0}. \qquad (4.6.77)$$

Substituting the value of $\dfrac{B^*(t-T(t))}{B^*(t-T_0)}$ from (4.6.62) in (4.6.77) gives the difference equation

$$B(t) = R(0)B(t-T_0) \exp\left(\left\{kt - \left(\frac{\sigma^2 k^2}{4T_0}\right) t + \left(\frac{\sigma^2 k^2}{2T_0^2}\right) t^2 + \frac{\sigma^2 k}{2T_0}\right\}\right)$$

$$\because \quad r \simeq \frac{kt}{T_0}, \text{ see (4.6.49).}$$

Taking logarithms on both sides

$$\log B(t) = \log R(0) + \log B(t-T_0) + kt - \frac{\sigma^2 k^2 t}{4T_0} + \frac{\sigma^2 k^2}{2T_0^2} t^2 + \frac{\sigma^2 k}{2T_0} \qquad (4.6.78)$$

Putting $Y(t) = \log B(t)$ and as before taking $R(0) = 1$, see after (4.6.35),

$$Y(t) - Y(t-T_0) = kt - \frac{\sigma^2 k^2 t}{4T_0} + \frac{\sigma^2 k^2}{2T_0^2} t^2 + \frac{\sigma^2 k}{2T_0} = a_0 + a_1 t + a_2 t^2 \text{ (say)}, \qquad (4.6.79)$$

$$\text{where} \quad a_0 = \frac{\sigma^2 k}{2T_0};\ a_1 = k\left(1 - \frac{\sigma^2 k}{4T_0}\right) \quad \text{and} \quad a_2 = \left[\frac{\sigma^2 k^2}{2T_0^2}\right]. \qquad (4.6.80)$$

As before we assume a solution of the form

$$Y(t) = b_0 + b_1 t + b_2 t^2 + b_3 t^3 \text{ of (4.6.79)}$$

determining the coefficients b_0, b_1, b_2 and b_3 as in the case of (4.6.40) we get

$$B(t) = B(0) \exp\left[\frac{k}{2} + \left(\frac{\sigma^2 k}{2T_0}\right)\left(\frac{1}{T_0} - \frac{k}{2}\right)\right] t + \frac{k}{2T_0}\left(1 + \frac{\sigma^2 k}{4T_0}\right)^2 t^2 + \left(\frac{\sigma^2 k^2}{T_0}\right) t^3, \qquad (4.6.81)$$

is the second improved solution of Coale (1976).

Remarks 4.7: For the usual value of σ^2 and T_0 found in actual human populations the expression (4.6.81) becomes useful only for relatively rapid changes in fertility and large values of t which is contradictory; as it is not possible to maintain a high rate of increase or decrease in the level of fertility over a long period without crossing the biological or socially possible limits otherwise the limit approaching zero. Hence the approximate solution given in (4.6.81) is perfectly adequate.

4.6.5 Characteristic of the Population with Changing Fertility

The approximate solution of (4.6.33) see also (4.6.35) and (4.6.44)) viz.

$$B(t) = B(0) \exp\left(\frac{k}{2} t + \left(\frac{k}{2T_0}\right) t^2\right) \qquad (4.6.82)$$

$$\Rightarrow \quad \log B(t) = \log B(0) + \frac{k}{2}t + \left(\frac{k}{2T_0}\right) t^2$$

$$\Rightarrow \quad \frac{d \log B(t)}{dt} = \frac{k}{2} + \frac{k}{2T_0}(2t) = \frac{k}{2} + \frac{k}{T_0}t = \frac{k}{2} + r(t), \text{ see (4.6.49).} \qquad (4.6.83)$$

Thus we see that the relative rate of increase of births in a population with changing fertility is the sum of the intrinsic rate of increase plus half the annual rate of change in fertility, see (4.6.30).

The effect on population is determined from (see (4.4.15))

$$P(t) = \int_0^{\omega} B(t-x)\, p(x)\, dx, \qquad (4.6.84)$$

where ω is the upper bound of the surviving age and may be replaced by infinity.

Now, (4.6.82) $\Rightarrow$ $B(t-x) = B(0) \exp\left(\frac{k}{2}(t-x) + \frac{k}{2T_0}(t-x)^2\right)$

$$(4.6.84) \Rightarrow P(t) = B(0) \int_0^{\omega} \left\{ \exp\left(\frac{k}{2}(t-x) + \frac{k}{2T_0}(t-x)^2\right)\right\} p(x)\, dx$$

$$= B(0) \exp\left(\frac{k}{2}t + \frac{k}{2T_0}t^2\right) \int_0^{\omega} \exp\left(-\frac{k}{2}x + \frac{k}{2T_0}x^2 - \frac{ktx}{T_0}\right) p(x)\, dx \qquad (4.6.85)$$

$$\Rightarrow \frac{dP(t)}{dt} = B(0) \exp\left(\frac{k}{2}t + \frac{k}{2T_0}t^2\right) \int_0^{\omega} \exp\left(-\frac{k}{2}x + \frac{k}{2T_0}x^2 - \frac{ktx}{T_0}\right)\left(-\frac{kx}{T_0}\right) p(x)\, dx$$

$$+ B(0) \exp\left(\frac{k}{2}t + \frac{k}{2T_0}t^2\right)\left(\frac{k}{2} + \frac{k}{T_0}t\right) \int_0^{\omega} \exp\left(-\frac{k}{2}x + \frac{k}{2T_0}x^2 - \frac{tkx}{T_0}\right) p(x)\, dx$$

$$= P(t)\left(\frac{k}{2} + \frac{k}{T_0}t\right) - B(0) \exp\left(\frac{k}{2}t + \frac{k}{2T_0}t^2\right) \frac{k}{T_0}$$

$$\cdot \int_0^{\omega} x \exp\left(-\frac{k}{2}x + \frac{k}{2T_0}x^2 - \frac{2tx}{T_0}\right) p(x)\, dx$$

$$= P(t)\left(\frac{k}{2} + \frac{k}{T_0}t\right)$$

$$- \frac{k}{T_0}\left[\frac{B(0) \exp\left(\frac{k}{2}t + \frac{k}{2T_0}t^2\right) \int_0^{\omega} x \exp\left(-\frac{k}{2}x + \frac{k}{2T_0}x^2 - \frac{ktx}{T_0}\right) \cdot p(x)\, dx}{B(0) \exp\left(\frac{k}{2} + \frac{k}{2T_0}t^2\right) \int_0^{\omega}\left(-\frac{k}{2}x + \frac{k}{2T_0}x^2 - \frac{ktx}{T_0}\right) p(x) dx}\right] \cdot P(t)$$

$$\Rightarrow \frac{dP(t)}{dt} = P(t)\left(\frac{k}{2} + \frac{k}{T_0}t\right) - \frac{k}{T_0}P(t) \frac{\int_0^{\omega} x \exp\left(-\frac{k}{2}x + \frac{k}{2T_0}x^2 - \frac{ktx}{T_0}\right) p(x)\, dx}{\int_0^{\omega} \exp\left(-\frac{k}{2}x + \frac{k}{2T_0}x^2 - \frac{ktx}{T_0}\right) p(x)\, dx}$$

$$= P(t)\left(\frac{k}{2} + \frac{k}{T_0}t\right) - \frac{k}{T_0}P(t) \frac{\exp\left(-\frac{k}{2}\xi + \frac{k}{2T_0}\xi^2\right) \int_0^{\omega} x \exp\left(-\frac{ktx}{T_0}\right) p(x)\, dx}{\exp\left(-\frac{k}{2}\xi' + \frac{k}{2T_0}\xi'^2\right) \int_0^{\omega} \exp\left(-\frac{ktx}{T_0}\right) p(x)\, dx}$$

by the M.V. Theorem (A.1.13) of integral calculus, where $0 \le \xi \le \omega$, $0 \le \xi' \le \omega$.

Assuming $\xi \simeq \xi'$, for a fairly good approximation,

$$\frac{dP(t)}{dt} \cong P(t)\left(\frac{k}{2}+\frac{k}{T_0}t\right)-\frac{k}{T_0}P(t)\frac{\int_0^\omega x\exp\left(-\frac{ktx}{T_0}\right)p(x)\,dx}{\int_0^\omega \exp\left(-\frac{ktx}{T_0}\right)p(x)\,dx}$$

Putting $\frac{kt}{T_0}=r(t)$, see (4.6.49),

$$\Rightarrow \frac{dP(t)}{dt}=P(t)\left(\frac{k}{2}+r(t)\right)-\frac{k}{T_0}P(t)\frac{\int_0^\omega x\exp(-r(t)x)\,p(x)\,dx}{\int_0^\omega \exp(-r(t)x)\,p(x)\,dx}$$

$$\frac{dP(t)}{dt}=P(t)\left(\frac{k}{2}+r(t)\right)-\frac{k}{T_0}P(t)\overline{X}(t),$$

$$\text{where,}\quad \overline{X}(t)=\frac{\int_0^\omega x\exp(-r(t)x)\,p(x)\,dx}{\int_0^\omega \exp(-r(t)x)\,p(x)\,dx}$$

$$\Rightarrow \frac{1}{P(t)}\frac{dP(t)}{dt}=\frac{k}{2}+r(t)-\frac{k}{T_0}\overline{X}(t)$$

$$\frac{d\log P(t)}{dt}=\frac{k}{2.}+r(t)-\frac{k}{T_0}\overline{X}(t)=\frac{k}{2}+r(t)\left(1-\frac{\overline{X}(t)}{t}\right). \tag{4.6.86}$$

Thus it follows that the relative rate of increase of the Population at any time is intrinsic rate of increase at time t less than $\frac{k}{T_0}=\frac{r(t)}{t}$ times the average age of the population ($\overline{X}(t)$) plus half the annual rate of change of its fertility.

If fertility is continuously decreasing the rate of increasing the birth rate is zero, then

$$(4.6.83) \Rightarrow r(t)=-\frac{k}{2} \Rightarrow \frac{kt}{T_0}=-\frac{k}{2}$$

$$\Rightarrow \frac{t}{T_0}=-\frac{1}{2} \Rightarrow t=-\frac{T_0}{2},$$

time is equal to half a generation before the population is replaced.

As shown by Coale (1976), it is possible to derive a different relationship between the rate of increase of the population with declining fertility and stable Population in the form

$$r_k=r_0+\left(\frac{1}{2}-\frac{\overline{X}k}{T_0}\right)k,$$

where r_k = growth rate with fertility changing by k, r_0 = intrinsic growth rate, k = annual rate change of fertility and $\overline{X}k$ is the mean age of the population with declining fertility.

Now, if $k<0$ and $\frac{\overline{X}}{T_0}>\frac{1}{2}$ then $\left(\frac{1}{2}-\frac{\overline{X}}{T_0}\right)k>0$. Thus the rate of growth of a population with a history of changing fertility is greater than intrinsic growth rate.

4.6.6 Comparison of the Age Distributions of a Stable Population with that of a Quasi Stable Population with Changing Fertility

In order to compare the relation between a stable age distribution and the age distribution of a Quasi Stable Population with changing Fertility Pattern, let us consider two Populations—one Stable and the other Quasi Stable characterized by changing Fertility Pattern, that have the same number of births at time t. The number of persons at age x in a stable population,

$$\begin{aligned} N_S(x,t) &= B(t-x)\ p(x), \text{ see (4.6.45)}, N_S(x;t) \equiv B(x,t) \\ &= B(t)\exp(-r(t)x)\ p(x), \text{ (from (4.4.26))}. \end{aligned} \tag{4.6.87}$$

Whereas in the corresponding Quasi Stable Population with Fertility changing at the rate k, the numbers of persons is given by,

$$N_k(x;t) = N_S(x,t)\exp\left(-\frac{k}{2}x + \frac{k}{2T_0}x^2\right), \quad \text{(see (4.6.44))} \tag{4.6.88}$$

A comparison of (4.6.87) and (4.6.88) reveals,

(i) The number of persons at age zero and at age T_0, the mean length of generation are same for both the populations. However, between 0 and T_0 population with declining fertility has more number of persons.

(ii) $\dfrac{N_k}{N_s}$ reaches a maximum at age equal to half the mean length of the generation.

(iii) The effect of a slight increase in number N_k from age zero to T_0 and a rapid diminution above age T_0 on the proportionate age distribution depends on the age distribution of the stable population itself.

(iv) If the total number of persons in each population were the same, the ratio of $\left(\frac{N_k}{N_s}\right)$ would be exactly $= \exp\left(-\left(\frac{k}{2}\right)x + \left(\frac{k}{2T_0}\right)x^2\right)\ \forall\, x$. However, this would be possible only if the small increase for ages below T offsets the larger decrease for ages above T. This could occur in a stable population that declines very steeply with age.

4.6.7 Expression for $\frac{b_k}{b_s}$ in Terms of the Moments of the Stable Age Distribution

We have, b_k = birth rate of a population with history of declining fertility at an annual rate k for t year and b_s = stable population birth rate.

We have from (4.6.50),

$$1 = \int_0^\omega C(x;t)dx = \int_0^\omega \underbrace{b(t)}_{=b_k} \exp\left(-\frac{k}{2}x + \frac{k}{2T_0}x^2\right)\exp(-r(t)x)\ p(x)\ dx$$

$$\Rightarrow \quad b(t) = b_k = \frac{1}{\displaystyle\int_0^\omega \exp\left(-\frac{k}{2}x + \frac{k}{2T_0}x^2\right)\exp(-r(t)x)\ p(x)\ dx} \tag{4.6.89}$$

Expanding b_k in Taylor's series about $k = 0$ (see (A.1.8)) and taking the first two terms as an approximation, we have

$$b_k = (b_k)_{k=0} + k\left(\frac{db_k}{dk}\right)_{k=0} \quad (4.6.90)$$

Now, $$\left(\frac{db_k}{dk}\right)_{k=0} = -\frac{1}{\left[\int_0^\omega \exp\left(-\frac{k}{2}x + \frac{k}{2T_0}x^2\right) e^{-r(t)x} p(x)dx\right]^2_{k=0}}$$

$$\times \int_0^\omega \left(-\frac{x}{2} + \frac{x^2}{2T_0}\right) \exp\left(-\frac{k}{2}x + \frac{k}{2T_0}x^2\right) e^{-r(t)x} p(x)dx\bigg|_{k=0}. \quad (4.6.91)$$

We have $b_k\big|_{k=0} = b_s$ (from assumptions). (4.6.92)

Let $\mu'_j = \int_0^\omega x^j e^{-r(t)x} p(x)\, dx \bigg/ \int_0^\omega e^{-r(t)x} P(x)\, dx$

$$\Rightarrow \int_0^\omega x^j e^{-r(t)x} p(x)\, dx = \frac{\mu'_j}{b_s}, \left[\because\ b_s = \frac{1}{\int_0^\omega e^{-y(t)x} p(x)\, dx} \text{ from (4.6.89) putting } k = 0.\right] \quad (4.6.93)$$

From (4.6.91) and (4.6.93) we have,

$$\left(\frac{db_k}{dk}\right)_{k=0} = -(b_k)^2_{k=0}\left[-\frac{\mu'_1}{2} + \frac{\mu'_2}{2T_0}\right]\frac{1}{b_s} = -b_s^2\left(-\frac{\mu'_1}{2} + \frac{\mu'_2}{2T_0}\right)\frac{1}{b_s}$$

$$= \frac{b_s}{2}\left(\mu'_1 - \frac{\mu'_2}{2T_0}\right). \quad (4.6.94)$$

Hence, substituting from (4.6.92) and (4.6.94) in (4.6.90) we get

$$b_k = b_s + k\frac{b_s}{2}\left(\mu'_1 - \frac{\mu'_2}{2T_0}\right), \text{ (form 4.6.94)}$$

$$= b_s\left[1 + \frac{k}{2}\left(\mu'_1 - \frac{\mu'_2}{2T_0}\right)\right] \quad (4.6.95)$$

Remarks 4.8:
$$\frac{b_k}{b_s} = \left(1 + \frac{k}{2}\left(\mu'_1 - \frac{\mu'_2}{2T_0}\right)\right)$$

is an exact relation which does not directly enable us to realise the properties of the stable age distribution affecting the ratio of $\frac{b_k}{b_s}$.

Empirically (see Coale (1976)) the relation given below provides a family of good estimates viz.

$$\left\{\frac{b_k}{b_s}\right\} = 1.0 + \left[3.807 - 1.039\frac{\overline{a_s}}{T} + 9.389\left(\frac{\overline{a_s}}{T}\right)^2\right] k + \left(66.0 - 149.0\frac{\overline{a_s}}{T}\right) k^2,$$

where $\frac{\overline{a_s}}{T}$ is the mean age of the stable population.

4.7 Population Growth Models Using Stochastic Processes

Population is a time dependent r.v., hence it can be better studied if it is taken as a stochastic process. In this section, we shall study some population growth models when population is a stochastic process.

4.7.1 Generalised Birth and Death Process Models

A stochastic version of the population growth model, based on time dependent birth and death parameters evolved by D.G. Kendall (1948) is presented in this section. This is a more generalised stochastic model, the particular case or the deterministic version of which may lead to Stable or Quasi stable Population models described in the earlier sections. The formulation of the general time dependent birth and death process is as follows:

Let an integer valued time dependent random variable n_t represent the size of the population measured at any time t and suppose in an infinitesimal element of time δt, the only possible transitions are

$$\begin{aligned} n_{t+\delta t} &= n_t + 1 \text{ with probability } n_t\lambda(t)\delta t + 0(\delta t) \\ &= n_t - 1 \text{ with probability } n_t\mu(t)\delta t + 0(\delta t) \\ &= n_t \text{ with probability } 1 - \{\lambda(t)\delta t + \mu(t)\delta t\}n_t + 0(\delta t) \end{aligned} \tag{4.7.1}$$

with the initial conditions

$$P_1(0) = 1,\ P_0(0) = 0,\ P_n(0) = 0 \text{ for } n \neq 1, \tag{4.7.2}$$

where $P_n(t)$ represents the probability that the population size is $n, n = 0, 1, 2, \ldots$ at any time t.

Denoting $\lambda(t) = \lambda$ and $\mu(t) = \mu$ we have the Kolmogorov equations as

$$\begin{aligned} P_n(t+\delta t) = P_n(t)[1 - n(\lambda\delta t + \mu\delta t) + 0(\delta t)] + P_{n-1}(t)[(n-1)\lambda\delta t + 0(\delta t)] \\ + P_{n+1}(t)[(n+1)\mu\delta t + 0(\delta t)], \text{ for } n > 0 \end{aligned} \tag{4.7.3}$$

$$\begin{aligned} \Rightarrow\ P_n(t+\delta t) - P_n(t) = -nP_n(t)(\lambda+\mu)\delta t + (n-1)P_{n-1}(t)\lambda\delta t \\ + (n+1)P_{n+1}(t)\mu\delta t + 0(\delta t) \end{aligned}$$

$$\Rightarrow\ \lim_{\delta t\to 0} \frac{P_n(t+\delta t) - P_n(t)}{\delta t} = -n(\lambda+\mu)P_n(t) + (n-1)\lambda P_{n-1}(t) + (n+1)\mu P_{n+1}(t)$$

$$\Rightarrow\ \frac{\partial P_n(t)}{\partial t} = -n(\lambda+\mu)P_n(t) + \lambda(n-1)P_{n-1}(t) + \mu(n+1)P_{n+1}(t),\ n > 0$$

For, $n = 0,\ P_0(t+\delta t) = P_1(t)[\mu\delta t + 0(\delta t)] + P_0(t)[1 - (\lambda+\mu)\delta t \cdot 0 + 0(\delta t)]$ (4.7.4)

$$\Rightarrow\ \frac{\partial P_0(t)}{\partial t} = \mu P_1(t)$$

We employ the method of probability generating function for the solution of the above differential equation. The entire analysis is given by Kendall (1948).

Let $\phi(t) = \sum_{n=0}^{\infty} P_n(t)z^n;\ |z| < 1$

$$\frac{\partial\phi(t)}{\partial t} = \sum_{n=0}^{\infty} \frac{\partial P_n(t)}{\partial t} z^n \tag{4.7.5}$$

$$\frac{\partial \phi(t)}{\partial z} = \sum_{n=0}^{\infty} n P_n(t) z^{n-1} \tag{4.7.6}$$

From (4.7.4) and (4.7.5), we have

$$\frac{\partial \phi(t)}{\partial t} = \sum_{n=0}^{\infty} [-n(\lambda+\mu)P_n(t) + \lambda(n-1)P_{n-1}(t) + \mu(n+1)P_{n+1}(t)] z^n$$

$$= -(\lambda+\mu)\sum_{n=0}^{\infty} nP_n(t)z^n + \lambda z \sum_{n=1}^{\infty}(n-1)P_{n-1}(t)z^{n-1} + \mu \sum_{n=0}^{\infty}(n+1)P_{n+1}(t)z^n$$

$$\frac{\partial \phi}{\partial t} = \mu \sum_{n=1}^{\infty} nP_n(t)z^{n-1} + \lambda z \sum_{n=0}^{\infty} nP_n(t)z^n - (\lambda+\mu)z\sum_{n=1}^{\infty} nP_n(t)z^{n-1}$$

$$= \mu \sum_{n=1}^{\infty} nP_n(t)z^{n-1} + \lambda z^2 \sum_{n=1}^{\infty} nP_n(t)z^{n-1} - (\lambda+\mu)z\sum_{n=1}^{\infty} nP_N(t)z^{n-1}$$

or $$\frac{\partial \phi}{\partial t} = \mu\frac{\partial \phi}{\partial z} + \lambda z^2 \frac{\partial \phi}{\partial z} - (\lambda+\mu)z\frac{\partial \phi}{\partial z}$$

or $$\frac{\partial \phi}{\partial t} = \{\mu - \lambda z^2 - (\lambda+\mu)z\}\frac{\partial \phi}{\partial z}$$

or $$\frac{\partial \phi}{\partial t} = \{\lambda z(z-1) - \mu(z-1)\}\frac{\partial \phi}{\partial z}$$

or $$\frac{\partial \phi}{\partial t} = (\lambda z - \mu)(z-1)\frac{\partial \phi}{\partial z} \tag{4.7.7}$$

This is a Lagrangian type of differential equation

Noting that $\dfrac{\partial \phi}{\partial t} = \dfrac{\partial \phi}{\partial z} \cdot \dfrac{\partial z}{\partial t}$

The above is reduced to an ordinary differential equation

$$\frac{dz}{dt} = (\lambda z - \mu)(z-1) \tag{4.7.8}$$

The solution is a homographic (Watson-Bessel function) (1962) of the form

$$z = \frac{f_1 + Cf_2}{f_3 + Cf_4} \tag{4.7.9}$$

where f_1, f_2, f_3 and f_4 are functions of t.

$\Rightarrow \quad zf_3 + zCf_4 = f_1 + Cf_2$

$\Rightarrow \quad C(zf_4 - f_2) = f_1 - zf_3$

$$\Rightarrow \quad C = \frac{zf_3 - f_1}{f_2 - zf_4} \tag{4.7.10}$$

Consider $$\phi(z,t) = \Phi\left(\frac{f_1 + Cf_2}{f_3 + Cf_4}, t\right) = \Phi\left(\frac{zf_3 - f_1}{f_2 - zf_4}\right) \tag{4.7.11}$$

and the solution or structure of $\phi(z,t)$ is to be obtained. Further, if we assume the initial conditions (4.7.2) then,

$$\phi(z,t)\Big|_{t=0} = \sum_{n=0}^{\infty} P_n(t)z^n\Big|_{t=0} = z \quad \Rightarrow \quad \phi(z,0) = \Phi\left(\frac{z[f_3]_0 - [f_1]_0}{[f_2]_0 - z[f_4]_0}\right) = z$$

where $[f_i]_0$ indicates the value of f_i at $t = 0,\ i = 1,2,3,4.$

$$\text{Let} \quad Z = \frac{z[f_3]_0 - [f_1]_0}{[f_2]_0 - z[f_4]_0} \qquad \therefore \quad \Phi(Z) = z$$

$$\text{Now} \quad Z = \frac{z[f_3]_0 - [f_1]_0}{[f_2]_0 - z[f_4]_0}$$

$$\Rightarrow \quad Z[f_2]_0 - Z\{z[f_4]_0\} = z[f_3]_0 - [f_1]_0 \quad \text{or} \quad z\{[f_3]_0 + Z[f_4]_0\} = [f_1]_0 + Z[f_2]_0$$

$$\Rightarrow \quad z = \frac{[f_1]_0 + Z[f_2]_0}{[f_3]_0 + Z[f_4]_0}$$

$$\Phi(Z) = z = \frac{Z[f_2]_0 + [f_1]_0}{Z[f_4]_0 + [f_3]_0} \tag{4.7.12}$$

To obtain the structure of $\Phi(z)$:

Also, $\phi(z,t) = \Phi\left(\dfrac{zf_3 - f_1}{f_2 - zf_4}\right)$, form (4.7.11)

$$= \frac{\left(\dfrac{zf_3 - f_1}{f_2 - zf_4}\right)[f_2]_0 + [f_1]_0}{\left(\dfrac{zf_3 - f_1}{f_2 - zf_4}\right)[f_4]_0 + [f_3]_0}, \text{ (as in (4.7.12))}$$

$$= \frac{(zf_3 - f_1)[f_2]_0 + (f_2 - zf_4)[f_1]_0}{(zf_3 - f_1)[f_4]_0 + (f_2 - zf_4)[f_3]_0}$$

$$= \frac{f_2[f_1]_0 - f_1[f_2]_0 + z\{f_3[f_2]_0 - f_4[f_1]_0\}}{f_2[f_3]_0 - f_1[f_4]_0 + z\{f_3[f_4]_0 - f_4[f_3]_0\}} = \frac{g_1(t) + zg_2(t)}{g_3(t) + zg_4(t)}, \tag{4.7.13}$$

where $g_1(t) = f_2[f_1]_0 - f_1[f_2]_0,\ g_2(t) = f_3[f_2]_0 - f_4[f_1]_0$
$g_3(t) = f_2[f_3]_0 - f_1[f_4]_0,\ g_4(t) = f_3[f_4]_0 - f_4[f_3]_0$

$$\therefore \quad \phi(z,t) = [g_1(t) + zg_2(t)]\left[g_3(t)\left(1 + z\frac{g_4(t)}{g_3(t)}\right)\right]^{-1}$$

$$= \left[\frac{g_1(t) + zg_2(t)}{g_3(t)}\right]\left[1 + z\frac{g_4(t)}{g_3(t)}\right]^{-1}$$

$$= \left[\frac{g_1(t) + zg_2(t)}{g_3(t)}\right]\left[1 - z\frac{g_4(t)}{g_3(t)} + z^2\left(\frac{g_4(t)}{g_3(t)}\right)^2 + \cdots\right.$$

$$\left. + z^{n-1}\left(\frac{g_4(t)}{g_3(t)}\right)^{n-1} - z^n\left(\frac{g_4(t)}{g_3(t)}\right)^n + \cdots\right], \tag{4.7.14}$$

where n is odd, provided $\left|z\dfrac{g_4(t)}{g_2(t)}\right| < 1.$

Again $\phi(z,t) = \sum_{n=0}^{\infty} P_n(t) z^n$ (4.7.15)

$$\Rightarrow \quad \phi(1,0) \left.\begin{array}{l} = 0 \text{ for } n \neq 1 \\ = 1 \text{ for } n = 1 \end{array}\right\}, \ \because \text{ from initial conditions (4.7.2)}$$

$P_0(t) =$ term without involving z and $P_n(t) =$ coefficient of z^n in (4.7.15). Therefore, from (4.7.14) on comparing coefficients of z^0 and z^n, we get

$$P_0(t) = \frac{g_1(t)}{g_3(t)} = \xi(t) = \xi_t = \xi \text{ (say)} \tag{4.7.16}$$

$$P_n(t) = \left\{ \frac{g_2(t)}{g,(t)} \left(\frac{g_4(t)}{g_3(t)}\right)^{n-1} - \frac{g_1(t)}{g_3(t)} \left(\frac{g_4(t)}{g_3(t)}\right)^{n} \right\}$$

$$= \left(\frac{g_4(t)}{g_2(t)}\right)^{n-1} \left\{ \frac{g_2(t)}{g_3(t)} - \frac{g_4(t)}{g_3(t)}\xi(t) \right\} \tag{4.7.17}$$

Now $\phi(1,t) = \dfrac{g_1(t) + g_2(t)}{g_3(t) + g_4(t)} = 1$, from (4.7.13)

$$\Rightarrow \quad \frac{\dfrac{g_1(t)}{g_3(t)} + \dfrac{g_2(t)}{g_3(t)}}{1 + \dfrac{g_4(t)}{g_3(t)}} = 1$$

Put $\dfrac{g_4(t)}{g_3(t)} = -\eta_t = \eta$ (say) (4.7.18)

$$\therefore \quad \frac{\xi + \dfrac{g_2(t)}{g_3(t)}}{1 - \eta} = 1 \text{ (from (4.7.16))} \Rightarrow \xi + \frac{g_2(t)}{g_3(t)} = 1 - \eta$$

$$\Rightarrow \quad \frac{g_2(t)}{g_3(t)} = 1 - \eta - \xi \tag{4.7.19}$$

Putting (4.7.18) and (4.7.19) in (4.7.17)

$$\Rightarrow \quad P_n(t) = (-\eta)^{n-1}(1 - \eta - \xi + \eta\xi) \text{ when } n \text{ is odd}$$

$$= (-\eta)^{n-1}(1-\eta)(1-\xi) \quad \because n \text{ is odd, so (n-1) is even}$$

$$= (-\eta_t^{n-1}(1 - P_0(t))(1 - \eta_t), \text{ from (4.7.16)}, \tag{4.7.20}$$

which provides the probability distribution of the r.v. n (the size of the population) at time t, when n is odd.

Similarly one can show that

$$P_n(t) = (\eta)^{n-1}(1 - \eta - \xi + \eta\xi), \text{ when } n \text{ is even}$$

$$= (\eta)^{n-1}(1-\eta)(1-\xi)$$

$$= (\eta_t)^{n-1}(1 - P_0(t)(1 - \eta_t), \text{ from (4.7.16)} \tag{4.7.21}$$

and $P_0(t) = \xi_t$, from (4.7.16) (4.7.22)

Combining (4.7.20) and (4.7.21) we have

$$P_n(t) = (\eta_t)^{n-1}(1-\eta_t)(1-\xi) \text{ and } P_0(t) = \xi_t \tag{4.7.23}$$

Now
$$\begin{aligned}\phi(z,t) &= P_0(t) + zP_1(t) + z^2P_2(t) + \cdots + z^{n-1}P_{n-1}(t) + \cdots \\ &= \xi + z(1-\eta-\xi+\eta\xi) + z^2\eta(1-\eta-\xi+\eta\xi) \\ &\quad + z^3\eta^2(1-\eta-\xi+\eta\xi) + \cdots + (-\eta)^{n-1}z^n(1-\eta-\xi+\eta\xi) + \cdots \\ &= \xi + z(1-\eta-\xi+\eta\xi)\{1+\eta z+\eta^2z^2+\cdots+\eta^nz^n+\cdots\} \\ &= \xi + z(1-\eta-\xi+\eta\xi)\frac{1}{1-\eta z}\cdot \text{provided } |\eta z| < 1 \\ &= \frac{\xi(1-\eta z)+z(1-\eta-\xi+\eta\xi)}{1-\eta z} = \frac{\xi-\xi\eta z+z-\eta z-\xi z+\eta\xi z}{1-\eta z}\end{aligned}$$

$$\therefore \quad \phi(z,t) = \frac{\xi-\eta z-\xi z+z}{1-\eta z} \tag{4.7.24}$$

Now we require to estimate the parameters $\xi = \xi(t)$ *and* $\eta = \eta(t)$

From (4.7.24)

$$\log\phi(z,t) = \log(\xi-\eta z-\xi z+z) - \log(1-\eta z) \tag{4.7.25}$$

$$\Rightarrow \quad \frac{1}{\phi}\frac{\partial\phi}{\partial z} = \frac{1-\eta-\xi}{(\xi-\eta z-\xi z+z)} - \frac{(-\eta)}{1-\eta z}$$

$$\Rightarrow \quad \frac{1}{\phi}\frac{\partial\phi}{\partial z} = \frac{(1-\eta-\xi+\eta\xi)}{(1-\eta z)(\xi-\eta z-\xi z+z)}$$

or
$$\begin{aligned}\frac{\partial\phi}{\partial z} &= \phi\cdot\frac{(1-\eta-\xi+\eta\xi)}{(1-\eta z)(\xi-\eta z-\xi z+z)} \\ &= \frac{(\xi-\eta z-\xi z+z)}{(1-\eta z)}\cdot\frac{(1-\eta-\xi+\eta\xi)}{(1-\eta z)(\xi-\eta z-\xi z-z)}\end{aligned}$$

$$\therefore \quad \frac{\partial\phi}{\partial z} = \frac{(1-\eta)(1-\xi)}{(1-\eta z)^2} \tag{4.7.26}$$

Now
$$\begin{aligned}\frac{\partial\phi}{\partial t} &= \frac{\partial\phi}{\partial z}\cdot\frac{\partial z}{\partial t} \\ &= \frac{(1-\eta)(1-\xi)}{(1-\eta z)^2}\cdot(\lambda z-\mu)(z-1) \text{ (from (4.7.7) and (4.7.26))} \\ &= \frac{(1-\xi)(1-\eta)}{(1-\eta z)^2}(\lambda z^2-(\lambda+\mu)z+\mu)\end{aligned} \tag{4.7.27}$$

Again, from (4.7.25)

$$\frac{1}{\phi}\cdot\frac{\partial\phi}{\partial t} = \frac{\xi'-\eta'z-\xi'z}{\xi-\eta z-\xi z+z} + \frac{\eta'z}{1-\eta z}$$

where $\xi' = \dfrac{\partial\xi}{\partial t}$, $\eta' = \dfrac{\partial\eta}{\partial t}$

$$\Rightarrow \quad \frac{\partial\phi}{\partial t} = \frac{\xi'-\eta'z-\xi'z}{1-\eta z} + \frac{\eta'z(\xi-\eta z-\xi z+z)}{(1-\eta z)^2}$$

$$= \frac{\xi' + (\eta'\xi - \xi' - \eta' - \eta\xi')z + (\eta' + \eta\xi' - \eta'\xi)z^2}{(1-\eta z)^2}, \text{ on simplification.} \tag{4.7.28}$$

Equating terms independent of z and coefficient of z, z^2 from (4.7.28) and (4.7.27) we get

$$\left.\begin{aligned} &\xi = \mu(1-\xi)(1-\eta) && \text{(i)}\\ &\eta'\xi - \xi'\eta - \xi' - \eta' = -(\lambda+\mu)(1-\xi)(1-\eta) && \text{(ii)}\\ &\eta\xi' - \eta'\xi + \eta' = \lambda(1-\xi)(1-\eta) && \text{(iii)} \end{aligned}\right\} \tag{4.7.29}$$

In equation (4.7.29), (i) and (iii) are the estimating equations for the parameters $\xi = \xi(t)$ and $\eta = \eta(t)$.

$$\left.\begin{aligned} &\text{Now put} && U = 1-\xi,\ V = 1-\eta\\ &\Rightarrow && 1-U = \xi \text{ and } 1-V = \eta\\ &\therefore && U' = \frac{dU}{dt} = -\xi', \text{ and } V' = \frac{dV}{dt} = -\eta' \end{aligned}\right\} \tag{4.7.30}$$

Now, (i)−(iii) $\Rightarrow \xi' - \eta\xi' + \eta'\xi - \eta' = (\mu-\lambda)(1-\xi)(1-\eta)$

$\Rightarrow \quad -U' - (1-V)(-U') + (-V')(1-U) - (-V') = (\mu-\lambda)UV$

or $\quad -U' + (1-V)U' - V'(1-U) + V' = (\mu-\lambda)UV$

Dividing by U both sides, we get

$$-\frac{U'}{U} + (1-V)\frac{U'}{U} - V'\frac{(1-U)}{U} + \frac{V'}{U} = (\mu-\lambda)V$$

or
$$-\frac{U'}{U} + \frac{U'}{U} - \frac{U'V}{U} - \frac{V'}{U} + \frac{UV'}{U} + \frac{V'}{U} = (\mu-\lambda)V$$

or
$$\left[V' - \frac{U'}{U}V\right] = (\mu-\lambda)V \quad \text{or} \quad V' + \mu V^2 = (\mu-\lambda)V,$$

since $\xi' = \mu(1-\xi)(1-\eta) \;\Rightarrow\; -U' = \mu UV \;\Rightarrow\; -\frac{U'}{U} = \mu V$ (4.7.31)

or
$$\frac{dV}{dt} - (\mu-\lambda)V = -\mu V^2 \tag{4.7.32}$$

This is a linear differential equation of the Bernoulli's form $\frac{dy}{dx} + Py = Qy^m$ with Integrating factor (I.F.) $= \exp\left(\int pdx\right)$.

Dividing both sides of (4.7.32) by $-V^2$, we get

$$-\frac{1}{V^2}\frac{dV}{dt} + (\mu-\lambda)\frac{1}{V} = \mu$$

Let
$$\frac{1}{V} = W \;\Rightarrow\; -\frac{1}{V^2}\frac{dV}{dt} = \frac{dW}{dt} \tag{4.7.33}$$

$$\frac{dW}{dt} + (\mu-\lambda)W = \mu \tag{4.7.34}$$

Integrating Factor (I.F.) $= \exp\left(\int(\mu(t)-\lambda(t))dt\right) = e^{\rho(t)}$,

where $\rho(t) = \int(\mu(t)-\lambda(t))dt$ (4.7.35)

$$\therefore \quad e^{\rho(t)}\frac{dW(t)}{dt} + e^{\rho(t)}(\mu - \lambda)W = \mu e^{\rho(t)} \quad \text{or} \quad \frac{d(We^{\rho(t)})}{dt} = \mu e^{\rho(t)}$$

Integrating from 0 to t, we get

$$We^{\rho(t)} = \int_0^t \mu e^{\rho(\tau)} d\tau + C \tag{4.7.36}$$

Initially, when $t = 0$, from (4.7.23)

$$\xi_t = P_0(t) \Rightarrow \xi_0 = P_0(0) = 0 \text{ (by assumption (4.7.2))}, \tag{4.7.37}$$

$$P_n(t) = (1 - \xi_t)(1 - \eta_t)(\eta_t)^{n-1} \text{ and, from (4.7.23)} \tag{4.7.38}$$

$$\Rightarrow \quad P_n(0) = (1 - \xi_0)(1 - \eta_0)(\eta_0)^{n-1}$$

$$\Rightarrow \quad P_n(0) = (1 - \eta_0)(\eta_0)^{n-1} \quad (\because \xi_0 = 0), \text{ from (4.7.37)} \tag{4.7.39}$$

For $n = 1$, from (4.7.39)

$$P_1(0) = (1 - \eta_0)$$

$$\Rightarrow \quad 1 = 1 - \eta_0, \text{ (from (4.7.2))} \Rightarrow \eta_0 = 0$$

$$\therefore \quad \xi_0 = 0, \ \eta_0 = 0 \tag{4.7.40}$$

$$\Rightarrow \quad \left.\begin{aligned} U_t|_{t=0} &= (1 - \xi_t)|_{t=0} = 1 - \xi_0 = 1 \\ V_t|_{t=0} &= (1 - \eta_t)|_{t=0} = 1 - \eta_0 = 1 \end{aligned}\right\}, \text{ from (4.7.30) and (4.7.39)}$$

$$\therefore \quad U = 1, \ V = 1 \text{ when } t = 0 \tag{4.7.41}$$

From (4.7.36) we have

$$W = Ce^{-\rho(t)} + e^{-\rho(t)} \int_0^t \mu(\tau) e^{\rho(\tau)} d\tau \tag{4.7.42}$$

Now consider

$$W_t|_{t=0} = \left.\frac{1}{V_t}\right|_{t=0} = \left.\frac{1}{1 - \eta_t}\right|_{t=0} = 1, \text{ form (4.7.33) and (4.7.39)}$$

$$\therefore \quad W(0) = 1$$

$$\rho(t) = \int_0^t (\mu(\tau) - \lambda(\tau))d\tau, \text{ (from (4.7.35))} \Rightarrow \rho(0) = 0 \tag{4.7.43}$$

From (4.7.42) we have

$$W(0) = Ce^{-\rho(0)} + \int_0^0 e^{\rho(\tau)} \mu(\tau) d\tau$$

$$\Rightarrow \quad C = W(0) = 1 \tag{4.7.44}$$

Hence, (4.7.42) becomes

$$W = e^{-\rho(t)} \left[1 - \int_0^t \mu(\tau) e^{\rho(\tau)} d\tau \right] \tag{4.7.45}$$

Next, we prove that

(A) $W(t) = 1 + e^{-\rho(t)} \int_0^t e^{\rho(\tau)} \lambda(\tau) d\tau$

$$\text{(B) } W(t) = \frac{1}{2}(1 + e^{-\rho(t)}) + \frac{1}{2}e^{-P(t)} \int_0^t e^{\rho(\tau)}\{\lambda(\tau) + \mu(\tau)\}d\tau$$

To prove (A), let us consider the integral on the R.H.S. of (4.7.45) viz.

$$\int_0^t \mu(\tau)e^{\rho(\tau)}d\tau$$

Integrating by parts

$$\Rightarrow \quad e^{\rho(\tau)}\left[\int_0^t \mu(\tau)d\tau\right]\Bigg|_0^t - \int_0^t \left\{\frac{d}{d\tau}[e^{\rho(\tau)}]\int \mu(\tau)d\tau\right\}d\tau,$$

$$[\because \rho(\tau) = \int \{\mu(\tau) - \lambda(\tau)\}d\tau \Rightarrow \int_0^t \{\mu(\tau) - \lambda(\tau)\}d\tau = \rho(\tau)\Big]_0^t = \rho(t) - \rho(0)]$$

$$\Rightarrow \quad \int_0^t \mu(\tau)d\tau = \rho(t) + \int_0^t \lambda(\tau)d\tau \quad (\because \rho(0) = 0), \text{ from (4.7.43)} \qquad (4.7.46)$$

$$\therefore \quad \int_0^t \mu(\tau)e^{\rho(\tau)}d\tau = e^{\rho(t)}\int_0^t \mu(\tau)d\tau - \int_0^t \left\{\frac{d}{d\tau}(e^{\rho(\tau)})\int \mu(\tau)d\tau\right\}d\tau$$

$$= e^{\rho(t)}\left[\rho(t) + \int_0^t \lambda(\tau)d\tau\right]$$

$$- \int_0^t \frac{d(e^{\rho(\tau)})}{d\tau}\left[\rho(t) + \int_0^t \lambda(\tau)d\tau\right]d\tau, \text{ using (4.7.46)} \qquad (4.7.47)$$

$$= e^{\rho(t)}\rho(t) + e^{\rho(t)}\int_0^t \lambda(\tau)d\tau - \int_0^t \left\{\frac{d}{d\tau}e^{\rho(\tau)}\right\}\rho(\tau)d\tau$$

$$- \int_0^t \left\{\frac{d}{d\tau}(e^{\rho(\tau)})\int_0^t \lambda(\tau)d\tau\right\}d\tau \qquad (4.7.48)$$

$$\text{Since} \quad \int_0^t e^{\rho(\tau)}\lambda(\tau)d\tau = e^{\rho(t)}\int_0^t \lambda(\tau)d\tau - \int_0^t \frac{d}{d\tau}(e^{\rho(\tau)})\left(\int_0^t \lambda(\tau)d\tau\right),$$

integrating by parts. (4.7.49)

$$\therefore \quad \int_0^t \mu(\tau)e^{\rho(\tau)}d\tau = e^{\rho(t)}\rho(t) + \int_0^t e^{\rho(\tau)}\lambda(\tau)d\tau - \int_0^t \left\{\frac{d}{d\tau}e^{\rho(\tau)}\right\}\rho(\tau)d\tau,$$

using (4.7.49) in (4.7.48) (4.7.50)

Now, let us consider $\int_0^t \left[\frac{d}{d\tau}e^{\rho(\tau)}\right]\rho(\tau)d\tau$

Integrating above by parts, we get

$$= \left|\rho(\tau)\int \frac{d}{d\tau}e^{\rho(\tau)}d\tau\right|_0^t - \int_0^t \left\{\frac{d\rho(\tau)}{d\tau}\int \frac{d}{d\tau}[e^{\rho(\tau)}]d\tau\right\}d\tau$$

$$= \rho(\tau)e^{\rho(\tau)}\Big]_0^t - \int_0^t \frac{d\rho(\tau)}{d\tau}e^{\rho(\tau)}d\tau$$

$$= \rho(t)e^{\rho(t)} - \int_0^t \frac{d}{d\tau}(e^{\rho(\tau)}d\tau), \quad (\because \rho(0) = 0, \text{ from (4.7.43)})$$

$$= \rho(t)e^{\rho(t)} - e^{\rho(\tau)}\Big|_0^t = \rho(t)e^{\rho(t)} - [e^{\rho(t)} - 1]$$

$$= 1 + \rho(t)e^{\rho(t)} - e^{\rho(t)} \tag{4.7.51}$$

Substituting from (4.7.51) in (4.7.50) we get

$$\int_0^t \mu(\tau)e^{\rho(\tau)}d\tau = \int_0^t e^{\rho(\tau)}\lambda(\tau)d\tau + e^{\rho(t)} - 1 \tag{4.7.52}$$

Now putting (4.7.52) in (4.7.45) we get

$$W = e^{-\rho(t)}\left[\int_0^t e^{\rho(\tau)}\lambda(\tau)d\tau + e^{\rho(t)}\right]$$

or $$W = e^{-\rho(t)}\left[\int_0^t e^{\rho(\tau)}\lambda(\tau)d\tau\right] + 1, \tag{4.7.53}$$

which establishes the result (A).

Similarly, from (4.7.45), we have

$$W(t) = e^{-\rho(t)}\left[1 + \int_0^t e^{\rho(\tau)}\mu(\tau)d\tau\right] \tag{4.7.54}$$

From (4.7.53) we have

$$W(t) = 1 + e^{-\rho(t)}\int_0^t e^{\rho(\tau)}\lambda(\tau)d\tau \tag{4.7.55}$$

Adding (4.7.54) and (4.7.55) we get

$$W(t) = \frac{1}{2}(1 + e^{-\rho(t)}) + \frac{e^{-\rho(t)}}{2}\int_0^t e^{\rho(\tau)}(\lambda(\tau) + \mu(\tau))d\tau; \tag{4.7.56}$$

which proves the result (B)

Next we recall (4.7.31)

$$\frac{U'}{U} = -\mu V = -\frac{\mu}{W} \tag{4.7.57}$$

and also the differential equation (4.7.34)

$$W' + (\mu - \lambda)W = \mu$$

with $\rho'(t) = (\mu(t) - \lambda(t))$, from (4.7.35)

$$\Rightarrow \quad W' + \rho'(t)W = \mu$$

$$\Rightarrow \quad -\frac{W'}{W} - \rho' = -\frac{\mu}{W}.$$

From (4.7.57) $\Rightarrow \dfrac{U'}{U} = -\dfrac{W'}{W} - \rho'$

Integrating both sides we get

$$\log U = -\log W - \rho + C$$

At $t = 0,\ U = V = W = 1$ and $\rho = 0$, from (4.7.41) and (4.7.44)

$$\Rightarrow \quad C = 0$$

$$\therefore \quad \log U = \log\frac{1}{W} - \rho \Rightarrow U = \frac{e^{-\rho}}{W}$$

$$\left.\begin{aligned} \therefore \text{ From (4.7.30)} \quad & \xi(t) = 1 - U = 1 - \frac{e^{-\rho(t)}}{W} \\ \text{and} \quad & \eta(t) = 1 - V = 1 - \frac{1}{W}, \text{ see (4.7.33)} \end{aligned}\right\} \tag{4.7.58}$$

Thus the results (4.7.58), (4.7.53) and (4.7.54) may completely determine the population process $P_n(t)$.

Mean and variance of $n(t)$ can be obtained immediately as:

From (4.7.58), $\quad \dfrac{1-\xi_t}{1-\eta_t} = e^{-\rho(t)}$

Now from (4.7.15)

$$E(n_t) = \phi'(1,t) = (1-\xi)/(1-\eta), \text{ from (4.7.26)}$$

$$\therefore \quad E(n_t) = e^{-\rho(t)} \tag{4.7.59}$$

Also $\quad P_0(t) = \xi_t \quad$ and $\quad P_n(t) = (1-\xi_t)(1-\eta_t)\eta_t^{n-1}$, from (4.7.37) and (4.7.38)

$$\therefore \quad E(n_t) = 0 \times \xi_t + \sum_{n_t=1}^{\infty} (1-\xi_t)(1-\eta_t)\eta_t^{n-1} n_t$$

$$\therefore \quad E(n_t) = (1-\xi_t)(1-\eta_t) \sum_{n_t=1}^{\infty} n_t \eta_t^{n-1} \tag{4.7.60}$$

Now $\quad \displaystyle\sum_{n=0}^{\infty} \eta_{nt} = \frac{1}{1-\eta_t}.$

Differentiating both sides time w.r.t η_t, we get

$$\sum_{n=0}^{\infty} n_t \eta_t^{n-1} = \frac{1}{(1-n_t)^2} \tag{4.7.61}$$

and $\quad \displaystyle\sum n_t(n_t-1)\eta_t^{n-2} = 2/(1-\eta_t)^3$

$$\Rightarrow \quad \sum n_t^2 \eta_t^{n-2} - \sum n_t \eta_t^{n-2} = 2/(1-\eta_t)^3$$

$$\Rightarrow \quad \sum n_t^2 \eta_t^{n-1} - \sum n_t \eta_t^{n-1} = 2\eta_t/(1-\eta_t)^3$$

$$\Rightarrow \quad \sum n_t^2 \eta_t^{n-1} = \frac{2\eta_t}{(1-\eta_t)^3} + \frac{1}{(1-\eta_t)^2} \tag{4.7.62}$$

Using (4.7.61),

$$E(n_t) = \frac{(1-\xi_t)(1-\eta_t)}{(1-\eta_t)^2} = \frac{(1-\xi_t)}{(1-\eta_t)}. \tag{4.7.63}$$

Further,Var $(n_t) = E(n_t^2) - [E(n_t)]^2$

Using (4.7.60)

$$\text{Var } (n_t) = (1-\xi_t)(1-\eta_t) \sum_{n=1}^{\infty} \eta_t^{n-1} n_t^2 - \left[\frac{(1-\xi_t)}{(1-\eta_t)}\right]^2$$

$$= (1-\xi_t)(1-\eta_t)\left[\frac{2\eta t}{(1-\eta t)^3}+\frac{1}{(1-\eta_t)^2}\right]-\frac{(1-\xi_t)^2}{(1-\eta_t)^2}, \text{ using (4.7.62)}$$

$$\therefore \quad \text{Var}(n_t) = \frac{(1-\xi_t)(\xi_t+\eta_t)}{(1-\eta_t)^2} \tag{4.7.64}$$

Also, from (4.7.58)

$$1-\xi_t = \frac{e^{-\rho(t)}}{W},\ \xi_t+\eta_t = 2-\frac{(1-e^{-\rho(t)})}{W} \quad \text{and} \quad (1-\eta_t)^2 = \frac{1}{W^2}$$

$$\text{Hence,} \quad \text{Var}(n_t) = \left(\frac{e^{-\rho}}{W}\right)\left(\frac{2W-1-e^{-\rho}}{W}\right)\Big/\frac{1}{W^2} \tag{4.7.65}$$

$$\Rightarrow \quad \text{Var}(n_t) = e^{-\rho}(2W-1-e^{-\rho}). \tag{4.7.66}$$

4.7.2 Some Demographic Applications of Kendall's Model of Birth and Death Process

An application of Kendall's model of birth and death process is to identify the process when the mean growth rate is given. Especially, sometimes the problem arises to achieve a certain stipulated mean growth rate and assuming that the population behaviour is described by a certain process (say Gaussian Process or Logistic process). There lies a problem of choosing birth and death parameters enabling to achieve the stipulated growth rate under the process. To be specific, let the mean growth rate follows a Logistic Law given by

$$\overline{n}_t = n_0\frac{\alpha}{1+(\alpha-1)e^{-\beta t}};\ \alpha>0, \beta>0, \tag{4.7.67}$$

where $n_0 = n_t|_{t=0} =$ the initial population size

$$\Rightarrow \quad \frac{d\log\overline{n}_t}{dt} = [\lambda(t)-\mu(t)], \text{ see (4.1.1)}$$

$$= \frac{(\alpha-1)\beta}{e^{\beta t}+(\alpha-1)} = \rho(t) \tag{4.7.68}$$

is the mean growth rate at time t, it is independent of the initial population size.

$$\Rightarrow \quad \lambda(t) = \frac{(\alpha-1)\beta}{e^{\beta t}+(\alpha-1)}+\mu(t). \tag{4.7.69}$$

Suppose $\mu(t) = \mu_0$ and $\lambda(t) = \lambda_0$ holds as $t\to\infty$, then

$$\lambda_0 = \left[\lim_{t\to\infty}\frac{(\alpha-1)\beta}{e^{\beta t}+(\alpha-1)}+\mu_0\right]$$

$$\Rightarrow \quad \lambda_0 = \mu_0 \quad \text{holds for very large } t. \tag{4.7.70}$$

Therefore, if the population is stable and the mean growth rate takes place in the logistic form, then the population is stationary also, i.e., $\lambda_0 = \mu_0$. On the other hand, if the population is quasi stable in the form viz. $\lambda(t)$ depends on t but $\mu(t) = \mu_0$ is independent of t; then we have

$$\lambda(t) = \left[(\alpha-1)\beta e^{-\beta t}\frac{\overline{n}_t}{n_0\alpha}+\mu_0\right]. \text{ by substituting (4.7.67) in (4.7.69)} \tag{4.7.71}$$

This gives an idea of the level of birth rate needed to maintain a mean or desirable size of the population given that its death parameter is stable.

For example, for population like India where birth parameter is time dependent whereas death rate can approximately be assumed to be more or less stable, the problem lies in deciding the level of birth rate (by suitable family limitation programme) for the maintenance of optimal population size.

4.7.3 A Bisexual Population Growth Model of Goodman

So far, we have considered the population growth of only one sex, usually female. L.A. Goodman (1953) presented a stochastic model of population growth based on both sexes. We begin with a population having two subpopulation, the F-subpopulation and M-subpopulation with the sex-specific birth and death rates.

Let there be M males and F females at time t with B_m and D_m are the birth and death rates of males and B_f and D_f are the birth and death rates of females.

Again, let

$FB_m\delta t =$ Prob [that a female gives birth to a male child in time δt when the number of females at time t is F]

$FB_f\delta t =$ Prob [that a female gives birth to a female child in time δt when the number of females at time t is F]

Then, $B_m = KD_m$, where K is the ratio of males to females at birth. (4.7.72)

Further,

$P_{00}(t) =$ Prob [of extinction of population at time t]

$P_{01}(t) =$ Prob [zero male and one female at time t]

$P_{ij}(t) =$ Prof [i males and j females at time t]

Now the problem is to find the probability distribution of number of males and females at the end of the period $(0, t+dt)$. Consider

$P_{m,f}(t+dt) =$ Probability that there will be M males and F females at the end of the period $(0, t+dt)$

$=$ Prob [M males and F females at time t and no births and no deaths in $(t, t+dt)$]

$+$ Prob [$(M-1)$ males and F females at time t and one male birth in $(t, t+dt)$]

$+$ Prob [$(M$ males and $F-1$ females at time t and one female birth in $(t, t+dt)$]

$+$ Prob [$M+1$ males and F females at time t and one male death in $(t, t+dt)$]

$+$ Prob [M males and $F+1$ females at time t and one female death in $(t, t+dt)$]

$$\begin{aligned} P_{m,f}(t+dt) &= P_{m,f}(t)[1-(MD_m+FD_f+MB_m+FB_f)dt] \\ &\quad + P_{m-1,f}(t)FB_m dt + P_{m,f-1}(t)(F-1)B_f dt \\ &\quad + P_{m+1,f}(t)(M+1)D_m dt + P_{m,f+1}(t)(F+1)D_f dt; \end{aligned}$$

$m, M = 0, 1, 2, \ldots; f, F = 0, 1, 2 \ldots$ (4.7.73)

Subtracting $P_{m,f}(t)$ from both sides of (4.7.73) and dividing throughout by dt, we obtain by taking limit

$$\lim_{dt\to 0}\frac{P_{m,f}(t+dt)-P_{m,f}(t)}{dt}$$

$$= -(MD_m + FD_f + MB_m + FB_f)P_{m,f}(t) + FB_mP_{m-1,f}(t) + (F-1)B_fP_{m,f-1}(t) + (M+1)D_mP_{m+1,f}(t) + (F+1)D_fP_{m,f+1}(t);$$

$$m, M = 0,1,2,\ldots \text{ and } f, F = 0,1,2,\ldots \quad (4.7.74)$$

$$\Rightarrow \frac{dP_{m,f}(t)}{dt} = -(MD_m + FD_f + MB_m + FB_f)P_{m,f}(t) + B_mFP_{m-1,f}(t) + B_f(F-1)P_{m,f-1}(t) + D_m(M+1)P_{m+1,f}(t) + D_f(F+1)P_{m,f+1}(t). \quad (4.7.75)$$

The infinite set of differential-difference equations (4.7.75) can be converted into a single equation in the probability generating function. Defining as usual the probability generating function $\phi(s_1, s_2, t)$ as

$$\phi(s_1,s_2,t) = \sum_{m,f=0}^{\infty} P_{m,f}(t)s_1^m s_2^f,\ |s_i| \le 1, i = 1,2,. \quad (4.7.76)$$

Multiplying (4.7.75) by $s_1^m s_2^f$ and summing over all m and f, we obtain by making use of (4.7.76)

$$\frac{\partial\phi}{\partial t} = -D_m s_1\frac{\partial\phi}{\partial s_1} - (D_f + B_m + B_f)s_2\frac{\partial\phi}{\partial s_2} + B_m s_1 s_2\frac{\partial\phi}{\partial s_2} + B_f s_2^2\frac{\partial\phi}{\partial s_2} + D_m\frac{\partial\phi}{\partial s_1} + D_f\frac{\partial\phi}{\partial s_2}$$

$$= D_m(1-s_1)\frac{\partial\phi}{\partial s_1} + [B_m s_1 s_2 + B_f s_2^2 + D_f - (D_f + B_m + B_f)s_2]\frac{\partial\phi}{\partial s_2}. \quad (4.7.77)$$

Differentiating (4.7.77) with respect to s_1,

$$\frac{\partial^2\phi}{\partial t\partial s_1} = -D_m\frac{\partial\phi}{\partial s_1} + D_m(1-s_1)\frac{\partial^2\phi}{\partial s_1^2} + B_m s_2\frac{\partial\phi}{\partial s_2} + [B_m s_1 s_2 + B_f s_2^2 + D_f - (D_f + B_m + B_f)s_2]\frac{\partial^2\phi}{\partial s_1\partial s_2}. \quad (4.7.78)$$

Putting $s_1 = s_2 = 1$, we get

$$\left.\frac{\partial^2\phi}{\partial t\partial s_1}\right|_{s_1=s_2=1} = -D_m\left.\frac{\partial\phi}{\partial s_1}\right|_{s_1=s_2=1} + B_m\left.\frac{\partial\phi}{\partial s_2}\right|_{s_1=s_2=1}$$

$$\frac{\partial\psi_m}{\partial t} = -D_m\psi_m + B_m\psi_f \quad (4.7.79)$$

where $\left.\frac{\partial\phi}{\partial s_1}\right|_{s_1=s_2=1} = \psi_m$ (say) and $\left.\frac{\partial\psi}{\partial s_2}\right|_{s_1=s_2=1} = \psi_f$ (say).

Thus ψ_m is the expected number or the first moment of males at time t and ψ_f that of females. Again differentiating (4.7.77) with respect to s_2

$$\frac{\partial^2\phi}{\partial t\partial s_2} = D_m(1-s_1)\frac{\partial^2\phi}{\partial s_1\partial s_2} + [B_m s_1 + 2B_f s_2 - (D_f + B_m + B_f)]\frac{\partial\phi}{\partial s_2} + [B_m s_1 s_2 + B_f s_2^2 + D_f - (D_f + B_m + B_f)s_2]\frac{\partial^2\phi}{\partial s_2^2} \quad (4.7.80)$$

$$\Rightarrow \quad \left.\frac{\partial^2 \phi}{\partial t \partial s_2}\right|_{s_1=s_2=1} = [B_m + 2B_f - (D_f + B_m + B_f)] \left.\frac{\partial \phi}{\partial s_2}\right|_{s_1=s_2=1}$$

$$\Rightarrow \quad \frac{\partial \psi_f}{\partial t} = (B_f - D_f)\psi_f \tag{4.7.81}$$

Similarly $\left.\frac{\partial \psi_m}{\partial s_1}\right|_{s_1=s_2=1} = E(M(M-1)), \left.\frac{\partial \psi_m}{\partial s_2}\right|_{s_1=s_2=1} = E(F(F-1)]$

$$\left.\frac{\partial \psi_m}{\partial s_1 \partial s_2}\right|_{s_1=s_2=1} = E(M, F)$$

where $E(M)$ and $E(F)$ represents the expected number of males and females at any time t.

$$\Rightarrow \quad E(M^2) = \left.\frac{\partial \psi_m}{\partial s_1}\right|_{s_1=s_2=1} + \psi_m$$

$$E(F^2) = \left.\frac{\partial \psi}{\partial s_2}\right|_{s_1=s_2=1} + \psi_f$$

In view of this we have

$$\frac{\partial E(M^2)}{\partial t} = -2D_m\psi_m^2 + 2B_m\psi_{mf} + D_m\psi_m + B_m\psi_f \tag{4.7.82}$$

$$\frac{\partial E(MF)}{\partial t} = (B_f - D_f - D_m)\psi_{mf} + B_m\psi_f^2 \tag{4.7.83}$$

$$\frac{\partial E(F^2)}{\partial t} = 2(B_f - D_f)\psi_f^2 + (B_f + D_f)\psi_f. \tag{4.7.84}$$

After solving these five equations (4.7.79), (4.7.81), (4.7.82), (4.7.83) and (4.7.84) we get,

$$\psi_m = \frac{B_m}{D_m + B_f - D_f} \exp[(B_f - D_f)t] - \exp(-D_m t) \tag{4.7.85}$$

$$\phi_f = \exp[(B_f - D_f)t], \tag{4.7.86}$$

with the initial condition of one female. If we start with F_0 females and M_0 males at the start we get

$$E(M) = \psi_m = M_0 \exp(-D_m t) + F_0 \frac{B_m}{D_m + B_f - D_f}[\exp[(B_f - D_f)t] - \exp(-D_m t)] \tag{4.7.87}$$

and $E(F) = \psi_f = F_0 \exp[B_f - D_f)t]$ (4.7.88)

4.7.4 Models of Population Growth Under Random Environment (A Stochastic Version of the Logistic Population Model)

It is evident that the deterministic theory of population growth outlined in some of the preceding sections relating to logistic, stable and quasi-stable population models is not really adequate for the description of all the phenomena one might wish to study; for it fails to take into account the role of chance fluctuations in the development of the process. Extensive work in this respect has been done by Feller

(1951, 1952). Feller's characterization of the 'Built in Boundaries' as simplified by Keilson (1965) and reproduced by Goel and Dyn (1979) is given in the following lines.

Let us take the stochastic version of Verhulst Deterministic Logistic model given in Sec. 4.1 and Sec. 4.2.1.

$$\frac{dN(t)}{dt} - rN(t)\left[1 - \left(\frac{N(t)}{M}\right)^x\right] \Big/ \alpha$$

For simplicity, let $\alpha = 1$ and $N(t) = X(t) = x$.

Then we have

$$\frac{dx}{dt} = rx\left[1 - \frac{x}{M}\right], \text{ where } N(t) = x \tag{4.7.89}$$

Consider a stochastic form of r, the growth rate under a fluctuating environmental condition, then let $r = \overline{r} + \sigma F(t)$, say,

where, $\overline{r} = E(r), \sigma = \sqrt{\text{Var}(r)}$ and $F(t)$ represents some random fluctuations.

$$\begin{aligned} \Rightarrow \quad \frac{dx}{dt} &= [\overline{r} + \sigma F(t)]x\left(1 - \frac{x}{M}\right) \\ &= \overline{r}x\left(1 - \frac{x}{M}\right) + \sigma x\left(1 - \frac{x}{M}\right)F(t) \\ &= \alpha(x) + \beta(x)\ F(t) \end{aligned} \tag{4.7.90}$$

$$\text{where} \quad \alpha(x) = \overline{r}x\left(1 - \frac{x}{M}\right) \tag{4.7.91}$$

$$\text{and} \quad \beta(x) = \sigma x\left(1 - \frac{x}{M}\right). \tag{4.7.92}$$

A process given by

$$\frac{dx}{dt} = \alpha(x) + \beta(x)\ F(t)$$

is called Gaussian Delta Continuous Process Goel and Dyn (1979).

Let $\quad a(x) = \alpha(x) + \dfrac{1}{4}\dfrac{\partial}{\partial x}\beta(x)^2$ and $b(x) = [\beta(x)]^2$

Then the Fokkar Planck equation (F.P. equation) (ibid.), is given by

$$\frac{\partial P}{\partial t} = -\frac{\partial}{\partial x}[a(x)P] + \frac{1}{2}\frac{\partial^2}{\partial x^2}[b(x)P] \tag{4.7.93}$$

where $P = \text{Prob}\ [x\ |y, t]$ = The probability that the r.v. will take the value of x at time t given that it takes a value y at time zero, for any y and t.

Given the F.P. equation, the stochastic differential equation of Gaussian delta continuous process is uniquely given by

$$\frac{\partial x(t)}{\partial t} = a(x) - \frac{1}{4}\frac{\partial b(x)}{\partial x} + [b(x)]^{-1/2}F(t) \tag{4.7.94}$$

The process which are completely determined by the coefficients $a(x)$ and $b(x)$ of the F.P. equation with no additional boundary conditions are known as Unrestricted Processes. Such processes have unlimited state spaces if $b(x) > 0$ and $a(x)$ is infinite. Again such processes may have some 'built in' boundaries in case (i) $b(x)$

vanishes at some point or (ii) $a(x)$ becomes infinite. These processes are continuous counterpart of the birth and death process. Then type of boundary+ is determined by the integrability of the following two functions.

$$h_1(x) = M(x)\int_{x_0}^{x}[b(\xi)M(\xi)]^{-1}d\xi \tag{4.7.95}$$

and $$h_2(x) = [b(x)M(x)]^{-1}\int_{x_0}^{x}M(\xi)d\xi \tag{4.7.96}$$

the function $M(x)$ is defined as

$$M(x) = \exp\left(-2\int_{x_0}^{x}\frac{a(\xi)}{b(\xi)}d\xi\right). \tag{4.7.97}$$

Over the interval $I = [x_0, r]$ where x_0 is some interior point of the state space of the process and r is the boundary.*

Further, for $y \le z \le x$

$$\mu_1(z|z_0) = 2\int_{x_0}^{z}h_1dx = \text{average time to reach } z \text{ starting from } x_0 \text{ where } z_0 \text{ is a reflecting boundary} \tag{4.7.98}$$

and $$\mu_2(x_0|z) = 2\int_{x_0}^{z}h_2dx = \text{average time to reach } x_0 \text{ starting from } z \text{ where } z \text{ is the reflecting boundary.} \tag{4.7.99}$$

We have in the case of Logistic process

$$b(x) = [\beta(x)]^2 = \sigma^2x^2\left(1-\frac{x}{M}\right)^2,\ b(0) = 0 \text{ and } b(M) = 0.$$

Hence the process is an unrestricted singular process. The boundaries $x = 0$, $x = M$ are therefore singular.

Also near $x = 0$.

$$\begin{aligned} a(x) &= \alpha(x) + \frac{1}{4}\frac{\partial}{\partial x}\beta(x)^2 \\ &= \bar{r}x\left(1-\frac{x}{M}\right) + \frac{1}{4}\frac{\partial}{\partial x}\sigma^2x^2\left(1-\frac{x}{M}\right)^2 \\ &= x\left(1-\frac{x}{M}\right)\left[\bar{r} + \frac{\sigma^2}{2}\left(1-\frac{2x}{M}\right)\right] \end{aligned}$$

For small x,

$$a(x) \simeq \left(\bar{r} + \frac{\sigma^2}{2}\right)x \tag{4.7.100}$$

and $$b(x) \simeq \sigma^2x^2 \tag{4.7.101}$$

Further, since $h_1(x)$ and $h_2(x)$ are both not integrable over I, it follows that the boundaries $x = 0$ and $x = M$ are inaccessible natural boundaries and cannot reach in finite time, a result in conformity with the deterministic process. The process may there fore describe a population which is far from extinction and which fluctuates about some mean value ($< M$) due to fluctuations in the net growth rate.

+ Inaccessible, absorbing or reflecting.

The steady state probability density function of the unrestricted process $P(x|y,\infty)$ is given by

$$P(x|y,\infty) = \frac{c}{b(x)} \exp\left(2\int_{-\infty}^{x} [a(\xi)/b(\xi)]d\xi\right), \tag{4.7.102}$$

while c is a normalizing constant; when no boundary is in exit boundary or a regular absorbing boundary. Again

$$P(x|y,\infty) = [b(x)M(x)]^{-1} \bigg/ \left[\int_r b(x)M(x)\right]^{-1} dx,$$

when boundaries are on exit or absorbing boundaries. Hence, the steady state probability density function of the process is given by

$$P(x|y,\infty) = cx^{[2\overline{r}/\sigma^2 - 1]}\left(1 - \frac{x}{M}\right)^{-\left(\frac{2x}{\sigma^2}+1\right)} \tag{4.7.103}$$

Particular Cases

(i) If $2\overline{r}/\sigma^2 < 1$, then the density function is U shaped indicating that the density approaches 0 or M

(ii) If $2\overline{r}/\sigma^2 > 1$, then the density function is monotonically increasing and in J-shaped concentration of population around π.

To derive the time dependent probability density of the process, we introduce the variable,

$$z = \frac{1}{\sigma}\log\left[x \bigg/ \left(1 - \frac{x}{M}\right)\right]$$

$$\Rightarrow \qquad dz = \left[\sigma x\left(1 - \frac{x}{M}\right)\right]^{-1} dx.$$

Then (4.7.90) becomes

$$\frac{dz}{dt} = \frac{\overline{r}}{\sigma} + F(t),$$

which is nothing but the stochastic differential equation for the unrestricted Weinner Process.

The Fokkar Plank equation satisfied by the conditional probability density of Z given by, say $g(z|z_0t)$ is

$$\frac{dy}{dt} = -\frac{\partial}{\partial z}[(\overline{r}/\sigma)g] + \frac{1}{2}\frac{\partial^2 g}{\partial z^2},$$

with the boundary condition

$$\lim_{z\to\pm\infty} g(z|z_0t) = 0,$$

corresponding to the inaccessible boundaries

$$x = 0 \text{ (i.e. } z = -\infty) \text{ and } x = M \text{ (i.e. } z = +\infty),$$

remember that y is the value of x at time $t = 0$ (see (4.7.93)).

Thus, the Gaussian distribution is given by

$$\text{Gau }(Z|z,t) = \frac{1}{(2\pi t)^{1/2}} \exp -\frac{1}{2t}\left(z - z_0 - \frac{\overline{r}t}{\sigma}\right)^2,$$

with peak at $z = z_0 + \frac{\overline{r}t}{\sigma}$, $Z = z$ at time t.

Eventually this gives

$$P(x|y,t) = \frac{1}{\sigma\sqrt{2\pi t x}\left(1 - \frac{x}{M}\right)}.$$

$$\exp\left[\left(-\frac{1}{2t}\left\{\frac{1}{\sigma}\log\frac{x}{y} - \frac{1}{\sigma}\right\}\log\left(1 - \frac{x}{M}\right)\Big/\left(1 - \frac{y}{M}\right) - \frac{\overline{r}t}{\sigma}\right)^2\right] \quad (4.7.104)$$

which is the density function determining the behaviour of the population.

Note that when $M = \infty$, either the population is far from saturation or the supply of resource is unlimited.

In this case,

$$E(X) = \int_0^M xP(x|y,t)dx = \int_{-\infty}^{\infty} \exp(\sigma z) g(z|z_0,t) dz \quad \because x = e^{\sigma z}$$

$$= \exp\left(\frac{\sigma^2 t}{2}\right)\exp(\sigma(z_0 + \overline{r}t/\sigma)) = ye^{rt}\left[\exp\left(\frac{\sigma^2 t}{2}\right)\right],$$

as compared to $x = ye^{rt}$ for the deterministic case is absence of random fluctuations.

$$\begin{aligned}\text{Var }(X) &= E(X^2) - [E(X)]^2 \\ &= E(Y^2)e^{2rt}e^{\sigma^2 t}[e^{\sigma^2 t} - 1] \\ &= [E(X)]^2[e^{\sigma^2 t} - 1],\end{aligned}$$

as compared to zero variance in the deterministic case. The coefficient of variation is given by

$$\frac{[\text{Var}(X)]^{1/2}}{E(X)} = (e^{\sigma 2t} - 1)^{1/2}. \quad (4.7.105)$$

As t increases the coefficient of variation is increasing, showing that the average rate cannot describe the growth of the population.

4.8 Conclusion

The discussion in respect of the deterministic and stochastic models unfortunately cannot have unlimited scope in view of the fact that further generalization of the model is naturally warranted with Mathematical complication and the consequent inflexibility of the models, so generalised, for making the same more realistic. For example, a population expert would like to generalise Leslie's original formulation and might reasonably suppose that the number of individuals born during the interval $(t, t+1)$ has the Poisson distribution with mean $\sum_{X=0}^{m} F_x n_{xt}$ and the number in the age group (x to $x+1$) at the end of the interval has a binomial distribution with probability p_{x-t} and the index parameter $n_{x-1,t}$. Such generalizations are useful in the context of population problems in India. Similarly Lotka's stable population and Coale's quasi stable population analysis are also capable of being generalized which will definitely be useful in the context of population problem. Again such generalizations would considerably cut down the flexibility of the models because of Mathematical complexities. However, the complexities arising out of these kind of complications may partially be overcome by employing computer solutions and modern simulation techniques.

References

1. Bartlett, M.S. (1960): Stochastic Population models—Mathuen Monograph Mathuen, London.
2. Biswas, S. (1986)—Population Dynamics and its perspective—An invited paper in the 73rd Session (Statistics Section) of All India Science Congress, Delhi, 1986
3. Coale, Ansley (1976): Growth and Structures of Human Population—Princeton University Press, N.J.
4. Feller, William (1949): An Introduction to the Probability theory and its application, Vol. II, John Wiley and Sons, N.Y.
5. Feller, William (1951): Diffusion Processes in genetics-Proceedings of the symposium in Mathematical Statistics and Probability-Second Berkeley Symposium, Berkeley, Calif.
6. Feller William (1952): The Parabolic differential equations and the Associated semi-group of transformations–Ann of Math. Vol. 55, p. 46.
7. Goel, N.S. and Dyn, Nira Richter (1979): Stochastic models in Biology. Academic Press. N.Y.
8. Goodman, L.A. (1953): Population growth of the sexes, Biometrika, Vol. 9 pp. 212–225.
9. Keilson, J. (1965): A review of transient behaviour in regular diffusion and birth and death processes Part I, J. Appl. Probability Vol. 1, p. 297.
10. Keilson, J. (1965): A review of transient behaviour in regular diffusion and birth and death processes, Part II, J. Appl. Probability Vol. 2, p. 405.
11. Kendall. D.G. (1948): A generalised birth and death Processes–Annals of Mathematical Statistics, Vol. 19, pp. 1–15.
12. Leslie, P.H. (1945): On the use of matrices in certain Population Mathematics, Biometrika, Vol. 33, pp. 183–212.
13. Leslie, P.H. (1948): Some further notes is Matrices in population Mathematics, Biometrika, Vol. 35, pp. 213–245.
14. Levy, P. (1954): Processus Semi-Markoviens'—Proc. Inter. Cong. Math. (Amsterdam) Vol. 3, pp. 416–426.
15. Lewis, E.G. (1942): On the generation and the growth of Population. Sankhya Vol. 36, pp. 93–95.
16. Lopez, Alvardo (1961): Problems in Stable Population theory, Princeton University Press, N.J.

17. Lotka, A.J., Dullin L.I. and Spiegelman, M.R. (1949): Length of life, Ronald Press, New York.

18. Moran, P.A.P. (1961): Statistical processes in evolutionary theory, Clarendon Press, Oxford, London.

19. Nicholson, A.J. (1957): The self adjustment of population to change, Cold Spring Harbor Symposium on Quantitative Biology, Vol. 22, pp. 551–598.

20. Nishlet, R.M. and Gurney, W.SC (1982): Modelling Fluctuating Population—John Wiley & Sons, N.Y.

21. Keyfitz, Nathan (1966): A life table that agrees with the data—Part I, Journal of American Statistical Association, Vol. 61, pp. 305–312.

22. Keyfitz, Nathan (1966): A life table that agrees with the data-Part II, Journal of American Statistical Association, Vol. 61, pp. 1253–68.

23. Keyfitz, N. (1977): Introduction to the Mathematics of Population with Revisions, Addison Wisley, London.

24. Rhodes, E.C. (1940): Populations, Mathematics–III J. Roy. Stat. Soc, Vol. 103, pp. 362–387.

25. Smith, W.L. (1954): Asymptotic Renewal Theorems; Proc. Royal. Society Edin A64, pp. 9–48.

26. Taka'cs, L. (1954): On the secondary process generated by Poisson Process and their application in Physics, Acta-Math-Acad-Science Vol. 5, pp. 203–236.

27. Taka'cs, L. (1960): Stochastic Process, Mathuen, London.

28. Wolfenden, H.H. (1954): Populations statistics and Their Compilations, The University of Chicago Press, Chicago.

29. Watson, G.N. (1962): Treatise on the Theory of Bessel Functions, Cambridge University Press, London.

5

Survival and Competing Risk Theory

5.0 Introduction

The concept of 'risk' is introduced in survival theory as follows: Let us consider any system, be it Mechanical or Biological. It is always observed that no sooner the system starts functioning the set of all risks, either of which could possibly stop the working of the system, start competing among themselves in an attempt to take away the life of the system or to cause the failure of the same. As a matter of fact, either of the risks may be responsible for the failure of the system. More specifically, one can reasonably assume that in the human biosystem, right from the day when a person is born all the risks which are capable of taking the life of an individual start competing with each other; and a particular risk ultimately succeeds in taking away the life of an individual either in presence of all other possible risks or in the presence of a subset of the set of risks, depending on the individual's particular set up. Quantitatively, the problems of competing risks can therefore be basically classified in three forms:

(i) How much intense a particular risk is while succeeding to take away the life of an individual during a given period of time in presence of all other risks;

(ii) Given that for an individual only a particular risk being viable, the problem of measuring the intensity of the same, or

(iii) Given that an individual is free from a particular risks the net risk with which an individual is exposed to or how much effective the risk of a particular kind on an individual when a correlated risk is off from the population.

These are the major items of consideration in competing risk theory.

Let us assume that a set of finite number of risks say $R_1, R_2, \ldots, R_\delta, \ldots, R_k$ are operative in the population. The following represents the analysis of the competing risk theory in the line of Chiang (1968).

5.1 Measurement of Competing Risks

Definition of Hazard Rate $\mu(t, \delta)$

We define the hazard rate due to risk R_δ as the conditional probability of dying in $(t, t + dt)$ given that the individual is surviving at t. It is denoted by $\mu(t, \delta)$. Obviously, under the orderliness condition i.e. probability of more than one failure in $(t, t + dt)$ is negligible, we have

$$\sum_{\delta=1}^{k} \mu(t, \delta) = \mu(t) \tag{5.1.5}$$

where $\mu(t)$ is the total hazard rate. (Elandt and Johnson (1975)).

To measure the various types of competing risks in introduce the following indices:

(i) Crude probability of death

(ii) Net probability (Type A and Type B) of death

(iii) Partially crude probability of death

Crude Probability of Death

It is defined as the probability of dying due to a risk, say, $R_\delta(\delta = 1, 2, \ldots, k)$ in an interval (t_i, t_{i+1}) in presence of all other risks in the population, given that the individual is surviving at $t = t_i$. It is denoted by $Q_{i\delta}$.

Net Probability of Death (Type A)

This is defined as the probability of dying in an interval $(t_i, t_{i+1}]$ by a risk R_δ when all other risks are off from the population, given that the individual is surviving at $t = t_i$. It is denoted by $q_{i\delta}$.

Net Probability of Death (Type B)

This is defined as the probability of dying in $(t_i, t_{i+1}]$, given that the person is surviving at $t = t_i$ when a particular risk R_δ is off from the population. It is denoted as $q_{i\cdot\delta}$.

Partially Crude Probability of Death

This is defined as the probability of dying in $(t_i, t_{i+1}]$ due to R_δ in presence of all other risks, when another risk say $R_\epsilon(R_\delta \neq R_\epsilon)$ is off from the population and given that the individual is surviving at $t = t_i$. It is denoted as $Q_{i\delta\cdot\epsilon}(\delta \neq \epsilon); \delta,\ \epsilon = 1, 2, 3, \ldots, k$.

5.2 Inter-Relations of the Death Probabilities

The probabilities $Q_{i\delta}, q_{i\delta}, q_{i\cdot\delta}, Q_{i\delta\cdot\epsilon}$ are inter-related. What follows we find the relations between them.

We have two basic assumptions viz.

$$q_t = Q_{i1} + Q_{i2} + \ldots + Q_{ik} \tag{5.2.1}$$

and $$\mu(t) = \mu(t;1) + \mu(t;2) + \ldots + \mu(t;k) \tag{5.2.2}$$

where $\mu(t;\delta)$ represents the hazard rate due to the risk R_δ.

5.2.1 Expression of Crude Probability of Death $Q_{i\delta}$

Probability of surviving in (t_i, t) where $t_i < t \leq t_{i+1}$ is given by

$$p_i = \exp\left(-\int_{t_i}^{t_{i+1}} \mu(t)dt\right) \tag{5.2.3}$$

where $\mu(t)$ is the hazard rate.

Again the probability of dying in $(t, t+dt)$ due to risk R_δ is

$$= \mu(t;\delta)dt \tag{5.2.4}$$

Therefore, the probability of surviving in $(t_i, t]$ and then dying in R_δ between $(t, t+dt)$, where $t_i < t \leq t_{i+1}$, is given by

$$\int_{t_i}^{t_{i+1}} \exp\left(-\int_{t_i}^{t} \mu(\tau)d\tau\right) \cdot \mu(t;\delta)dt$$

$$\text{Hence,}\quad Q_{i\delta} = \int_{t_i}^{t_{i+1}} \exp\left(-\int_{t_i}^{t} \mu(\tau)d\tau\right)\mu(t;\delta)dt \tag{5.2.5}$$

$$= \int_{t_i}^{t_{i+1}} \left(-\int_{t_i}^{t} \mu(\tau)d\tau\right)\left[\frac{\mu(t;\delta)}{\mu(t)}\right]\cdot \mu(t)dt \tag{5.2.6}$$

Now, we have

$$\mu(t;1) + \mu(t;2) + \ldots + \mu(t;k) = \mu(t), \text{ see (5.2.2)}$$

since the risks $R_2, \ldots, R_k$ are independent.

Chiang (1968) assumed that although $\mu(t;\delta), (\delta = 1, 2, \ldots, k)$ depends on t but $\dfrac{\mu(t;\delta)}{\mu(t)}$, $(\delta = 1, 2, \ldots, k)$ is taken as independent of t, in some interval $(t_i, t_{i+1}]$, (5.2.7)

which is known as *Chiang's proportionality assumption.*

In view of Chiang's proportionality assumption $\dfrac{\mu(t;\delta)}{\mu(t)}$ in (5.2.6) may be taken as independent of time and we can rewrite (5.2.6) as:

$$Q_{i\delta} = \frac{\mu(t;\delta)}{\mu(t)} \int_{t_i}^{t_{i+1}} \exp\left(-\int_{t_i}^{t} \mu(\tau)d\tau\right)\mu(t)dt$$

$$= -\frac{\mu(t;\delta)}{\mu(t)} \int_{t_i}^{t_{i+1}} \frac{d}{dt}\left(\exp\left(-\int_{t_i}^{t} \mu(\tau)d\tau\right)\right) dt, \text{ using (A.1.15)}$$

$$= -\frac{\mu(t;\delta)}{\mu(t)}\left[\exp\left(-\int_{t_i}^{t_{i+1}} \mu(\tau)d\tau\right) - \exp\left(-\int_{t_i}^{t_i} \mu(\tau)d\tau\right)\right],$$

integration and differentiation are opposite to each other.

$$= -\frac{\mu(t;\delta)}{\mu(t)}\left[\exp\left(-\int_{t_i}^{t_{i+1}} \mu(\tau)d\tau\right) - 1\right]$$

$$\Rightarrow\quad Q_{i\delta} = \frac{\mu(t;\delta)}{\mu(t)}\left[1 - \exp\left(-\int_{t_i}^{t_{i+1}} \mu(\tau)d\tau\right)\right] \tag{5.2.8}$$

Let $\quad \exp\left(-\displaystyle\int_{t_i}^{t_{i+1}} \mu(\tau)d\tau\right) = p_i =$ Probability of surviving upto t_{i+1}, given that the person has survived upto t_i.

Then, (5.2.8)

$$\Rightarrow\qquad Q_{i\delta} = q_i \cdot \frac{\mu(t,\delta)}{\mu(t)}, \quad \text{where}\quad q_i = 1 - p_i$$

$$\Rightarrow\qquad \frac{Q_{i\delta}}{q_i} = \frac{\mu(t;\delta)}{\mu(t)} \tag{5.2.9}$$

$$\text{Similarly,}\quad \frac{Q_{i\epsilon}}{q_i} = \frac{\mu(t;\epsilon)}{\mu(t)} \tag{5.2.10}$$

$$\frac{(5.2.9)}{(5.2.10)} \quad\Rightarrow\quad \frac{Q_{i\delta}}{Q_{i\epsilon}} = \frac{\mu(t;\delta)}{\mu(t;\epsilon)}. \tag{5.2.11}$$

Thus the probabilities are proportional to hazard rates.

5.2.2 Relationship between Crude Probability of Death and Net Probability of Death (Type A)

Probabilities of surviving in $(t_i, t_{i+1}]$ given that R_δ is the only risk, is given by

$$p_{i\delta}(\text{say}) = \exp\left(-\int_{t_i}^{t_{i+1}} \mu(t;\delta)dt\right), \text{ where } p_{i\delta} = 1 - q_{i\delta} \tag{5.2.12}$$

$\therefore$ Probability of dying in $(t_i, t_{i+1}]$ due to the risk R_δ is given by

$$q_{i\delta} = 1 - \exp\left(-\int_{t_i}^{t_{i+1}} \mu(t;\delta)dt\right) = 1 - \exp\left(-\int_{t_i}^{t_{i+1}} \left(\frac{\mu(t;\delta)}{\mu(t)}\right)\mu(t)dt\right)$$

$$= 1 - \exp\left(-\frac{\mu(t;\delta)}{\mu(t)}\int_{t_i}^{t_{i+1}} \mu(t)dt\right),$$

by Chiang's proportionality assumption (5.2.7)

$$= 1 - \left[\exp\left(-\int_{t_i}^{t_{i+1}} \mu(t)dt\right)\right]^{\frac{\mu(t,\delta)}{\mu(t)}}$$

$$= 1 - (p_t)^{\frac{\mu(t;\delta)}{\mu(t)}}, \text{ from (5.2.3)} \tag{5.2.13}$$

$$\Rightarrow \quad q_{i\delta} = 1 - (p_t)^{\frac{Q_{i\delta}}{q_i}}, \text{ using (5.2.8)}, \tag{5.2.14}$$

which gives relationship between Net probability of death (type A) $q_{i\delta}$ and crude probability of death $Q_{i\delta}$.

5.2.3 Relationship between Net Probability of Death (type B) ($q_{i\cdot\delta}$) and Crude Probability of Death $Q_{i\delta}$

Probability of dying in $(t_i, t_{i+1}]$ where R_δ is off from the population, given that the person is surviving at $t = t_i$ i.e. $q_{i\cdot\delta}$ is given by

$$q_{i\cdot\delta} = 1 - \exp\left(-\int_{t_i}^{t_{i+1}} [\mu(t) - \mu(t;\delta)]dt\right), \text{ from (5.2.2)}$$

$$= 1 - \exp\left(-\int_{t_i}^{t_{i+1}} \mu(t)\left[1 - \frac{\mu(t;\delta)}{\mu(t)}\right]dt\right)$$

$$= 1 - \left[\exp\left(-\int_{t_i}^{t_{i+1}} \mu(\tau)dt\right)\right]^{\left(1-\frac{\mu(t;\delta)}{\mu(t)}\right)}, \text{ from (5.2.7)}$$

$$\Rightarrow \quad q_{i\cdot\delta} = 1 - (p_i)^{1-\frac{Q_{i\delta}}{q_i}}, \text{ (by using (5.2.9)} \tag{5.2.15}$$

$$\Rightarrow \quad q_{i\cdot\delta} = 1 - (p_i)^{\frac{q_i - Q_{i\delta}}{q_i}} \tag{5.2.16}$$

This gives a relationship between crude probability of death $Q_{i\delta}$ and Net probability of death (type B) $q_{i\cdot\delta}$.

We can rewrite (5.2.16),

$$\left.\begin{aligned} q_{i\cdot\delta} &= 1-(p_i)^{\left(\sum\limits_{j=1}^{k} Q_{ij}-Q_{i\delta}/q_i\right)}, \text{ using (5.2.1)} \\ &= 1-(p_i)^{\left(\sum\limits_{j\neq\delta}^{k} Q_{ij}/\sum\limits_{j=1}^{k} Q_{ij}\right)} \end{aligned}\right\} \tag{5.2.17}$$

5.2.4 Relationship between Partially Crude Probability of Death ($Q_{i\delta\cdot\epsilon}$) $\delta \neq \epsilon$ and Crude Probability of Death ($Q_{i\delta}$)

Let us consider $Q_{i\delta\cdot 1}(\delta \neq 1)$. This is the probability of dying in $(t_i, t_{i+1}]$ in risk R_δ, given that the risk R_1 is off from the population. The probability of surviving in $(t_i t]$ where $t_i < t \le t_{i+1}$, when R_1 is off from the population, is

$$\exp\left(-\int_{t_i}^{t}[\mu(\tau)-\mu(\tau;1)]d\tau\right), \text{ from (5.2.3) and (5.2.2).} \tag{5.2.18}$$

The probability of dying in $[t, t+dt)$, given the materialisation of (5.2.18) is

$$\mu(t;\delta)dt, \quad \text{where} \quad \delta \neq 1 \quad \text{and} \quad t_i < t \le t_{i+1} \tag{5.2.19}$$

Therefore, the probability of simultaneous materialization of (5.2.18) and (5.2.19) when $t_i < t \le t_{i+1}$, is

$$Q_{i\delta\cdot 1} = \int_{t_i}^{t_{i+1}}\left\{\exp\left(-\int_{t_i}^{t}[\mu(\tau)-\mu(\tau;1)]d\tau\right)\right\}\cdot\mu(t;\delta)dt$$

$$= \int_{t_i}^{t_{i+1}}\left\{\exp\left(-\int_{t_i}^{t}[\mu(\tau)-\mu(\tau;1)]d\tau\right)\right\}\frac{\mu(t;\delta)}{\mu(t)-\mu(t;1)}[\mu(t)-\mu(t;1)]dt$$

Therefore, from (5.2.7)

$$Q_{i\delta\cdot 1} = \frac{\left[\dfrac{\mu(t;\delta)}{\mu(t)}\right]}{\left[1-\dfrac{\mu(t;1)}{\mu(t)}\right]}\int_{t_i}^{t_{i+1}}\exp\left(-\int_{t_i}^{t}[\mu(\tau)-\mu(\tau;1)]d\tau\right)[\mu(t)-\mu(t;1)]dt$$

$$= \frac{\left(\frac{Q_{i\delta}}{q_i}\right)}{\left(1-\frac{Q_{i1}}{q_i}\right)}\int_{t_i}^{t_{i+1}}\exp\left(-\int_{t_i}^{t}[\mu(\tau)-\mu(\tau;1)]d\tau\right)[\mu(t)-\mu(t;1)]dt, \text{ from (5.2.9)}$$

$$= -\frac{Q_{i\delta}}{q_i-Q_{i1}}\int_{t_i}^{t_{i+1}}\frac{d}{dt}\left[\exp\left(-\int_{t_i}^{t}[\mu(\tau)-\mu(\tau;1)]d\tau\right)dt\right], \text{ using (A.1.15)}$$

$$= \frac{Q_{i\delta}}{q_i-Q_{i1}}\left[-\exp\left(-\int_{t_i}^{t}[\mu(\tau)-\mu(\tau;1)]d\tau\right)\right]_{t=t_i}^{t=t_{i+1}}$$

$$Q_{i\delta\cdot 1} = \frac{Q_{i\delta}}{q_i-Q_{i_1}}\left[1-\exp\left(-\int_{t_i}^{t_{i+1}}[\mu(\tau)-\mu(\tau;1)]d\tau\right)\right].$$

Further, we have

$$\exp\left(-\int_{t_i}^{t_{i+1}}[\mu(\tau)-\mu(\tau;1)]d\tau\right) = \text{Probability of surviving in } (t_i, t_{i+1}], \text{ when } R_1 \text{ is off from the population}$$

$$= \exp\left(-\int_{t_i}^{t_{i+1}}[\mu(\tau;2)+\mu(\tau;3)+\ldots+\mu(\tau;k)]d\tau\right)$$

$$= p_{i2}+p_{i3}+\ldots+p_{ik}, \quad \text{where } \; p_{i\delta} = 1-q_{i\delta}$$

$$\Rightarrow \quad 1-\exp\left(-\int_{t_i}^{t_{i+1}}[\mu(\tau;2)+\mu(\tau;3)+\ldots+\mu(\tau;k)]d\tau\right)$$

$$= 1-(p_{i2}+p_{i3}+\ldots+p_{ik}) = 1-(p_i-p_{i1}) = 1-p_{i\cdot 1}$$

$$= q_{i\cdot 1} = 1-(p_i)^{\left(\frac{q_i-Q_{i_1}}{q_i}\right)}, \text{ from (5.2.16)}$$

Thus $$Q_{i\delta\cdot 1} = \frac{Q_{i_\delta}}{q_i-Q_{i_1}}[1-p_i^{(q_i-Q_{i1})/q_i}]. \tag{5.2.20}$$

This gives the relationship between Partially Crude probability of death and crude probability of death.

In general, (from (5.2.20))

$$Q_{i\delta\cdot\epsilon} = \frac{Q_{i\delta}}{q_i-Q_{i\epsilon}}\left[1-(p_i)^{\frac{(q_i-Q_{i\epsilon})}{q_i}}\right] \tag{5.2.21}$$

Example 5.1: Show that $q_{i\delta} > Q_{i_t}$

Proof: We have $q_{i\delta} = 1-(p_i)^{\frac{Q_{i\delta}}{q_i}}$, from (5.2.14)

$$= 1-(p_i)^{\frac{\mu(t;\delta)}{\mu(t)}}, \text{ (by (5.2.9))}$$

$$= 1-(1-q_i)^{\frac{\mu(t,\delta)}{\mu(t)}}$$

$$= 1-\left(1-\frac{\mu(t;\delta)}{\mu(t)}q_i+\frac{\mu(t;\delta)}{\mu(t)}\left(\frac{\mu(t;\delta)}{\mu(t)}-1\right)\frac{q_i^2}{2!}+\cdots\right)$$

$$\Rightarrow \quad q_{i\delta} = \frac{\mu(t;\delta)}{\mu(t)}q_i-\frac{\mu(t;\delta)}{\mu(t)}\left[\frac{\mu(t;\delta)}{\mu(t)}-1\right]\frac{q_i^2}{2!}+\cdots$$

$$\Rightarrow \quad q_{i\delta} = \frac{\mu(t;\delta)}{\mu(t)}q_i+\left(1-\frac{\mu(t;\delta)}{\mu(t)}\right)\frac{\mu(t;\delta)}{\mu(t)}\frac{q_i^2}{2!}+\cdots$$

Since all the terms on the right hand side are positive, $\because \frac{\mu(t,\delta)}{\mu(t)} < 1$, it follows that

$$q_{i\delta} > \frac{\mu(t;\delta)}{\mu(t)}\cdot q_i = Q_{i\delta}, \text{ from (5.2.9)} \quad //$$

Example 5.2: Show that $Q_{i\delta} > Q_{i\epsilon} \Rightarrow q_{i\delta} > q_{i\epsilon}$ and $q_{i\cdot\delta} < q_{i\cdot\epsilon}$

Proof: We have $q_{i\delta} = 1-(p_i)^{\frac{Q_{i\delta}}{q_i}}$ from (5.2.14)

Given $Q_{i\delta} > Q_{i\epsilon}$

$$\Rightarrow \quad \frac{Q_{i\delta}}{q_i} > \frac{Q_{i\epsilon}}{q_i} \quad \text{and} \quad 1 - \frac{Q_{i\delta}}{q_i} < 1 - \frac{Q_{i\epsilon}}{q_i}$$

$$\Rightarrow \quad (p_i)^{\frac{Q_{i\delta}}{q_i}} < (p_i)^{\frac{Q_{i\epsilon}}{q_i}} \quad \text{and} \quad p_i^{1-\frac{Q_{i\delta}}{q_i}} > p_i^{1-\frac{Q_{i\epsilon}}{q_i}}$$

$$\Rightarrow \quad 1 - (p_i)^{\frac{Q_{i\delta}}{q_i}} > 1 - (p_i)^{\frac{Q_{i\epsilon}}{q_i}} \quad \text{and} \quad 1 - p_i^{1-\frac{Q_{i\delta}}{q_i}} < 1 - p_i^{1-\frac{Q_{i\epsilon}}{q_i}}$$

$\Rightarrow \quad q_{i\delta} > q_{i\epsilon}$ and $q_{i\cdot\delta} < q_{i\cdot\epsilon}$, from (5.2.14) and (5.2.15). //

5.2.5 Expressions for Crude, Net and Partially Crude Probabilities in an Interval (a,b)

Let $Y_\delta, (\delta = 1, 2, \ldots, k)$ be an individual's length of life under the risk R_δ (assuming that R_δ is the only operating risk factor), $Y_\delta \geq 0$.

Let $P_\delta(x) = P\{Y_\delta \leq x\}$ (5.2.22)

and $p_\delta(x)dx = P\{x \leq Y_\delta \leq x + dx\}$ (5.2.23)

i.e. $P_\delta(x)$ is the c.d.f. and $p_\delta(x)$ is the pdf of Y_δ.

If we denote,

$$Z = \min(Y_1, Y_2, \ldots, Y_k) = \operatorname*{Min}_\delta Y_\delta, \tag{5.2.24}$$

then we may note that Z is observable whereas Y_δ's, $(\delta = 1, 2, \ldots, k)$ are not observable.

Also, $Z > x \Rightarrow Y_\delta > x \; \forall \; \delta$ and we denote

$$1 - F_z(x) = \overline{F}_z(x) = P(Z > x) = P[Y_1 > x, Y_2 > x, \ldots, Y_k > x], \tag{5.2.25}$$

$\overline{F}_z(x)$ is called the survival function of the r.v. Z.

Also, if we define

$h_z(x)dx = P\{$Minimum of k theoretical life times $Y_\delta(\delta = 1, 2, \ldots, k)$ i.e. Z will lie between x to $x + dx/$ no individual dies before $x\}$

$$= \frac{f_z(x)dx}{\overline{F}_z(x)},$$

where $f_z(x) = -\frac{d}{dx}\overline{F}_z(x)$, is the density function of Z.

$$\Rightarrow \quad h_z(x) = \frac{f_z(x)}{\overline{F}_z(x)}, \text{ see (A.5.3)}, \tag{5.2.26}$$

then $h_z(x)$ is known as Force of decrement, Instantaneous force of mortality, Hazard rate or Failure rate.

Let $g_\delta(x)dx, (\delta = 1, 2, \ldots, k)$ denote the probability of failure due to the risk R_δ in an interval (x to $x + dx$) in the presence of all other $(k - 1)$ risks, given that the individual is alive at x, then $g_\delta(x)$ is a measure of crude probability of death.

Assuming that not more than one failure (or death) is possible in $(x, x + dx)$ then

$$h_z(x) = \sum_{\delta=1}^{k} g_\delta(x), \tag{5.2.27}$$

since $h_z(x)dx$ is the probability of failure (or death) from either of the causes R_δ, $(\delta = 1, 2, \ldots, k)$ in $(x, x + dx)$.

Now, the expressions for Crude, Net and Partially crude probabilities of death in an interval (a, b), which are denoted respectively as $Q_\delta(a, b)$, $q_\delta(a, b)$ and $Q_{\substack{\delta.j \\ \delta \neq j}}(ab)$, following David and Moeschberger (1978), are obtained as follows:

1. Crude Probability of Death $Q_\delta(a, b)$

We have,

$$\exp\left\{-\int_a^x h_z(t)dt\right\} = P\ \{\text{Surviving (all the risks) in } (a, x] \text{ having survived up to a}\}$$

$$= \overline{F}(t), \text{ (see (A.5.4),} \tag{5.2.28}$$

where $h_z(t)$ is the hazard rate when all risks are present.

$$\Rightarrow \int_a^b \exp\left\{-\int_a^x h_z(t)dt\right\} g_\delta(x)dx = P\ \{\text{Failure (or death) due to the risk } R_\delta \text{ in the presence of all other risks}\},$$

where $g_\delta(x)dx$ is the probability of failure due to R_δ in the interval $(x, x + dx)$.

$$= Q_\delta(a, b), \text{ by definition.} \tag{5.2.29}$$

Further, from (5.2.28), using (A.1.15) $h_z(t) = -\dfrac{d}{dt}\log \overline{F}_z(t)$

$$\Rightarrow \int_a^x h_z(t)dt = \int_a^x -\frac{d}{dt}\log \overline{F}_z(t)dt = -\log\frac{\overline{F}_z(x)}{\overline{F}_z(a)} \tag{5.2.30}$$

Hence, by putting (5.2.30) in (5.2.29), we get

$$Q_\delta(a, b) = \frac{1}{\overline{F}_z(a)}\int_a^b \overline{F}_z(x) g_\delta(x)dx. \tag{5.2.31}$$

(5.2.28) and (5.2.30) are the required expressions for $Q_\delta(a, b)$.

2. Net Probability of Death $q_\delta(a, b)$

As in (5.2.30), $$\int_a^b h_\delta(t)dt = -\log\frac{\overline{P}_\delta(b)}{\overline{P}_\delta(a)}, \tag{5.2.32}$$

where $\overline{P}_\delta(\cdot) = 1 - P_\delta(\cdot)$ and $P_\delta(\cdot)$ is defined in (5.2.22).

$$\Rightarrow \exp\left\{-\int_a^b h_\delta(t)dt\right\} = \frac{\overline{P}_\delta(b)}{\overline{P}_\delta(a)},\ \delta = 1, 2, \ldots, k \tag{5.2.33}$$

$$= 1 - \frac{P_\delta(b) - P_\delta(a)}{\overline{P}_\delta(a)}$$

$= 1 - P$ [Failure (death) in (a, b) when R_δ is the only risk], from definition of $P_\delta(\cdot)$ in (5.2.22)

$= 1-$ Net probability (Type A) of death in (a, b) due to R_δ

$= 1 - q_\delta(a, b)$

$$\Rightarrow q_\delta(a, b) = 1 - \exp\left\{-\int_a^b h_\delta(t)dt\right\}, \text{ is the required expression.} \tag{5.2.34}$$

3. Partially Crude Probability of Death $Q_{\substack{\delta\cdot\epsilon \\ \delta\neq\epsilon}}(a,b)$

As in 1. above, we have

$$\exp\left\{-\int_a^x h_z^{(-j)}(t)dt\right\} = P \text{ [Surviving all the risks, when } R_j \text{ is absent, in } (a,x) \text{, having survived up to } a],$$

where $h_z^{(-j)}(t)$ is the hazard rate after elimination of R_j.

$$\Rightarrow \int_a^b \left[\exp\left\{-\int_a^x h_z^{(-j)}(t)dt\right\}\right] g_\delta^{(-j)}(x)dx$$

$= P$ [Death (or failure) from R_δ ($R_\delta \neq R_\epsilon$) in (a,b) in presence of all risks, except R_ϵ],

where $g_\delta^{(-j)}(d)dx$ represents the probability of death (or failure) in $(x, x+dx)$ due to R_δ when R_ϵ is off.

$=$ Partially crude probability of death due to R_δ in (a,b) when R_j is off from the population.

$= Q_{\substack{\delta\cdot\epsilon \\ \delta\neq\epsilon}}(a,b)$.

Hence, partially crude probability of death in (a,b) is,

$$Q_{\substack{\delta\cdot\epsilon \\ \delta\neq\epsilon}}(a,b) = \int_a^b \left[\exp\left\{-\int_a^x h_z^{(-j)}(t)dt\right\}\right] g_\delta^{(-j)}(x)dx \qquad (5.2.35)$$

5.3 Estimation of Crude, Net and Partially Crude Probabilities of Death

Here we obtain the MLE of $Q_{i\delta}$, p_i and $Q_{i\delta\cdot\epsilon}$. We also obtain the Minimum Chi Square estimators of $Q_{i\delta}$ and p_i.

5.3.1 Estimation of Crude Probability $Q_{i\delta}$ and Net Probability p_i

Consider the joint distributions of deaths due to various risks $R_\delta(\delta = 1, 2, \ldots, k)$ as well as the survivors in the age group (t_i, t_{i+1}) as follows:

Table 5.1

	Deaths due various Risks							Survivors	Total
	R_1	R_2	$\ldots$	R_δ	$\ldots$	R_k			
Frequency	d_{i1}	d_{i2}	$\ldots$	$d_{i\delta}$	$\ldots$	d_{ik}		l_{i+1}	l_i
Probability	Q_{i1}	Q_{i2}	$\ldots$	$Q_{i\delta}$	$\ldots$	Q_{i_k}		p_i	1

Since in the above set up

$$l_i = l_{i+1} + d_{i1} + d_{i2} + \ldots + d_{ik}, \qquad (5.3.1)$$

where $l_i - l_{i+1} = d_i = \sum_{j=1}^{k} d_{ij}$

$$1 = p_i + Q_{i1} + Q_{i2} + \ldots + Q_{ik} \qquad (5.3.2)$$

A. Maximum Likelihood Estimators for p_i and $Q_{i\delta}$

It follows that the joint distribution of $l_{i+1}, d_{i1}, d_{i2}, \ldots d_{ik}$ given l_i is multinomial, given by

$$L = \frac{l_i!}{l_{i+1}!d_{i1}!d_{i2}!\ldots d_{ik}!}(p_i)^{l_{i+1}}(Q_{i1})^{d_{i1}}(Q_{i2})^{d_{i2}}\ldots(Q_{ik})^{d_{ik}} \tag{5.3.3}$$

$$\Rightarrow \quad \log L = \log C + \sum_{\delta=1}^{k} d_{i\delta}\log Q_{i\delta} + l_{i+1}\log p_i,$$

where C is a constant, independent of the parameters p_i, Q_{ij}, $j=1,\ldots,k$, given by

$$C = \frac{l_i!}{l_{i+1}!d_{i1}!\ldots d_{ik}}.$$

The likelihood equation is

$$\frac{\partial \log L}{\partial Q_{i\delta}} = \frac{d_{i\delta}}{Q_{i\delta}} + \frac{l_{i+1}}{p_i}\frac{\partial p_i}{\partial Q_{i\delta}} = 0$$

$$\Rightarrow \quad \left[\frac{d_{i\delta}}{Q_{i\delta}} - \frac{l_{i+1}}{p_i}\right] = 0,$$

Since $\sum Q_{i\delta} = q_i$, from (5.2.1) $= (1 - p_i)$,

$$\Rightarrow \quad \frac{\partial p_i}{\partial Q_{i\delta}} = -1 \tag{5.3.4}$$

$$\Rightarrow \quad \frac{l_{i+1}}{p_i} = \frac{d_{i\delta}}{Q_{i\delta}} \tag{5.3.5}$$

$$= \frac{\sum d_{i\delta}}{\sum Q_{i\delta}} = \frac{d_i}{q_i}, \text{ from (5.3.1) and (5.3.2)} \tag{5.3.6}$$

$$\Rightarrow \quad \frac{\hat{p}_i}{\hat{q}_i} = \frac{l_{i+1}}{d_i}, \text{ from (5.3.5) and (5.3.6)} \tag{5.3.7}$$

$$\Rightarrow \quad \frac{\hat{p}_i}{1-\hat{p}_i} = \frac{l_{i+1}}{d_i} \Rightarrow \frac{1}{1-\hat{p}_i} = \frac{d_i + l_{i+1}}{d_i}$$

$$\Rightarrow \quad \hat{p}_i = \frac{l_{i+1}}{d_i + l_{i+1}}$$

$$\Rightarrow \quad \hat{p}_i = \frac{l_{i+1}}{l_i}, \text{ from (5.3.1) and (5.3.6)} \tag{5.3.8}$$

$$\Rightarrow \quad \hat{q}_i = 1 - \frac{l_{i+1}}{l_i} = \frac{d_i}{l_i}. \tag{5.3.9}$$

Further, from (5.3.5),

$$\frac{Q_{i\delta}}{d_{i\delta}} = \frac{p_i}{l_{i+1}}$$

$$\Rightarrow \quad \hat{Q}_{i\delta} = \hat{p}_i \cdot \frac{d_{i\delta}}{l_{i+1}} = \frac{l_{i+1}}{l_i} \cdot \frac{d_{i\delta}}{l_{i+1}} \text{ (by using (5.3.5))} = \frac{d_{i\delta}}{l_i}. \tag{5.3.10}$$

(5.3.8) and (5.3.10) give MLEs of p_i and $Q_{i\delta}$, respectively.

Properties of MLEs $\hat{p}_i$ and $\hat{Q}_{i\delta}$

Since $d_{i\delta}, l_{i+1}$ are multinomial variates

$$\therefore \left.\begin{array}{l} E(d_{i\delta}/l_i) = l_iQ_{i\delta}, \text{Var}\ (d_{i\delta}/l_i) = l_iQ_{i\delta}(1 - Q_{i\delta}) \\ \text{Cov}\ (d_{i\delta}, d_{i\epsilon}/l_i) = -l_iQ_{i\delta}Q_{i\epsilon} \\ E(l_{i+1}/l_i) = l_ip_i, \text{Var}\ (l_{i+1}/l_i) = l_ip_i(1 - p_i) \end{array}\right\} \quad (5.3.11)$$

We can check that the Maximum Likelihood estimators are unbiased as,

$$E(\hat{Q}_{i\delta}) = E\left(\frac{d_{i\delta}}{l_i}\right), \text{ (from (5.3.10))} = E\left(\frac{1}{l_i}E(d_{i\delta}|l_i)\right)$$

$$= E\left(\frac{1}{l_i}l_iQ_{i\delta}\right), \text{ (from (5.3.11))} = Q_{i\delta} \quad (5.2.12)$$

$$E(\hat{p}_i) = E\left(\frac{l_{i+1}}{l_i}\right) = E\left(\frac{1}{l_i}E(l_{i+1}|l_i)\right)$$

$$= E\left(\frac{1}{l_i}\cdot l_ip_i\right), \text{ (from (5.3.11))} = p_i \quad (5.2.13)$$

Thus both $\hat{p}$ and $\hat{Q}_{i\delta}$ are unbiased for p_i and $Q_{i\delta}$ respectively.

Further, Var $(\hat{Q}_{i\delta}) = E(\hat{Q}^2_{i\delta}) - [E(\hat{Q}_{i\delta})]^2$

$$\text{Now,}\quad E(\hat{Q}^2_{i\delta}) = E\left(\frac{d^2_{i\delta}}{l^2_i}\right) = E\left(\frac{1}{l_i}E\left(\frac{d^2_{i\delta}}{l_i}|l_i\right)\right)$$

$$= E\left(\frac{1}{l_i}\frac{(l_iQ_{i\delta}(1 - Q_{i\delta}) + l_i^2Q^2_{i\delta})}{l_i}\right), \text{ from (5.3.11)}$$

$$= E\left[\frac{1}{l_i}\{Q_{i\delta}(1 - Q_{i\delta}) + l_iQ^2_{i\delta}\right]$$

$$= E\left(\frac{1}{l_i}\right)Q_{i\delta}(1 - Q_{i\delta}) + Q^2_{i\delta}$$

$$\therefore \quad \text{Var}(\hat{Q}_{i\delta}) = E\left(\frac{1}{l_i}\right)Q_{i\delta}(1 - Q_{i\delta}) + Q^2_{i\delta} - Q^2_{i\delta}$$

$$= E\left(\frac{1}{l_i}\right)Q_{i\delta}(1 - Q_{i\delta}) \quad (5.3.14)$$

$$\text{Cov}\ (\hat{Q}_{i\delta}, \hat{Q}_{i\epsilon}) = E(\hat{Q}_{i\delta}\hat{Q}_{i\epsilon}) - E(\hat{Q}_{i\delta})E(\hat{Q}_{i\epsilon}), (\delta \neq \epsilon)$$

$$= E\left(\frac{d_{i\delta}}{l_i}\cdot\frac{d_{i\epsilon}}{l_i}\right) - Q_{i\delta}Q_{i\epsilon}, \text{ from (5.3.10) and (5.3.12)}$$

$$= E\left(\frac{1}{l_i}E\left(\frac{d_{i\delta}d_{i\epsilon}}{l_i}\bigg|l_i\right)\right) - Q_{i\delta}Q_{i\epsilon}$$

$$= E\left[\frac{1}{l_i}\text{Cov}\left(\frac{d_{i\delta}, d_{i\epsilon}}{l_i}\bigg|l_i\right)\right] + E\left(\frac{d_{i\delta}}{l_i}\bigg|l_i\right)E\left(\frac{d_{i\epsilon}}{l_i}\bigg|l_i\right) - Q_{i\delta}Q_{i\epsilon}$$

$$= E\left(\frac{1}{l_i}\left(-\frac{l_iQ_{i\delta}Q_{i\epsilon}}{l_i}\right)\right) + \left(\frac{l_iQ_{i\delta}}{l_i}\right)\left(\frac{l_iQ_{i\epsilon}}{l_i}\right) - Q_{i\delta}Q_{i\epsilon}$$

$$= -E\left(\frac{1}{l_i}\right) Q_{i\delta}Q_{i\epsilon} + Q_{i\delta}Q_{i\epsilon} - Q_{i\delta}Q_{i\epsilon}$$

$$= -E\left(\frac{1}{l_i}\right) Q_{i\delta}Q_{i\epsilon}, \ (\delta \neq \epsilon). \tag{5.3.15}$$

Finally, we can show

$$\text{Cov}\,(\hat{Q}_{i\delta}, \hat{Q}_{j\epsilon}) = 0 \quad \text{for} \quad i \neq j, \delta\ \epsilon.$$

Since

$$\text{Cov}\,(\hat{Q}_{i\delta}, \hat{Q}_{j\epsilon}) = E(\hat{Q}_{i\delta}\hat{Q}_{j\epsilon}) - E(\hat{Q}_{i\delta})E(\hat{Q}_{j\epsilon})$$

$$= E\left(\frac{d_{i\delta}}{l_i} \cdot \frac{d_{j\epsilon}}{l_j}\right) - Q_{i\delta}Q_{j\epsilon}, \text{ from (5.3.10) and (5.3.12)}$$

$$= E\left(\frac{d_{i\delta}}{l_i}\right) E\left(\frac{d_{j\epsilon}}{l_j}\right) - Q_{i\delta}Q_{j\epsilon}, \tag{5.3.16}$$

$\Big($since $\frac{d_{i\delta}}{l_i}$, $\frac{d_{j\epsilon}}{l_j}$ are independent random variables as they correspond to two multinomial populations, given by

$$\sum_{\delta} d_{i\delta} + l_{i+1} = l_i \quad \text{and} \quad \sum_{\delta} Q_{i\delta} = q_i \Rightarrow \sum_{\delta} Q_{i\delta} + p_i = 1 \tag{5.3.17}$$

$$\text{and} \quad \sum d_{j\epsilon} + l_{j+1} = l_j \quad \text{and} \quad \sum_{\epsilon} Q_{j\epsilon} = q_j \Rightarrow \sum_{\epsilon} Q_{j\epsilon} + p_j = 1.\Big) \tag{5.3.18}$$

(5.3.16) gives

$$\text{Cov}\,(\hat{Q}_{i\delta}, \hat{Q}_{j\epsilon}) = Q_{i\delta}Q_{j\epsilon} - Q_{j\delta}Q_{j\epsilon}$$

$$= 0 \quad \text{for} \ i \neq j, \delta \neq \epsilon, \text{ from (5.3.11)}. \tag{5.3.19}$$

B. Neyman's Modified Minimum χ^2 Estimators of p_i and $Q_{i\delta}$

Neyman's modified χ^2 is given by

$$\chi^2 = \sum_j \left(\frac{O_j - E_j}{O_j}\right)^2, \ j = 1, 2, \ldots, k \text{ and } i+1, \tag{5.3.20}$$

where O_j and E_j represent the observed and the expected frequencies at the jth cell of the multinomial Table 5.1, given by

$$\left.\begin{aligned} &O_j = d_{ij}, E_j = l_iQ_{ij} \quad \text{for} \ \ j = 1, 2, \ldots, k, \ \text{from (5.3.11)} \\ &\text{and for} \quad j = i+1, O_{i+1} = l_{i+1}, \ E_{i+1} = l_ip_i \ \text{from (5.3.11)}. \end{aligned}\right\} \tag{5.3.21}$$

From (5.3.19) and (5.3.20)

$$\chi^2 = \sum_{\delta=1}^{k} \frac{(d_{i\delta} - l_iQ_{i\delta})^2}{d_{i\delta}} + \frac{(l_{i+1} - l_ip_i)^2}{l_{i+1}} \tag{5.3.22}$$

$$\frac{\partial \chi^2}{\partial Q_{i\delta}} = 0 \Rightarrow \frac{2(d_{i\delta} - l_iQ_{i\delta})(-l_i)}{d_{i\delta}} + \frac{2(l_{i+1} - l_ip_i)}{l_{i+1}}(-l_i)\frac{\partial p_i}{\partial Q_{i\delta}} = 0$$

$$\Rightarrow \frac{d_{i\delta} - l_iQ_{i\delta}}{d_{i\delta}} = \frac{l_{i+1} - l_ip_i}{l_{i+1}} \left(\text{since } \frac{\partial p_i}{\partial Q_{i\delta}} = -1, \text{ from (5.3.4)}\right)$$

$$\Rightarrow \ 1 - l_i \cdot \frac{Q_{i\delta}}{d_{i\delta}} = 1 - \frac{l_i}{l_{i+1}}p_i$$

$$\Rightarrow \quad \frac{Q_{i\delta}}{d_{i\delta}} = \frac{p_i}{l_{i+1}}$$

$$\text{or} \quad \widetilde{Q}_{i\delta} = d_{i\delta}\frac{\widetilde{p}_i}{l_{i+1}}, \tag{5.3.23}$$

where $\widetilde{Q}_{i\delta}$ and $\widetilde{p}_i$ are the modified minimum χ^2 estimators of $Q_{i\delta}$ and p_i.

To obtain $\widetilde{p}_i$ we proceed as follows:

From (5.3.21)

$$\frac{\partial \chi^2}{\partial p_i} = \frac{2(d_{i\delta} - l_i Q_{i\delta})(-l_i)}{d_{i\delta}}\frac{\partial Q_{i\delta}}{\partial p_i} + \frac{2(l_{i+1} - l_i p_i)}{l_{i+1}}(-l_i) = 0$$

$$\Rightarrow \quad \frac{d_{i\delta} - l_i Q_{i\delta}}{d_{i\delta}} = \frac{l_{i+1} - l_i p_i}{l_{i+1}} \left(\text{since } \frac{\partial Q_{i\delta}}{\partial p_i} = -1, \text{ from (5.3.4)}\right)$$

$$\Rightarrow \quad \frac{Q_{i\delta}}{d_{i\delta}} = \frac{p_i}{l_{i+1}}$$

$$\Rightarrow \quad \frac{\sum_\delta Q_{i\delta}}{\sum_\delta d_{i\delta}} = \frac{p_i}{l_{i+1}}$$

$$\Rightarrow \quad \frac{q_i}{d_i} = \frac{p_i}{l_{i+1}}. \quad \left(\because \sum_\delta Q_{i\delta} = q_i \text{ and } \sum_\delta d_{i\delta} = d_j, \text{ see (5.3.2) and (5.3.1)}\right)$$

$$\Rightarrow \quad \frac{p_i}{1-p_i} = \frac{l_{i+1}}{d_i}$$

$$\Rightarrow \widetilde{p}_i = \frac{l_{i+1}}{l_{i+1} + d_i} = \frac{l_{i+1}}{l_i}, \text{ from (5.3.1)}, \tag{5.3.24}$$

is the modified χ^2 estimator of $\widetilde{p}_i$ which is same as $\hat{p}_i$, the maximum likelihood estimator, see (5.3.8).

Putting (5.2.24) in (5.2.23)

$$\widetilde{Q}_{i\delta} = \frac{l_{i+1}}{l_i} \cdot \frac{d_{i\delta}}{l_{i+1}} = \frac{d_{i\delta}}{l_i}, \tag{5.3.25}$$

which again makes $\widetilde{Q}_{i\delta}$, the same as $\hat{Q}_{i\delta}$, the maximum likelihood estimator in (5.3.10).

Thus the maximum likelihood and Modified Minimum χ^2 provide one and the same result.

5.3.2 MLE of Partially Crude Probability, $Q_{i\delta\cdot j}$

Here, we obtain MLE of $Q_{i\delta\cdot j}$. We have,

$$Q_{i\delta\cdot j} = \frac{Q_{i\delta}}{q_i - Q_{ij}}\left[1 - (p_i)^{\frac{(q_i - Q_{ij})}{q_i}}\right], \ \forall\ \delta \neq j, \delta = 1, 2, \ldots, k \text{ from (5.2.20)}.$$

By the invariance property of Maximum Likelihood estimator (see Sec. A.2.6) the m.l.e. of $Q_{i\delta\cdot\epsilon}$ is given by

$$\hat{Q}_{i\delta\cdot\epsilon} = \frac{\hat{Q}_{i\delta}}{(\hat{q}_i - \hat{Q}_{i\epsilon})}\left[1 - (\hat{p}_i)^{\frac{(\hat{q}_i - \hat{Q}_{i\delta})}{\hat{q}_i}}\right]$$

$$= \frac{\left(\frac{d_{i\delta}}{l_i}\right)}{\cdot\left(\frac{d_i}{l_i} - \frac{d_{i\epsilon}}{l_i}\right)}\left[1 - \left(\frac{l_{i+1}}{l_i}\right)^{\left(\frac{d_i}{l_i} - \frac{d_{i\delta}}{l_i}\right)\cdot\frac{l_i}{d_i}}\right], \text{ from (5.3.8), (5.3.9) and (5.3.10)}$$

$$= \frac{d_{i\delta}}{d_i - d_{i\epsilon}}\left[1 - \left(\frac{l_{i+1}}{l_i}\right)^{\frac{d_i - d_{i\delta}}{d_i}}\right]. \tag{5.3.26}$$

Similarly the maximum likelihood estimators of $q_{i\delta}$ and $q_{i\cdot\delta}$ may be obtained.

5.4 Independent and Dependent Risks

Let us again consider the k competing risks $R_\delta(\delta = 1, 2, \ldots, k)$ and the corresponding length of life Y_δ (a r.v.) under the risk R_δ. Now, it may happen that the presence of one risk may enhance or reduce the effect of other risk (s) i.e. the risks are dependent. On the other hand, the presence or absence of a risk does not affect the intensity of other risk (s) i.e. the risks are independent. We shall discuss these two situations separately.

5.4.1 Theory of Independent Risks

We assume that the risks to which each individual is exposed act independently.

Further, let

$$X_i = \underset{l}{\text{Min}}\, Y_l = \text{Observed life time of the individual given that the death is due to risk } R_i (i = 1, 2, \ldots, k) \tag{5.4.1}$$

$$\left.\begin{aligned} p_i(x)dx &= P\{x \le Y_i \le x + dx\} \\ P_i(x) &= P(Y_i \le x) \end{aligned}\right\} \tag{5.4.2}$$

and $\pi_i = P\{\text{the death is due to risk } R_i\}$

$$= P\{X_i = \underset{l}{\text{Min}}\, Y_l\}, \text{ where, } \pi_i > 0, \ \sum_{i=1}^{k} \pi_i = 1. \tag{5.4.3}$$

Here we can, without loss of generality, omit the causes for which $\pi_i = 0$ and we can assume that for every risk there is a non zero probability of failure (death).

Then the p.d.f. $f_i(x)$ of X_i, assuming the independence of Y_l's, is given by

$$f_i(x)dx = P\{\text{of death in } (x, x + dx)/R_i\}$$

$$\times P\{\text{surviving after time } x \text{ due to the risks } R_1, R_2, \ldots R_{i-1}, R_{i+1}, \ldots, R_k\}$$

$$= \frac{p_i(x)dx}{\pi i} P\{Y_1 > x, Y_2 > x, \ldots, Y_{i-1} > x, Y_{i+1} > x, \ldots, Y_k > x\},$$

$$\text{from (5.4.2) and (5.4.3)} \tag{5.4.4}$$

$$= \frac{p_i(x)dx}{\pi_i} \prod_{\substack{l=1 \\ l \ne i}}^{k} P\{Y_l > x\}, \text{ since } Y_l\text{'s are independent r.v.'s} \tag{5.4.5}$$

$$\therefore \quad f_i(x) = \frac{p_i(x)}{\pi_i} \prod_{\substack{l=1 \\ l \neq i}}^{k} \overline{P}_l(x), \quad \text{where } \overline{P}_l(\cdot) = 1 - P_l(\cdot), \text{ from (5.4.2).} \quad (5.4.6)$$

$f_l(x)$ is called the death density function, i.e., $f_i(x)dx$ represents the conditional probability of dying in $(x, x+dx)$ due to cause R_i in the presence all other risks in the population, given that death will occur due to risk R_i.

$$\text{Then,} \quad f_i(x) = \frac{p_i(x)}{\pi_i} \frac{\prod_{l=1}^{k} \overline{P}_l(x)}{\overline{P}_l(x)} = \frac{1}{\pi_i} \frac{p_i(x)}{\overline{P}_i(x)} \prod_{l=1}^{k} \overline{P}(x) \quad (5.4.7)$$

$$\text{or} \quad f_i(x) = \frac{1}{\pi_i} h_i(x) \overline{F}_X(x) \quad (5.4.8)$$

$$\text{where} \quad h_i(x) = \frac{p_i(x)}{\overline{P}_i(x)} \quad (5.4.9)$$

$$\text{and} \quad \overline{F}_x(x) = P\{X > x\} \quad (5.4.10)$$

$$= P\{\min Y_l > x\}$$

$$= P\{Y_1 > x \cap Y_2 > x \cap \ldots \cap Y_k > x\} \quad (5.4.11)$$

As defined in (5.4.6) death density function due to the risk R_i i.e. $f_i(x)$ can also be written as

$$f_i(x)dx = P[\text{dying in } (x, x+dx), \text{ cause of death } R_i]$$

$$= \frac{S(x)h_i(x)dx}{\int_0^\infty S(x)h_i(x)dx}, \text{ from (5.4.9)} \quad (5.4.12)$$

where $S(x)$ = probability of surviving at least upto $x = P[X \geq x]$

Now under Chiang's proportionality assumption (5.2.7), we have

$$h_i(x) = C_i h(t) \quad \text{where } h(x) = \sum_{i=1}^{k} h_i(x) \text{ and } C_i \text{ is independent of } x.$$

$$\text{Therefore,} \quad f_i(x)dx = \frac{S(x)C_i h(x)dx}{C_i \int_0^\infty S(x)\lambda(x)dx} = \frac{S(x)h(x)dx}{\int_0^\infty S(x)h(x)dx} \quad (5.4.13)$$

$$\text{Now,} \quad \int_0^\infty f_i(x)dx = 1 \Rightarrow \frac{\int_0^\infty S(x)h(x)dx}{\int_0^\infty S(x)h(x)dx} = 1.$$

$$\text{Note that} \quad \int_0^\infty S(x)h_i(x)dx \neq 1, \text{ (see the example below),} \quad (5.4.14)$$

$$\text{but} \quad \int_0^\infty S(x)h(x)dx = 1. \quad (5.4.15)$$

Example 5.3: Let $\lambda_1, \lambda_2 \ldots, \lambda_k$ be the death intensities corresponding to the risks $R_1, R_2, \ldots, R_k$ respectively, then the probability of dying due to $R_i (i = 1, 2, \ldots, k)$ is $\frac{\lambda_i}{\lambda}$ where, $\lambda = \sum_{i=1}^{k} \lambda_i$.

Solution: The hazard rate corresponding to the risk R_i is λ_i

$\Rightarrow$ survival function $= \exp(-\lambda_i t) = \overline{P}_i(t)$

and density function $= p_i(t) = \lambda_i \exp(-\lambda_i(t))$

$$\therefore \quad f_i(t) = \frac{p_i(t)}{\pi_i} \prod_{\substack{j=1 \\ j \neq i}}^{k} \overline{P}_j(t), \text{ from (5.4.7), where } \pi_i = P \text{ (dying in } R_i)$$

$$= \frac{\lambda_i e^{-\lambda_i t}}{\pi_i} \prod_{\substack{j=1 \\ j \neq i}}^{k} e^{-\lambda j t} = \frac{\lambda_i e^{-\lambda_i t}}{\pi_i} \exp\left(-t \sum_{j=1}^{k} \lambda_j\right) \Big/ e^{-\lambda_i t}$$

$$\Rightarrow \quad f_i(t) = \frac{\lambda_i}{\pi_i} e^{-\lambda t}, \quad \text{where } \lambda = \sum_{j=1}^{k} \lambda j$$

$$\therefore \quad 1 = \int_0^\infty f_i(t)dt = \frac{\lambda_i}{\pi_i} \int_0^\infty e^{-\lambda t} dt = \frac{\lambda_i}{\lambda_i} \cdot \frac{1}{\lambda}$$

$$\Rightarrow \quad \frac{\lambda_i}{\lambda} \frac{1}{\pi_i} = 1 \quad \Rightarrow \quad \frac{\lambda_i}{\lambda} = \pi_i = P[\text{dying in } R_i], \ (i = 1, 2, \ldots, k).$$

5.4.2 Dependent Risks

If the risks an independent, $f_i(x)$ is given by (5.4.7) If, however, the risks are dependent, then we get from (5.4.4)

$$\Rightarrow f_i(x) = \frac{p_i(x)}{\pi} \int_x^\infty \int_x^\infty \cdots$$

$$\int_x^\infty p(y_1, y_2, \ldots, y_{i-1}, y_{i+1}, \ldots, y_k | y_i = x) dy_1 dy_2 \ldots dy_{i-1} dy_{i+1} \ldots dy_k \tag{5.4.16}$$

Bivariate Dependent Risks

In the bivariate case i.e. when $k = 2$, from (5.4.16), we have

$$f_1(x) = \frac{p_1(x)}{\pi_1} \int_x^\infty p(y_2 | Y_1 = x) dy_2, \tag{5.4.17}$$

and

$$f_2(x) = \frac{p_2(x)}{\pi_2} \int_x^\infty p(y_1 | Y_2 = x) dy_1. \tag{5.4.18}$$

Let us assume that $(Y_1, Y_2) \sim$ BVN $(\mu_2, \mu_2 \sigma_1^2, \sigma_2^2, \rho)$, (BVN = Bivariate normal distribution). Then

$$\pi_1 = P[\text{dying in } R_1] = P[Y_2 - Y_1 > 0]$$

Now, $Y_2 - Y_1 \sim N(\mu_2 - \mu_1, \sigma_2^2 + \sigma_1^2 - 2\rho\sigma_1\sigma_2)$, see Biswas (1991).

$$\therefore \quad \pi_1 = \frac{1}{\sqrt{2\pi}\sqrt{\sigma_1^2 + \sigma_2^2 - 2\rho\sigma_1\sigma_2}} \int_0^\infty \exp\left(-\frac{1}{2}\left[\frac{\tau - (\mu_2 - \mu_1)}{\sqrt{\sigma_1^2 + \sigma_2^2 - 2\rho\sigma_1\sigma_2}}\right]^2\right) d\tau$$

$$\text{Put} \quad \frac{\tau - (\mu_2 - \mu_1)}{\sqrt{\sigma_1^2 + \sigma_2^2 - 2\rho\sigma_1\sigma_2}} = \xi \quad \text{and} \quad \eta = \frac{\mu_1 - \mu_2}{\sqrt{\sigma_1^2 + \sigma_2^2 - 2\rho\sigma_1\sigma_2}} \tag{5.4.19}$$

$$\text{Then} \quad \pi_1 \frac{1}{\sqrt{2\pi}} \int_{\frac{\mu_1 - \mu_2}{\sqrt{\sigma_1^2 + \sigma_2^2 - 2\rho\sigma_1\sigma_2}}}^\infty \exp(-\xi^2/2) d\xi = 1 - \Phi(\eta) = \overline{\Phi}(\eta) \tag{5.4.20}$$

$$\text{where} \quad \eta = \frac{\mu_1 - \mu_2}{\sigma_1^2 + \sigma_2^2 - 2\rho\sigma_1\sigma_2}$$

$$\text{Next consider} \quad f_1(x) = \frac{p_1(x)}{\pi} \int_x^\infty p(y_2|Y_1 = x) dy_2 \tag{5.4.21}$$

Since the conditional distribution of Y_2 given $Y_1 = x$ is

$$N\left(\mu_2 + \rho\frac{\sigma_2}{\sigma_1}(x - \mu_1)\sigma_2\sqrt{1 - \rho^2}\right), \text{ see Biswas (1991)},$$

$$\therefore \quad f_1(x) = \frac{e^{-\frac{1}{2}\left(\frac{x - \mu_1}{\sigma_1}\right)^2}}{\pi_1\sigma_1} \frac{1}{\sqrt{2\pi\sigma_2^2(1 - \rho^2)}}$$

$$\int_x^\infty \exp\left[-\frac{1}{2}\left\{\frac{y_2 - \left(\mu_2 + \rho\frac{\sigma_2}{\sigma_1}(x - \mu_1)\right)}{\sigma_2\sqrt{1 - \rho^2}}\right\}^2\right] dy_2 \tag{5.4.22}$$

Case I: Suppose $\rho\sigma_2 \neq \sigma_1$.

From (5.4.22), we have

$$f_1(x) = \frac{1}{\pi_1} \frac{1}{\sigma_1\sqrt{2\pi}} \exp\left(-\frac{1}{2}\left(\frac{x - \mu_1}{\sigma_1}\right)^2\right)$$

$$\frac{1}{\sigma_2\sqrt{1 - \rho^2}\sqrt{2\pi}} \int_x^\infty \exp\left[-\frac{\left\{y_2 - \left(\mu_2 + \rho\frac{\sigma_2}{\sigma_1}(x - \mu_1)\right)\right\}^2}{2\sigma_2^2(1 - \rho^2)}\right] dy_2$$

$$= \frac{\phi\left(\frac{x - \mu_1}{\sigma_1}\right)}{\pi_1\sigma_1} \frac{1}{\sqrt{2\pi}} \int_{\frac{x - \mu_1'}{\sigma_1'}}^\infty \exp\left(-\frac{z^2}{2}\right) dz, \text{ where, } z = \frac{y_2 - \mu_1'}{\sigma_1'},$$

$$\mu_1' = \mu_2 + \rho\frac{\sigma_2}{\sigma_1}(x - \mu_1) \quad \text{and} \quad \sigma_1' = \sigma_2\sqrt{1 - \rho^2}.$$

$$\therefore \quad f_1(x) = \frac{\phi\left(\frac{x - \mu_1}{\sigma_1}\right)}{\pi_1\sigma_1} \overline{\Phi}\left(\frac{x - \mu_1'}{\sigma_1'}\right), \tag{5.4.23}$$

where ϕ and Φ are standardized normal density function and distribution function, respectively.

Case II: Next suppose $\rho\sigma_2 = \sigma_1$, so that $\sigma_2^2 - \sigma_2^2\rho^2 = \sigma_2^2 - \sigma_1^2$, then (5.4.22) gives

$$f_1(x) = \frac{\frac{1}{\sigma_1\sqrt{2\pi}}\exp\left[-\frac{1}{2}\left(\frac{x-\mu_1}{\sigma_1}\right)^2\right]}{\pi_1} \frac{1}{\sqrt{\sigma_2^2-\sigma_1^2}} \int_x^\infty \exp\left(-\frac{(y-(\mu_2+(x-\mu_1)))^2}{\sigma_2^2-\sigma_1^2}\right) dy$$

Putting $\{(y-\mu_2)-(x-\mu_1)\}/(\sigma_1^2-\sigma_2^2)^{1/2} = \xi$, we have

$$f_1(x) = \frac{\phi\left(\frac{x-\mu_1}{\sigma_1}\right)}{\pi_1\sigma_1}\left\{\frac{1}{\sqrt{2\pi}}\int_{\frac{\mu_1-\mu_2}{(\sigma_1^2-\sigma_2^2)^{1/2}}}^\infty \exp(-\xi^2/2)d\xi\right\}. \tag{5.4.24}$$

Again, by putting $\rho\sigma_2 = \sigma_1$ in (5.4.19), we get

$$\pi_1 = \frac{1}{\sqrt{2\pi}}\int_{\frac{\mu_1-\mu_2}{\sqrt{\sigma_1^2-\sigma_2^2}}}^\infty \exp(-\xi^2/2)d\xi. \tag{5.4.25}$$

Hence, from (5.4.24) and (5.4.25), we have

$$f_1(x) = \frac{\phi\left(\frac{x-\mu_1}{\sigma_1}\right)}{\pi_1\sigma_1}\cdot\pi_1 = \frac{\Phi\left(\frac{x-\mu_1}{\sigma_1}\right)}{\sigma_1}. \tag{5.4.26}$$

5.5 Censoring

A distinct feature of the survival data is that the data is often censored in the sense that only partial information is available in respect of the r.v. of interest. The censoring may be adopted to save time or money or it may occur in natural way i.e. it is inherent in the nature of the experiment.

Let us suppose that n units (or individuals) are put under observation. Units are observed till they fail (or die) and their failure times are noted. If $T_{(1)}, T_{(2)}, \ldots, T_{(n)}$ are the life times of the n units, in the order they fail, then $T_{(1)} < T_{(2)} < \ldots < T_{(n)}$. These experiments are called Life Testing Experiments. In Uncensored experiments we have the observation from all the n units but in Censored experiments we do not have observations from all the units, by design or otherwise. Bellow we discuss different types of censoring in Life Testing Experiments.

1. Type I Censoring (or Time Censoring)

Let t_0 be some fixed (pre-assigned) time to stop taking observations. The units, whose life times exceed t_0 is a r.v. In fact, here instead of observing T_1, $T_2, \ldots, T_n$, which may be the life times of n independent units, virtually, we observe $Y_1, T_2, \ldots, Y_n$, where

$$\left.\begin{aligned} Y_i &= T_i \quad \text{if} \quad T_i \le t_0 \\ &= t_0 \quad \text{if} \quad T_i > t_0 \end{aligned}\right\} \tag{5.5.1}$$

We may note that the distribution of Y_i has a non-zero probability at $Y_i = t_0$, given by

$$P[Y_i = t_0] = P[T_i > t_0]. \tag{5.5.2}$$

2. Type II: Censoring (or Failure Censoring)

In Type II censoring the experiment stops as soon as a fixed (preassigned) number of observations, say $r(< n)$ is obtained. Here, we observe only $T_{(1)}, T_{(2)}, \ldots, T_{(r)}$, $(r < n)$, and the time of censoring is a r.v.

Suppose, we are testing the longevities of n bulbs. The ordered observations corresponding to n bulbs can be denoted as $T_{(1)}$, $T_{(2)}$, $\ldots, T_{(n)}$, but if we stop at the failure of the rth bulb in order $(r < n)$ i.e. we are able to observe only $T_{(1)}$, $T_{(2)}, \ldots T_{(r)}$, then it is a Type II censoring. We may require to obtain the best estimate of the mean longevity of the bulbs using the censored data [Feller (1968)]

3. Joint Censoring

This is a combination of 1. and 2. above i.e. the experiment stops at time t_0 (fixed) or at rth (fixed) failure, whichever occurs first.

The above censoring techniques are suitable for controlled experiments but in natural experiments these are not feasible. In Biological or Medical applications in Clinical trails another type of censoring, called Random Censoring, is used.

4. Random Censoring (or Progressive Censoring)

In a clinical trial, patients usually enter in the trial at different times and then each patient is given a therapy; in terminal diseases a life extending therapy.

The problem lies in estimating the mean life time of patients based on progressively censored data which may, usually, occur in the following form:

(i) Loss to follow up: It is due to non-availability of the patient to respond.

(ii) Drop outs: It may be due to some adverse effect of the experiment (treatment) and the patient may prefer to withdraw himself from the experiment

(iii) Termination of the study: The patient may remain alive even at the end of the study.

The random censoring also may be either time censoring or failure censoring.

5.5.1 Analysis of Censored Data

We assume that the distribution of life time (T) is exponential with density

$$\left.\begin{aligned} f(t) &= \lambda e^{-\lambda \cdot t}, \lambda > 0,\ t \geq 0 \\ \text{or}\quad f(t) &= \frac{1}{\mu} e^{-\frac{t}{\mu}}, \mu > 0, t \geq 0 \end{aligned}\right\} \tag{5.5.3}$$

μ is the mean of the distribution.

What follows, we consider different types of censored data from this population.

1. Progressively Censored Type I Data

Let us assume that for the ith patient T_i is the time point at which the observation is censored due to any of the three reasons (i), (ii) or (iii) discussed above, if the patient is not dead before T_i whereas t_i is the uncensored observation i.e. the ith patient died at t_i.

We are to estimate the parameter λ from the progressively censored data (t_i, T_i), $i = 1, 2, \ldots, n$.

The contribution to the likelihood of the sample due to the ith individual is

$$\left.\begin{aligned} f(t_i) &= \lambda e^{-\lambda \cdot t_i}, \quad \text{if} \quad 0 \le t_i \le T_i \\ R(t_i) &= e^{-\lambda \cdot T_i}, \quad \text{if} \quad t_i > T_i \end{aligned}\right\} \tag{5.5.4}$$

Thus the likelihood function of the sample is

$$L(\lambda) = \prod_{i=1}^{n} [\{f(t_i)\}^{\delta_i} \{R(T_i)^{1-\delta_i}\}]$$

$$= \prod_{i=1}^{n} [\{\lambda e^{-\lambda \cdot t_i}\}^{\delta_i} \{e^{-\lambda \cdot T_i}\}^{1-\delta_i}], \text{ from (5.5.4)}, \tag{5.5.5}$$

$$\text{where} \quad \delta_i = \begin{cases} 1, & \text{if} \quad 0 \le t_i \le T_i \text{ i.e. } t_i \text{ is uncensored observation, its probability is } (1 - e^{-\lambda T_i}) \\ 0, & \text{if} \quad t_i > T_i \text{ i.e. } T_i \text{ is censored observation, with prob } e^{-\lambda t_i} \end{cases} \tag{5.5.6}$$

$$(5.5.5) \Rightarrow \log L(\lambda) = \sum_{i=1}^{n} [\delta_i (\log \lambda - \lambda t_i) + (1 - \delta_i)(-\lambda T_i)] \tag{5.5.7}$$

$$\Rightarrow \frac{\partial \log L}{\partial \lambda} = \sum_{i=1}^{n} [\{(\delta_i/\lambda) - \delta_i t_i\} - (1 - \delta_i) T_i] \tag{5.5.8}$$

Then the solution of the likelihood equation $\dfrac{\partial \log L}{\partial \lambda} = 0$ is

$$\hat{\lambda} = \frac{d}{\sum_{i=1}^{n} [\delta_i t_i + (1 - \delta_i) T_i]} \tag{5.5.9}$$

$$\text{where} \quad d = \sum_{i=1}^{n} \delta_i = \text{Number of deaths} \tag{5.5.10}$$

Further, $\dfrac{1}{\lambda} = \mu =$ Mean of the population.

The estimate of μ is given by the invariance property of MLE (see Sec. A.2.6 (a)), as

$$\hat{\mu} = \frac{\sum_{i=1}^{n} [\delta_i t_i + (1 - \delta_i) T_i]}{d} \tag{5.5.11}$$

The sampling variance of $\hat{\lambda}$ is obtained as

$$\text{Var}(\hat{\lambda}) = -1/E\left[\frac{\partial^2 \log L}{\partial \lambda^2}\right], \text{ from (A.2.29)}$$

$$= -1/E_d\left[E_{\hat{\lambda}}\left(\frac{\partial^2 \log L}{\partial \lambda^2}/d\right)\right], \text{ (see (A.2.5))}.$$

$$\text{From (5.5.8),} \quad \frac{\partial^2 \log L}{\partial \lambda^2} = -\frac{1}{\lambda^2} \sum \delta_i = -\frac{d}{\lambda^2} \tag{5.5.12}$$

$$\text{Hence,} \quad -1/E_d\left[\frac{\partial^2 \log L}{\partial \lambda^2}\right] = 1/E_d\left[\frac{d}{\lambda^2}\right] = \frac{\lambda^2}{E[d]}. \tag{5.5.13}$$

Now, from (5.5.6)

$$E(\delta_i) = 1 \cdot (1 - e^{-\lambda T_i}) + 0 \cdot (e^{-\lambda T_i}) = 1 - e^{-\lambda T_i} = Q_i \text{ (say)} \tag{5.5.14}$$

$$\Rightarrow \quad E(d) = \sum_{i=1}^{n} E(\delta_i) = \sum_{i=1}^{n} Q_i. \tag{5.5.15}$$

$$\text{Therefore,} \quad \text{Var}(\hat{\lambda}) = \frac{\lambda^2}{\sum_{i=1}^{n} Q_i} \tag{5.5.16}$$

$$\text{Similarly, it can be shown that} \quad \text{Var}(\hat{\mu}) = \frac{\mu^2}{\sum_{i=1}^{n} Q_i} \tag{5.5.17}$$

2. Type II Censored Data

Here we assume that all the n patient have the same exponential life time distribution and have the same point of entry into the study. The study terminates after the death of the dth patient ($d \leq n$). Here n and d both are fixed. This is the case of Type II censoring. The observations are $t_{(I)} < t_{(2)} < \ldots < t_{(d)}$.

A. Maximum Likelihood Estimator (MLE)

To obtain MLE of λ we proceed as follows:

If we would have taken the complete sample $0 < t_{(1)} < t_{(2)} < \ldots < t_{(n)}$ then the likelihood function of the sample is given by (since $t_{(i)}$'s are i.i.d. r.v.'s),

$$L(t_{(1)}, t_{(2)}, \ldots, t_{(n)};\ \lambda) = n!\lambda^n \exp\left(\lambda \sum_{i=1}^{n} t_{(i)}\right) \tag{5.5.18}$$

Since, in the experiment here the test is terminated at the observation $t_{(d)}$, to get the likelihood function of the censored sample $t_{(1)} < t_{(2)} < \ldots < t_{(d)}$, the expression (5.5.18) is integrated w.r.t. $t_{(d+1)} < t_{(d+2)} < \ldots < t_{(n)}$, which gives

$$L(t_{(1)}, t_{(2)}, \ldots, t_{(d)};\ \lambda) = \frac{n!\lambda^d}{(n-d)!} \exp\left[-\lambda \sum_{i=1}^{d} t_i\right] [\exp(-\lambda t_{(d)})]^{n-d}. \tag{5.5.19}$$

By standard procedure, the MLE of λ is

$$\hat{\lambda} = d \Big/ \left[\sum_{i=1}^{d} t_{(i)} + (n-d)t_{(d)}\right]. \tag{5.5.20}$$

Then, the MLE of $\mu = 1/\lambda$ (by the invariance property of MLE, see Sec. A.2.6(a)) is

$$\hat{\mu} = \left[\sum_{i=1}^{d} t_{(i)} + (n-d)t_{(d)}\right] \Big/ d. \tag{5.5.21}$$

B. Best Linear Unbiased Estimator of μ

We define a linear function of the form

$$U = a_1 t_{(1)} + a_2(t_{(2)} - t_{(1)}) + \ldots + a_d(t_{(d)} - t_{(d-1)}), \tag{5.5.22}$$

where $t_{(1)}, (t_{(2)} - t_{(1)}), \ldots, (t_{(d)} - t_{(d-1)})$ are mutually independent exponential variates, from the memory less property of exponential distribution, Feller (1968).

Now, in $t_{(1)} < t_{(2)} < \ldots < t_{(n)}$, $t_{(i)}$'s $i = 1, 2 \ldots, n$ are independent.

$$\therefore \quad P[t_{(1)} > t] = P[t_{(1)} > t, t_{(2)} > t, \ldots, t_{(n)} > t]$$
$$= (e^{-\lambda \cdot t})^n = e^{-n \cdot \lambda \cdot t} \tag{5.5.23}$$

Similarly, $P[(t_{(2)} - t_{(1)}) > t] = P[(t_{(2)} - t_{(1)}) > t, (t_{(3)} - t_{(1)}) > t,$
$$\ldots, (t_{(d)} - t_{(1)}) > t]$$
$$= (e^{-\lambda \cdot t})^{n-1} = e^{-(n-1) \cdot \lambda \cdot t} \tag{5.5.24}$$

In general, $P[(t_{(k)} - t_{(k-1)} > t)] = \cdot e^{-(n-k+1) \cdot \lambda \cdot t}$ (5.5.25)

Now, (5.5.23) shows that $t_{(1)}$ is exponential with parameter $n\lambda$

i.e. $f(t_{(1)}) = n\lambda e^{-n \cdot \lambda \cdot t_{(1)}}$

$\therefore \quad E[t_{(1)}] = 1/n\lambda \quad \text{and} \quad var[t_{(1)}] = 1/(n\lambda)^2$ (5.5.26)

Similarly, from (5.5.24), $(t_{(2)} - t_{(1)})$ is exponential with mean and variance as

$E[t_{(2)} - t_{(1)}] = 1/\{(n-1)\lambda\}, \text{Var}\ [t_{(2)} - t_{(1)}] = \{(n-1)\lambda\}^2$ (5.5.27)

From (5.5.25), $[t_{(k)} - t_{(k-1)}]$ is exponential with

$E[(t_{(k)} - t_{(k-1)}] = 1/[(n-k+1)\lambda]$ and $\text{Var}[t_{(k)} - t_{(k-1)}] = 1/[(n-k+1)\lambda]^2.$ (5.5.28)

To obtain the minimum value of Var (U) for the choice of a_i's, $i = 1, 2, \ldots, n$ we proceed as follows:

Let us unconditionally minimize (using Lagrange's method of multipliers, see Landau (1960)),

$$\text{Var}\ (U) - \eta[E(U) - (1/\lambda)] = \phi \text{ (say)}$$

$$\Rightarrow \left\{\frac{a_1^2}{[n\lambda]^2} + \frac{a_2^2}{[(n-1)\lambda]^2} + \cdots + \frac{a_k^2}{[(n-k+1)\lambda]^2} + \cdots + \frac{a_d^2}{[(n-d+1)\lambda]^2}\right\}$$
$$-\eta\left\{\frac{a_1}{[n\lambda]} + \frac{a_2}{[(n-1)\lambda]} + \cdots + \frac{a_k}{[(n-k+1)\lambda]} + \cdots + \frac{a_d}{[(n-d+1)\lambda]}\right\} = \phi \tag{5.5.29}$$

from (5.5.22) and (5.5.26) to (5.5.28), where η is the Lagrangian multiplier.

$$\text{Then,} \quad \frac{\partial \phi}{\partial a_k} = 0 \Rightarrow \frac{2a_k}{[(n-k+1)\lambda]^2} - \frac{\eta}{(n-k+1)\lambda} = 0$$
$$\Rightarrow a_k = \eta\lambda(n-k+1)/2, k = 1, 2, \ldots, d. \tag{5.5.30}$$

Further, from (5.5.29)

$$\frac{\partial \phi}{\partial \eta} = 0 \Rightarrow \frac{a_1}{n} + \frac{a_2}{n-1} + \cdots + \frac{a_k}{n-k+1} = 1$$

$$\Rightarrow \quad \frac{\eta\lambda}{2} + \frac{\eta\lambda}{2} + \cdots + \frac{\eta\lambda}{2} = 1, \text{ using (5.5.30)}$$

$$\Rightarrow \quad d\frac{\eta\lambda}{2} = 1 \Rightarrow \eta = \frac{2}{d \cdot \lambda} \tag{5.5.31}$$

$\therefore$ From (5.5.30) and (5.5.31) we get

$$a_k = \frac{n-k+1}{d}, k = 1, 2, \ldots, d. \tag{5.5.32}$$

Now, substituting the values of a_k's from (5.5.32) in (5.5.22) we get

$$U = \frac{n}{d}t_{(1)} + \frac{n-1}{d}(t_{(2)} - t_{(1)}) + \cdots + \frac{n-k+2}{d}(t_{(d-1)} - t_{(d-2)}) + \frac{n-k+1}{d}(t_{(d)} - t_{(d-1)}) \tag{5.5.33}$$

$$= \frac{1}{d}t_{(1)} + \frac{1}{d}t_{(2)} + \cdots + \frac{1}{d}t_{(d-1)} + \frac{n-d+1}{d}t_{(d)}. \tag{5.5.34}$$

Now, from (5.5.33) and (5.5.26) to (5.5.28) we get,

$$E(U) = \frac{n}{d}\left[\frac{1}{n \cdot \lambda}\right] + \frac{n-1}{d}\left[\frac{1}{(n-1)\lambda}\right] + \cdots + \frac{n-d+1}{d}\left[\frac{1}{(n-d+1)\lambda}\right]$$

$$= \frac{1}{d \cdot \lambda} + \frac{1}{d \cdot \lambda} + \cdots + \frac{1}{d \cdot \lambda} = d\frac{1}{d \cdot \lambda} = \frac{1}{\lambda} = \mu. \tag{5.5.35}$$

i.e. U is unbiased for μ and its variance is minimum.

Hence, the BLUE of μ, say $\widetilde{\mu}$, is given by (5.5.33) (by putting $U = \widetilde{\mu}$) as

$$\widetilde{\mu} = \sum_{k=1}^{d} \frac{n-k+1}{d}\{t_{(k)} - t_{(k-1)}\} \tag{5.5.36}$$

or from (5.5.34), $$\widetilde{\mu} = \left[\sum_{k=1}^{d} t_{(k)} + (n-d+1)t_{(d)}\right]/d \tag{5.5.37}$$

Distribution of $\widetilde{\mu}$:

Let $z_k = (n-k+1)[t_{(k)} - t_{(k-1)}]$ (5.5.38)

From (5.5.28)

$$E[z_k] = \frac{1}{\lambda} = \mu, \ k = 1, 2, \ldots, d. \tag{5.5.39}$$

i.e. z_k's are i.i.d. exponential variate with mean μ.

$\therefore \ \dfrac{z_k}{\mu}$ is a gamma variate with scale parameter '1' and degrees of freedom '1' i.e. $\gamma(1)$.

$\therefore$ From the reproductive property of gamma distribution $\sum_{k=1}^{d} \dfrac{z_k}{d}$ is a $\gamma(d)$ variate.

Now, from (5.5.36) and (5.5.38),

$$\widetilde{\mu} = \sum_{k=1}^{d} \frac{z_k}{d} \Rightarrow \frac{d \cdot \widetilde{\mu}}{\mu} = \sum_{k=1}^{d} \frac{z_k}{\mu} \text{ follows a } \gamma(d) \text{ distribution} \tag{5.5.40}$$

Its density function is given by

$$f\left(\frac{d \cdot \widetilde{\mu}}{\mu}\right) = (\Gamma d)^{-1} \exp\left[\frac{d \cdot \widetilde{\mu}}{\mu}\right]\left(\frac{d \cdot \widetilde{\mu}}{\mu}\right)^{d-1} \frac{d}{\mu} \tag{5.5.41}$$

$$f(\widetilde{\mu}) = (\Gamma d)^{-1} \exp\left[\frac{d \cdot \widetilde{\mu}}{\mu}\right]\left(\frac{d}{\mu}\right)^{d} (\widetilde{\mu})^{d-1} \tag{5.5.42}$$

It can be easily seen that

$$E[\widetilde{\mu}] = \mu,\ E[\widetilde{\mu}]^2 = \left(\frac{\mu}{d}\right) d(d+1) \tag{5.5.43}$$

and $$\operatorname{Var}[\widetilde{\mu}] = \frac{\mu^2}{d}. \tag{5.5.44}$$

The Rao–Cramer lower bound for the variance of the unbiased estimator $\widetilde{\mu}$ is given by, (see (A.2.29)),

$$\frac{1}{E\left[\left(\dfrac{\partial \log f}{\partial \widetilde{\mu}}\right)^2\right]}, \text{ where } f \text{ is given by (5.5.42).}$$

Now, $$E\left[\left(\frac{\partial \log f}{\partial \widetilde{\mu}}\right)^2\right] = E\left[\left(\frac{d}{\mu} - \frac{d}{\mu^2}\widetilde{\mu}\right)^2\right] = \frac{d^2}{\mu^2} + \frac{d^2}{\mu^4}E(\widetilde{\mu})^2 - \frac{2d^2}{\mu^3}E(\widetilde{\mu})$$

$$= \frac{d^2}{\mu^2} + \frac{d^2}{\mu^4}\left(\frac{\mu}{d}\right)^2 d(d+1) - \frac{2d^2}{\mu^3}\mu, \text{ from (5.5.43)}$$

$$= \frac{d}{\mu^2}$$

$$\therefore \frac{1}{E\left[\left(\dfrac{\partial \log f}{\partial \widetilde{\mu}}\right)^2\right]} = \frac{\mu^2}{d} = \operatorname{Var}(\widetilde{\mu}), \text{ from (5.5.44)} \tag{5.5.45}$$

Thus $\widetilde{\mu}$ is a Minimum Variance Bound Estimator (MVBE).

5.6 Parametric Survival Models

In Sec. 5.5 we have considered the analysis of Exponential Models. There are many other distributions which are also used as survival models. All these models are called Parametric Survival Models. The characteristics of a few important such models are given in this section.

Let T (a r.v.) represent the Life Time (or Time-to failure) of an individual (or device). We denote (see Sec. A.5)

$$\left.\begin{aligned} f(t) &= \text{Density function, } F(t) = \text{c.d.f, } R(t) = 1 - F(t) \\ &= \text{Survival function,} \\ h(t) &= f(t)/R(t) = \text{Hazard function (or Failure rate), corresponding to } T \end{aligned}\right\} \tag{5.6.1}$$

Further, $R(t) = \exp\left[-\int_0^t h(u)du\right]$.

1. Exponential Model (or Distribution)

For this model we have

$$\left.\begin{aligned} f(t) &= \lambda e^{-\lambda\cdot t},\ F(t) = 1 - e^{-\lambda\cdot t},\ R(t) = e^{-\lambda\cdot t} \\ h(t) &= E[T] = 1/\lambda,\ \operatorname{Var}[T] = 1/(\lambda)^2 \end{aligned}\right\} \tag{5.6.2}$$

2. Gamma Model

Sum of i.i.d. exponential variables follows a gamma distribution. Here we have

$$f(t) = \frac{\lambda^n}{\Gamma(n)} t^{n-1} e^{-\lambda \cdot t}, \ \lambda > 0, n > 0 \tag{5.6.3}$$

Other characteristics of this model e.g. $R(t)$, $h(t)$ etc. do not come in closed form and can be obtained from the relations given in (5.6.1)

3. Weibull Model

If instead of T, T^α is Exponentially distributed, then the distribution of T is Weibull. Here,

$$f(t) = \lambda \alpha t^{\alpha-1} \exp[-\lambda t^\alpha], \lambda > 0. \ F(t) = 1 - \exp[-\lambda t^\alpha] \tag{5.6.4}$$

$$R(t) = \exp[-\lambda t^\alpha] \tag{5.6.5}$$

$$h(t) = \lambda \alpha (\lambda t)^{\alpha-1} \tag{5.6.6}$$

4. Rayleigh Model

If the hazard function is of the form

$$h(t) = \lambda_0 + \lambda_1 t, \lambda_0 > 0, \lambda_1 > 0, t \geq 0 \tag{5.6.7}$$

then corresponding distribution is Rayleigh. Here

$$\left.\begin{aligned} f(t) &= (\lambda_0 + \lambda_1 t) \exp[-(\lambda_0 t + \lambda_1 t^2)], \\ \text{and} \quad R(t) &= \exp[-(\lambda_0 t + \lambda_1 t^2)] \end{aligned}\right\} \tag{5.6.8}$$

A generalized Rayleigh model is given by the hazard function,

$$h(t) = \sum_{i=0}^{k} \lambda_i t^i \tag{5.6.9}$$

5. Log-Normal Model

If $\log T$ is $N(\mu, \sigma^2)$ then the distribution of T is log-normal.

Here, we have,

$$f(t) = \frac{1}{t\sigma\sqrt{2\pi}} \exp\left[-\frac{(\log t - \mu)^2}{2\sigma^2}\right], \sigma > 0, -\infty \leq \mu \leq \infty, t \geq 0 \tag{5.6.10}$$

$$R(t) = 1 - \Phi\left(\frac{\log t - \mu}{\sigma}\right). \tag{5.6.11}$$

where $\Phi(\cdot)$ is the c.d.f. of $N(0, 1)$.

Other characteristics can be written similarly but they are not in closed form.

5.7 Non-Parametric Survival Models

The Classical methods of estimating the survival probability $S(t)$ at time t are based on Actuarial approaches depending on Life tables. Below we represent an outline of the same.

Let us divide the interval $(0, t)$ into k half open sub-intervals $I_j = (\tau_{j-1}, \tau_j]$, $j = 1, 2, \ldots, k$.

These sub-intervals are, mostly, of equal length, generally of one year duration.

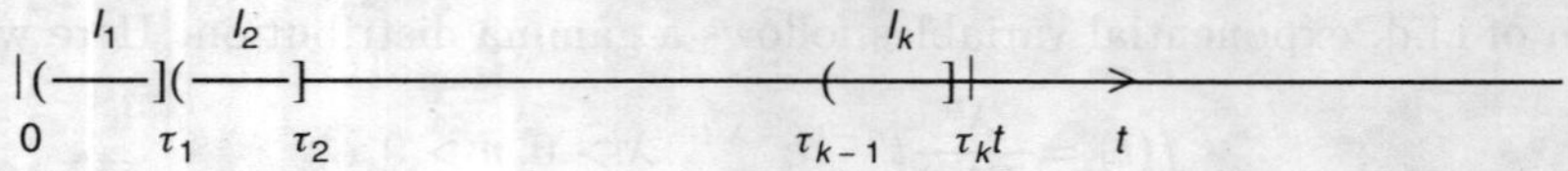

Let us denote,

$$\left.\begin{aligned} n_j &= \text{Number of survivors at the beginning of } I_j \\ d_j &= \text{Number of persons died during } I_j \\ l_j &= \text{Number of persons lost to follow up during } I_j \\ w_j &= \text{Number of persons withdrew during } I_j \end{aligned}\right\}. \tag{5.7.1}$$

In what follows, we discuss here some Non-parametric methods of estimating the survival probability.

5.7.1 Reduced Sample Method

Reduced sample is the old name for Censored sample. Here, the sample (n) consists of only those persons which are exposed to the risk of death during the entire length of study $(0, \tau_k]$ i.e.

$$n = n_1 - \sum_{j=1}^{k} l_j - \sum_{j=1}^{k} w_j. \tag{5.7.2}$$

Let $d =$ Total number of persons died in $(0, \tau_k] = \sum_{j=1}^{k} d_j$. (5.7.3)

Then, an estimate of survival probability is

$$\hat{S}(\tau_k) = \hat{P}[T > \tau_k] = \left(1 - \frac{d}{n}\right). \tag{5.7.4}$$

As the number of persons lost to follow up and number of persons withdrawn are excluded from the sample, hence this is called reduced sample method.

5.7.2 Acturial Method

With the same setup as above, we have

$$S(\tau_k) = P[T > \tau_k] = P[T > \tau_k | T > \tau_{k-1}] P[T > \tau_{k-1}]. \tag{5.7.5}$$

Using (5.7.5) again and again, finally, we get

$$S(\tau_k) = P[T > \tau_1] P[T > \tau_2 | T > \tau_1] \ldots P[T > \tau_k / T > \tau_{k-1}], \tag{5.7.6}$$

$$\tau_0 = 0 \quad \text{and} \quad P(T > 0) = 1$$

Let, $P[T > \tau_j | T > \tau_{j-1}] = p_{j,j-1} = p(j)$, (say). (5.7.7)

Then, $S(\tau_k) = \prod_{j=1}^{k} p(j)$ (5.7.8)

If l_j and w_j are zero then estimate of $p(j)$ will be given by

$$\hat{p}(j) = \left(1 - \frac{d_j}{n_j}\right). \tag{5.7.9}$$

Then, $\hat{S}(\tau_k) = \hat{P}[T > \tau_k] = \prod_{j=1}^{k} \hat{p}(j) = \prod_{j=1}^{k} \left(1 - \frac{d_j}{n_j}\right)$ (5.7.10)

But if l_j and w_j are non-zero, we assume that these persons were at the risk of dying for half of the sub-interval I_j. Thus, the sample size may be taken as

$$n'_j = n_j - \frac{1}{2}(l_j + w_j) \tag{5.7.11}$$

$$\left.\begin{aligned} &\text{Then,} \quad \hat{p}(j) = \left(1 - \frac{d_j}{n'_j}\right), \text{ and the estimate of } S(t) \text{ is,} \\ &\hat{S}'(\tau_k) = \prod_{j=1}^{k}\left(1 - \frac{d_j}{n'_j}\right) \end{aligned}\right\}. \tag{5.7.12}$$

This is called a Product Estimator. It is Actuarial Estimate of the Survival function from life table. It gives the estimate only at the end of the interval.

The Variance of the Actuarial Estimator $\hat{S}(\tau_k)$

From (5.7.10)

$$\log \hat{S}(\tau_k) = \sum_{j=1}^{k} \log \hat{p}(j) \tag{5.7.13}$$

Using Delta method for obtaining large sample variance of an estimator (see (A.2.21)), we get

$$\begin{aligned} \text{Var}[\log \hat{p}(j)] &\cong \text{var}[\hat{p}(j)] \left\{\frac{d \log \hat{p}(j)}{d\hat{p}(j)}\right\}^2_{\hat{p}(j)=p(j)} \\ &= \frac{p(j)[1-p(j)]}{n_j}\left[\frac{1}{p(j)}\right]^2 = \frac{q(j)}{n_j p(j)}, \quad \text{where } q(j) = 1 - p(j) \\ &= \frac{d_j}{n_j(n_j - d_j)}, \text{ from (5.7.9).} \end{aligned} \tag{5.7.14}$$

Assuming that all $\hat{p}(j)$ and hence $\log \hat{p}(j)$ are independent,

$$\text{Var}\,[\log \hat{S}(\tau_k)] \doteq \sum_{j=1}^{k} \frac{q(j)}{n_j p(j)} \tag{5.7.15}$$

To get Var $[\hat{S}(\tau_k)]$ we gain use (A.2.21) to get

$$\begin{aligned} \text{Var}\,[\hat{S}(\tau_k)] &\cong \text{Var}[\log \hat{S}(\tau_k)]\left[\frac{\partial \hat{S}(\tau_k)}{\partial \log \hat{S}(\tau_k)}\right]^2 = \text{Var}[\log \hat{S}(\tau_k)]\left[\frac{\partial \log \hat{S}(\tau_k)}{\partial \hat{S}(\tau_k)}\right]^{-2} \\ &= \left[\sum_{j=1}^{k} \frac{q(j)}{n_j p(j)}\right][\hat{S}(\tau_k)]^2, \text{ from (5.7.15).} \end{aligned} \tag{5.7.16}$$

$$\text{Then,} \quad \text{Var}[\hat{S}(\tau_k)] \cong [\hat{S}(\tau_k)]^2 \sum_{j=1}^{k} \frac{d_j}{n_j(n_j - d_j)}, \text{ from (5.7.14).} \tag{5.7.17}$$

Remarks 5.1: If l_j and w_j are non-zero then (5.7.12) will be the estimate of $S(t)$ and as such variance of the estimate will be given by

$$\text{Var}[\hat{S}'(\tau_k)] \cong [\hat{S}'(\tau_k)]^2 \sum_{j=1}^{k} \frac{d_j}{n'_j(n'_j - d_j)} \tag{5.7.18}$$

5.7.3 Product Limit Estimator

If the width of (equal) intervals is reduced, the number of deaths in each interval will change and as such the estimator (5.7.10) or (5.7.12) may change. If we go on reducing the width of the intervals till in each interval there is either one death or no death, then we have reached a limit, because if we further reduce the width there will be no change in the product estimator of survival probability. This estimator is a **product limit estimator**. Kaplan-Meier estimator is a product limit estimator which we discuss next.

1. Kaplan-Meier Estimator of the Survival Probability (With Random Time Censoring)

Kaplan and Meier (1958) evolved a generalization of the Actuarial estimator in the sense that the intervals (I_i) unlike the Actuarial estimator need not be equal. The estimator is known as the product limit (PL) estimator.

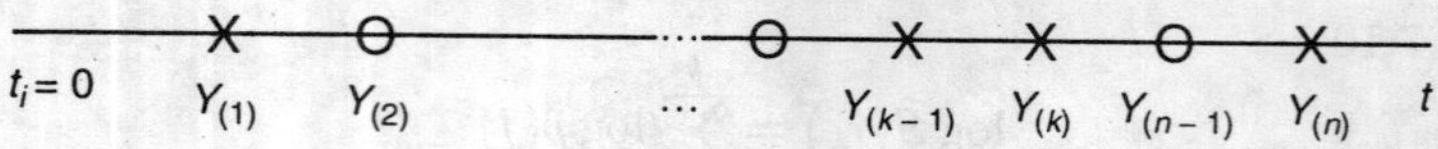

Fig. 5.2

We denote by $O \equiv$ Censored observation and by $X \equiv$ Uncensored observation.

Let $C_1, C_2, \ldots, C_n$ be the random censoring times of n patients having a common c.d.f $G(\cdot)$; C_i's are thus i.i.d.r.v's. We denote by $T_1, T_2, \ldots, T_n$ as the actual life times. Define a r.v. $Y_i (i = 1, 2, \ldots, n)$ such that

$$\begin{aligned} Y_i &= T_i \quad \text{if} \quad C_i \geq T_i \\ &= C_i \quad \text{if} \quad C_i < T_i \end{aligned}$$

and let $Y_{(1)} < Y_{(2)} < \ldots < Y_{(n)}$, be n ordered observations.

Note that $Y_i = \min(T_i, C_i) = T_i \Lambda C_i$ and

$$\begin{aligned} \delta_i = I(T_i \leq C_i) &= 1, \quad \text{if} \quad T_i \leq C_i, \\ &= 0, \quad \text{if} \quad T_i > C_i. \end{aligned}$$

Next we consider the pairs

$$(Y_{(1)}, \delta_{(2)}), (Y_{(2)}, \delta_{(2)}), \ldots, (Y_{(n)}, \delta_{(n)}),$$

where $\delta_{(i)}$ is the value of δ_i corresponding to $Y_{(i)}$.

Note that $\delta_{(i)}$'s are not ordered but placed in the above from by the abuse of notation.

Assuming no ties, we have the Kaplan Meier estimator of the Survival probability given by

$$\hat{R}(t) = \hat{P}(T > t) = \prod_{Y_{(i)} < t} \hat{p}_i,$$

where $p_i = P[T > Y_{(i)} | T > Y_{(i-1)}]$ where T is the longevity of an individual.

$$\text{or} \qquad \hat{R}(t) = \prod_{u: Y_{(i)} \leq t} \left(1 - \frac{1}{n_i}\right), \tag{5.7.19}$$

where the product runs over all uncensored observations.

$$\Rightarrow \qquad \hat{R}(t) = \prod_{Y_{(i)} \le t} \left(1 - \frac{1}{n_i}\right)^{\delta_{(i)}}, \tag{5.7.20}$$

where n_i = number of persons in the risk set.

The risk set is denoted by $N(y_{(i)})$ = number of persons alive upto time $y_{(i)-0}$.

Also $\quad n_1 = n, n_2 = n - 1, \ldots, n_i = n - i + 1.$

$$\text{Therefore, } \hat{R}(t) = \prod_{Y_{(i)} \le t} \left(\frac{n-i}{n-i+1}\right)^{\delta_{(i)}}, \tag{5.7.21}$$

is the Kaplan Meier estiamtor of the survival probability upto time $T = t$, based on progressively censored sample.

2. Variance of the Kaplan Meier Estimator

Using the procedure of Delta Method (see (A.2.21)) we cam obtain the variance of Kaplane-Meier estimator as follows:

From (5.7.19),

$$\hat{R}(t) = P(T > t) = \prod_{Y_{(i)} \le t} \hat{p}_i = \prod_{u:Y_{(i)} \le t} \left(1 - \frac{1}{n_i}\right) \tag{5.7.22}$$

$$\Rightarrow \quad \log \hat{R}(t) = \sum_{u:Y_{(i)} \le t} \log \hat{p}_i.$$

$$\therefore \quad \frac{1}{\hat{R}(t)} \frac{\partial \hat{R}(t)}{\partial p_i} = \frac{1}{\hat{p}_i}$$

$$\Rightarrow \quad \frac{\partial \hat{R}(t)}{\partial \hat{p}_i} = \hat{R}(t) \cdot \frac{1}{\hat{p}_i} \tag{5.7.23}$$

$$\text{Var}(\hat{R}(t)) = \sum_{u:Y_{(i)} \le t} \left(\frac{\partial \hat{R}(t)}{\partial \hat{p}_i}\right)^2_{\hat{p}_i = p_i} \text{Var}(\hat{p}_i), \text{ (vide (A.2.21))} \tag{5.7.24}$$

$$= \sum_{u:Y_{(i)} \le t} \left(\frac{\hat{R}(t)}{\hat{p}_i}\right)^2_{\hat{p}_i = p_i} \frac{p_i q_i}{n_i}$$

$$= \sum_{u:Y_{(i)} \le t} \frac{(R(t))^2}{p_i} \frac{q_i}{n_i} \tag{5.7.25}$$

$$\therefore \quad \text{Var }(\hat{R}(t)) = (\hat{R}(t))^2 \sum_{u:Y_{(i)} \le t} \frac{\left(\frac{1}{n-i+1}\right)}{\left(1 - \frac{1}{n-i+1}\right)} \frac{1}{(n-i+1)},$$

from (5.7.21), for $\delta_{(i)}$'s = 1 only. (5.7.26)

$$= (\hat{R}(t))^2 \sum_{u:Y_{(i)} \le t} \frac{1}{(n-i)} \frac{1}{(n-i+1)}$$

$$\text{Var }(\hat{R}(t)) = (\hat{R}(t))^2 \sum_{Y_{(i)} \le t} \frac{\delta_{(i)}}{(n-i)(n-i+1)},$$

considering all observations. (5.7.27)

For Kaplan-Meier Estimator with tied data and its standard error, the reader is referred to Miller (1981).

5.8 Self Consistency

An estimator of Survival function is said to be self consistent if the proportion estimated to survive beyond t is equal to the proportion of subjects observed to survive past t; plus the sum for all individuals censored before t of the estimated conditional probability of surviving past 't' given survival to be the censoring time (Leurgans et. al. (1982)).

That is, an estimator $R(t)$ is daid to be self consistent (denoted by $\hat{S}_c(t)$) if

$$\hat{S}_c(t) = \frac{1}{n}\left[\sum_{i=1}^{n} 1 \cdot I(T_i > t) + \sum_{i=1}^{n} 0 \cdot I(Y_i \le t, \delta_i = 1) + \sum_{i=1}^{n} \frac{\hat{S}_c(t)}{\hat{S}_c Y_{(i)}} I(Y_i \le t, \delta_i = 0)\right] \tag{5.8.1}$$

Efron (1967) has shown that a self consistent estimator, as defined above, gives rise to Kaplan-Meier estimator. Hence method of self consistency gives an easy way to construct Kaplan-Meier estimator.

5.8.1 'Self Consistency' in Bivariate Survival Estimator

Munoz (1980a, 1980b) has extended the definition of 'self-consistency' for a bivariate survival function estimator for a randomly censored sample as follows:

$$\begin{aligned} R(t_1, t_2) &= P(T_1 > t_1, T_2 > t_2) \\ &= \frac{1}{n}\Bigg[\sum_{i=1}^{n} I(T_{1i} > t_1, T_{2i} > t_2) + \sum_{\delta_i = (1,0)} \frac{\hat{R}(t_1 - 0, t_2) - \hat{R}(t_1, t_2)}{\hat{R}(t_1 - 0, y_{2i}) - \hat{R}(t_1, y_{2i})} \quad \text{if } t_1 \le y_{1i} = y_{2i} < t_2 \\ &\quad + \sum_{\delta_i = (1,0)} \frac{\hat{R}(t_1 t_2 - 0) - \hat{R}(t_1 t_2)}{\hat{R}(y_{1i}, t_2 - 0) - \hat{R}(y_{1i,t_2})} + \sum_{\delta_i = (1,0)} \frac{\hat{R}(t_1, t_2)}{\hat{R}(y_{1i,} y_{2i})}\Bigg], \\ &\quad t_2 < y_{2i} = y_{1i} < t_1 \qquad \text{if } y_{1i} < t_1, y_{2i} > t_2, \text{ and } y_{1i} = y_{2i}. \end{aligned} \tag{5.8.2}$$

where $y_{1i} = \min(T_{1i}, C_i), y_{2i} = \min(T_{2i}, C_i), C_i$ being the random censoring times; T_{1i} and T_{2i} being the life times of first and second components respectively and $\delta_i = (\delta_{1i}, \delta_{2i}), (i = 1, 2, \ldots, n)$ where

$$\begin{aligned} \delta_{1i} &= 1, \quad \text{if } T_{1i} \Lambda C_i = T_{1i}, \\ &= 0, \quad \text{if } T_{1i} \Lambda C_i = C_i. \end{aligned}$$

Similarly for δ_{2i} also.

It may be noted that the second summand on the right hand side of (5.8.2) estimates

$$P(T_{1i} > t_1, T_{2i} > t_2 | T_{1i} = t_1, T_{2i} > y_{2i}),$$

the third summand estimates

$$P(T_{1i} > t_1, T_{2i} > t_2 | T_{1i} = y_{1i}, T_{2i} = t_2)$$

and the last one estimates

$$P(T_{1i} > t_1, T_{2i} > t_2 | T_{1i} = y_{1i}, T_{2i} > y_{2i}).$$

It follows that, an estimator in order to be self consistent has to assign positive probability to 'lines' associated with 'one component censored observations' and to 'regions' associated with 'two component censored observations'.

There is also another way of looking at the self consistency which may be given as follows:

Given an arbitrary survival function estimator say $\hat{R}^{(0)}$, one can use (5.8.2) to obtain a new estimator $\hat{R}^{(1)}$ from the left hand side of (5.8.2) as a function of $\hat{R}^{(0)}$. One can continue with this iteration scheme and stop at the nth stage when

$$\hat{R}^{(m-1)} = \hat{R}^{(m)}$$

and $\hat{R}^{(m)}$ becomes a self consistent estimator. Again a bivariate self consistent survival function estimator is a Bivariate Kaplan-Meier estimator (Efron (1967)) and this gives us a simple way of constructing Bivariate Kaplan-Meier Estimator.

5.8.2 Generalised Maximum Likelihood Estimator (G.M.L.E)

Kaplan-Meier estimator apart from having the property of 'self consistency' is a Generalised Maximum Likelihood estimator (G.M.L.E) which reduces to 'Ordinary Maximum likelihood estimator' if the family of the probability measures corresponding to the joint distribution of failure times is dominated. However, in general the family has no 'dominated' character, therefore, we define a G.M.L.E. due to Keifer and Wolfwitz (1956) as follows:

Let $\underset{\sim}{X}$ be the observation vector and P be the class of probability measures. Let $\hat{P}$ be the estimate of P. Suppose we define

$$f(x, \hat{P}, P) = \frac{d\hat{P}(x)}{d(\hat{P} + P)} \tag{5.8.3}$$

as Radon Nikodym derivative.

$$\Rightarrow \quad f(x, P, \hat{P}) = \frac{dP(x)}{d(\hat{P} + P)}, \quad \text{if} \quad \hat{P}(x) > P(x)$$

$$\Rightarrow \quad f(x, \hat{P}, P) > f(x, P, \hat{P}). \tag{5.8.4}$$

Then $\hat{P}$ is called G.M.L.E. of P.

Johansen (1978) noted that if the family is dominated, the definition of G.M.L.E. reduces to a classical one and what is more relevant in the present situation is that if $\hat{P}$ gives positive probability to the data then

$$\hat{P}_{(i)}(x) > P(x) \forall\ P \in \mathbb{P} \text{ such that } P(x) > 0.$$

In the next place, we prove a specific result viz. G.M.L.E. is the Kaplan Meier Estimator.

Theorem 5.8.1: A G.M.L.E. is a Kaplan-Meier Estimator.

Proof: We prove the result for univariate survival function only.

Let Y_i's and δ_i's are defined as earlier and $Y_{(1)} < T_{(2)} \cdots < Y_{(n)}$ are ordered observations and $\delta_{(i)}$'s correspond to $Y_{(i)}$'s.

Then define

$$L = P[(y_{(1)}, \delta_{(1)}), (y_{(2)}, \delta_{(2)}), \ldots, (y_{(n)}, \delta_{(n)})]$$

$$= \prod_{i=1}^{n} P(T_i = y_{(i)})^{\delta_{(i)}} P(T_i > y_{(i)})^{1-\delta_{(i)}}$$

Let $\quad P(T_i = y_{(i)}) = p_i \quad \text{if} \quad \delta_{(i)} = 1$

and $\quad P(y_{(\hat{j})} < T_i < y_{(j+1)}) = p_j \quad \text{if} \quad \delta_{(i)} = 0, j = i, i+1, \ldots, n.$

Then, $\displaystyle L = \prod_{i=1}^{n} \left\{ pi^{\delta_{(i)}} \left(\sum_{j=i}^{n} p_j \right)^{1-\delta_{(i)}} \right\}$

$$= \prod_{i=1}^{n} \left(\lambda_i \sum_{j=i}^{n} p_j \right)^{\delta_{(i)}} \left((1 - \lambda_i) \sum_{j=i}^{n} p_j) \right)^{1-\delta_{(i)}}$$

Putting $\displaystyle \lambda_i = \frac{p_i}{\sum_{j=i}^{n} p_j} \Rightarrow 1 - \lambda_i = \left\{ \sum_{j=i+1}^{n} p_j \Big/ \sum_{j=i}^{n} p_j \right\} \quad \text{and} \quad \lambda_n = 1.$

$$\Rightarrow \quad (1 - \lambda_i) \sum_{j=i}^{n} p_j = \sum_{j=i+1}^{n} p_j, \because \lambda = \frac{p_n}{\sum_{j=i}^{n} p_j} = \frac{p_n}{p_n} = 1$$

Then we have $\displaystyle L = \prod_{i=1}^{n} \lambda_i^{\delta_{(i)}} (1 - \lambda_i)^{1-\delta_{(i)}}$

$$\because \prod_{j=1}^{i-1} (1 - \lambda_i) = \left[\frac{\sum_{j=2}^{n} p_j \sum_{j=3}^{n} p_j \cdots \sum_{j=i}^{n} p_j}{\sum_{j=1}^{n} p_j \sum_{j=2}^{n} p_j \cdots \sum_{j=i-1}^{n} p_j} \right] = \sum_{j=i}^{n} p_j \Big/ \sum_{j=1}^{n} p_j$$

$$= \sum_{j=i}^{n} p_j, \quad \because \sum_{j=1}^{n} p_j = 1$$

and therefore

$$L = \left(\prod_{j=1}^{n} \lambda_i \sum_{j=i}^{n} p_j \right)^{\delta_{(i)}} \left(\sum_{j=i}^{n} p_j \right)^{1-\delta_{(i)}}$$

$$= \prod_{i=1}^{n} \lambda_i^{\delta_{(i)}} \left(\sum_{j=1}^{n} p_j \right)^{\delta_{(i)}} \left(\prod_{j=1}^{i-1} (1 - \lambda_j) \right)^{-\delta_{(i)}} \left(\prod_{j=1}^{i-1} (1 - \lambda_j) \right),$$

$$\because \sum_{j=i}^{n} p_j = \prod_{j=1}^{i-1} (1 - \lambda_j).$$

$$= \prod_{i=1}^{n} \lambda_i^{\delta(i)} \left\{ \prod_{j=1}^{i-1} (1-\lambda_j)^{\delta(i)} \prod_{j=1}^{i-1} (1-\lambda_j)^{-\delta(i)} \right\} \prod_{j=1}^{i-1} (1-\lambda_j)$$

$$= \prod_{i=1}^{n} (\lambda_i)^{\delta(i)} \prod_{j=1}^{i-1} (1-\lambda_j) \qquad (5.8.5)$$

$$= (\lambda_1^{\delta(1)} \cdot 1)(\lambda_2^{\delta(2)}(1-\lambda_1)) \ldots (\lambda_n^{\delta(n)}(1-\lambda_1)\ldots(1-\lambda_{n-1}))$$

$$= \lambda_1^{\delta(1)} \lambda_2^{\delta(2)} \ldots \lambda_n^{\delta(n)} (1-\lambda_1)^{n-1}(1-\lambda_2)^{n-2} \ldots (1-\lambda_{n-2})^2 (1-\lambda_{n-1})$$

$$= \prod_{i=1}^{n} \lambda_i^{\delta(i)} (1-\lambda_i)^{n-i}$$

$$\Rightarrow \quad \log L = \sum_{i=1}^{n} \delta_{(i)} \log \lambda_i + \sum_{i=1}^{n} (n-i) \log(1-\lambda_i)$$

Now setting

$$\frac{\partial \log L}{\partial \lambda_i} = 0 \Rightarrow \frac{\partial \log L}{\delta \lambda_i} = \sum_{i=1}^{n} \frac{\delta_{(i)}}{\lambda_i} \sum_{i=1}^{n} (n-i) \frac{1}{\lambda_i}$$

$$\Rightarrow \hat{\lambda}_i = \frac{\delta_{(i)}}{\delta_{(i)} + (n-i)} = \frac{\delta_{(i)}}{n-i+1},$$

which corresponds to the maximization of each product in (5.8.5)

It gives $\hat{p}_i = \dfrac{\delta_{(i)}}{\delta_{(i)} + (n-i)} \left\{ \prod_{j=1}^{i-1} (1-\lambda_j) \right\} = \dfrac{\delta_{(i)}}{n-i+1} \prod_{j=1}^{i-1} \left(1 - \dfrac{\delta_{(j)}}{n-j+1} \right),$

which corresponds to the self consistent estimator in the univariate case. //

References

1. Biswas S. (1991): Topics in Statistical Methodology-Wiley Eastern limited.
2. Breslow, N. and Crowley, J. (1974): A large sample study of the life table and product limit estimates under random censorship, Annals of statistics, Vol. 2, pp. 437–453.
3. Chiang, C.L. (1968): Introduction to Stochastic Processes in Biostatistics, John Wiley, New York.
4. Cox, D.R. (1972): Regression models and life tables, Journal of the Royal Statistics Society, series B, Vol. 34, pp. 187–220.
5. Cox, D.R. and Isham, V. (1980): Point Processes Monograph on Statistics and Applied Probability, Chapman and Hall, London.
6. Cox, D.R. and Oakes, D. (1982): Analysis of Survival Data. Monograph on Statistics and Applied Probability. Chapman and Hall, London.
7. David, H.A. and Moeschberger, M.L. (1978): The Theory of Competing Risks, Monograph No. 39, Charles Griffin and Co. Ltd. London.

8. Efron, B. (1967): The two sample problem with censored data. Proceedings of the fifth Berkeley Symposium on Mathematical Statistics and Probability, Vol. 25, pp. 831–853, University of California Press, Berkeley, California.

9. Elandt, R.C. and Johnson, N.L. (1975): Conditional Fairune Time Distributions under Competing Risks Theory with Dependent Failure Times and Proportional Hazard Rates. Institute of Statistics, Mimeo Series No. 1105, University of North Carolina.

10. Elandt, R.C. and Johnson, N.L. (1980): Survival Models and Data Analysis, John Wiley and Sons., New York.

11. Feller, W. (1968): An Introduction to Probability Theory and its Applications, Vol. II, John Wiley, New York.

12. Gross, A.J. and Clark, V.A. (1975): Survival distributions: Reliability Application in Biomedical Science, John Wiley, New York.

13. Johansen, S. (1978): The Product Limit Estimator of a Maximum likelihood estimator, Scandinavian Journal of Statistics, Vol. 5, pp. 195–199.

14. Kalbfleisch, J. and Prentice, R.L. (1979): Hazard rate models with Covariates Biometrics, Vol. 35, pp. 25–39.

15. Kalbfleisch, J. and Prentice, R.L. (1980): The Statistical Analysis of Failure Time Data, John Wiley and Sons, New York.

16. Kaplan, E.L. and Meier, P. (1958): Non-Parametric Estimation from Incomplete Observations, Journal of the American Statistical Association, Vol. 53, pp. 457–481.

17. Keifer, J. and Wolfwitz (1956): Consistency of the maximum likelihood estimator in the presence of infinitely many incidental parameters, Annals of Mathematical Statistics, Vol. 27, pp. 887–906.

18. Leurgans's, Tsai, Eei-Yanss and Crowley, J. (1984): Freund's Bivariate Exponential Distribution and Censoring, IMS Lecture Notes, Monograph Series, Vol. 2, pp. 230–242.

19. Miller, Rupert Jr. (1981): Survival Analysis, Wiley series in Probability and Mathematical Statistics, Applied Probability and Statistics, John Wiley and Sons, New York.

20. Munoz, A. (1980a): Non-Parametric Estimation from Censored Bivariate Observations. Technical report No. 60, August 1980, Division of Biostatistics, Stanford University, Stanford, California.

21. Munoz, A. (1980b): Consistency of the Self Consistent Estimator of the Distribution Function from Censored Operations. Technical report No. 61, August 1980, Division of Biostatistics, Stanford University, Stanford, California.

6

Some Miscellaneous Items in Indian Demography

6.0 Introduction

Basic data for Indian Population are available from (i) Census (ii) Registration or Vital Statistics (iii) Periodic Surveys Like National Sample Survey and through other collecting agencies like Central Statistical Organizations, special surveys conducted by Government authorities or private agencies etc. In this chapter, we shall consider the techniques of constructing vital rates and indices using census or registration data, keeping in view of the limitations of these data.

The registration system of India started as early as 1864–69, with the official vital registration system implemented in Bengal followed by the same in Madras in 1885. In 1889 in all the provinces of British India enactment of the voluntary registration of births, deaths and marriages was made thereafter. The first census started operating from 1882 and thereafter in 1891: and since then in every ten years complete censuses are being taken with the exception of 1941. In 1941, Government undertook a 1% sample census (known as Y-sample, after the name of Frank Yates, Statistician who gave the sampling design of this census). Now, if the provision of the vital registration act was implemented effectively, it could have concurrently with the census data provided a wealth of Demographic information. Unfortunately perhaps at no time of the history, the registration department of any of the states of India functioned satisfactorily to provide vital statistics of a reasonable degree of accuracy because of two basic reasons:

(a) The general apathy of the public towards registration of vital events.

(b) The lack of a permanent government organisation whose whole time responsibility is the registration of the vital events. There is, therefore, a general despondency prevailing among Indian Demographers on account of huge gap in Indian Demographic data created by the lack of reliable vital statistics. For the purpose of the estimation of vital rates such as birth, death, infant mortality and life tables rates, census authorities had to depend mainly on their own returns.

But as vital statistical data are available in continuous form, however defective they are, a section of Demographers have thought it worthwhile to make estimates through the vital registration system by special modeling while making real and valid assumptions about the nature of the errors in the data. We shall, in the following section, present methods of census and vital statistics for estimating the vital rates. The census data also are defective in a number of aspects such as age returns, underenumeration of mortality figures, birth figures etc. etc. Therefore, a life table which is constructed based on the mortality experience of Inter-census period is also subject to various baffling problems. In this chapter, we shall also present a method which was adopted by census authority to construct Indian life tables.

6.1 Techniques of Construction of Vital Rates (Birth and Death Rates) From Census Data—(Reverse Survival and Differencing Methods)

6.1.1 Reverse Survival Method

The population in the age-group 0–10 enumerated in the census 1981 have been born in 1971–1981. They are the survivors of those births which occurred during the decade 1971–1981 in absence of migration.

Denoting by

${}_9P_0^Z$ = Population in the age group (0 to 10) in the census year Z (say $Z = 1981$)

The number of births which occurred during $(Z-10, Z)$

$$= {}_9P_0^Z \times \frac{10 \cdot l_0}{{}_9L_0},$$

where l_0 and ${}_9L_0$ are the life table entries for a life table appropriate for the mortality situation prevailing in $(Z-10, Z)$. Hence the mean birth rate for the decade is

$$B_{(Z-10,Z)} = \left\{ {}_{10}l_0 \cdot \frac{{}_9P_0^Z}{{}_9L_0 \cdot 10} \div \frac{1}{2}(P^Z + P^{Z-10}) \right\} \times K. \tag{6.1.1}$$

This method is known as '*Reverse Survival Method*' given the census mean growth rate $G_{(Z-19,Z)}$ (per K persons).

We have $B_{(Z-10,Z)} - G_{(Z-10,Z)} = D_{(Z-10,Z)}$, (6.1.2)

assuming no migration or a complete balance in the emigration and migration.

6.1.2 Hardy's Differencing Method

This method was introduced by Census Actuary, Hardy, using 1901 census data. The method enables to estimate the death rate using Census data and Registration data as follows:

Let P^Z = Total population as enumerated in the census year Z.

P_{10+}^{Z+10} = Population in the age group 10 and above enumerated on the censers date $(Z+10)$

Then assuming balance in emigration and migration,

$(P^Z - P_{10+}^{Z+10})$ = Number of deaths which occurred during Z to $Z+10$ to those births of the persons who were born after Z and died before the census date $(Z+10)$.

Hardy assumed the approximate equation

$$(P^Z - P_{10+}^{Z+10}) \simeq \Delta D_{0,5} + (D_{5+} - \Delta D_{5+}), \tag{6.1.3}$$

where $\Delta D_{0,5}$ represents a very negligible fraction of the deaths in the age sector $(0,5)$ which is included in the differences $(P^Z - P_{10+}^{Z+10}), D_{5+}$ represents all the deaths in the age sector 5 above and ΔD_{5+} represents a fraction of death in the age sector 5 and above. In other words, according to Hardy $(P^Z - P_{10+}^{Z+10})$ includes most of the deaths in the age sector 5 and above and excludes most of the deaths in the age sector $(0, 5)$ during Z to $Z + 10$.

In fact exclusion is much heavier than inclusion. But assuming

$$\Delta D_{0,5} \simeq \Delta D_{5+} \tag{6.1.4}$$

$$P^Z - P_{10+}^{Z+10} = \Delta D_{0,5} + (D_{5+} - \Delta D_{5+}) \tag{6.1.5}$$

$$P^Z - P_{10}^{Z+10} \simeq D_{5+} \tag{6.1.6}$$

Hardy next suggested to use λ viz. the ratio of all deaths and ratio of deaths in the age sector 5 and above. He maintained that although both numerator and denominator in the ratio λ are grossly under enumerated due to fault in the registration system but their ratio is sufficiently accurate.

Thus

$$D_{(Z,Z+10)} = \left[\frac{(P^Z - P_{10+}^{Z+10})}{10 \cdot \dfrac{(P^Z + P^{Z+10})}{2}} \cdot \lambda\right] \cdot K, \tag{6.1.7}$$

provides the estimated death rate during Z to $Z + 10$ per K persons in the population.

Again $\quad G_{(Z,Z+10)} + D_{(Z,Z+10)} = B_{(Z,Z+10)},$ (6.1.8)

gives the estimated mean birth rate for the decade

Remarks 6.1: (i) Reverse survival method over-estimates the mean birth rate, because reverse survival make assumptions that the present population in the age group (0–9) are the survivors of the births which occurred during the previous decade. Since the survival proportion is estimated from a current life table which shows improvement in the survival proportion more than what is pertinent for the decade so the estimated birth rate has an upward bias.

(ii) Differencing method under estimates the mean death rate, because by similar arguments as above, the estimated death rate from it has a downward bias.

6.2 Methods of Estimating Vital Rates from Registration Data

6.2.1 Chandrasekharan and Deming's Method

Chandrasekharan and Deming (1949) have evolved a Mathematical technique which when applied to a comparison of the registrar's list of births and deaths with that of a list obtained from a house to house canvass provide estimates of the total number of events over the area in a specified period. The application of the theory which is to be developed requires a comparison of:

(1) The Registrar's list (referred to as R),

(2) The result of a complete house to house canvass carried out by interviewers (referred to as I) and the classification of the entries on these lists into the following four exhaustive groups or categories of events.

I C: The number of events which are recorded in I as well as in R and therefore assumed to be correct without verification.

II N_1: Entries recorded in R but not in I and after investigation found to be correct.

III N_2: Entries recorded only in I but not in R and after investigation found to be correct.

IV X: Entries recorded on one list or the other (but not in both) and found incorrect after verification.

If N represents the total number of events then Chandrasekharan and Deming (1949) estimate N as

$$\hat{N} = C + N_1 + N_2 + \hat{Y}, \tag{6.2.1}$$

where $\hat{Y}$ represents the estimate of the number of events which are missed by both R and I which is given by (6.2.1).

Theorem 6.2.1: The estimate of Y is given by

$$\hat{Y} = \frac{N_1 N_2}{C}, \tag{6.2.2}$$

assuming the independence of reporting of R and I.

Proof: Let p_1 and p_2 be the probability that an event is detected by R and I, respectively. Then,

$$p_1 = \frac{C+N}{N} \quad \text{and} \quad p_2 = \frac{C+N}{N}. \tag{6.2.3}$$

Total of all categories of events $= N = (C + N_1 + N_2 + Y)$.

Then, the number of events missed by both i.e.

$$Y = (C + N_1 + N_2 + Y)(1 - p_1)(1 - p_2).$$

Asymptotically unbiased extimates of p_1 and p_2 are given by [ibid.]

$$\hat{p}_1 = \frac{C}{C+N_2} \quad \text{and} \quad \hat{p}_2 = \frac{C}{C+N_1}. \tag{6.2.5}$$

Replacing, p_1, p_2 and Y in (6.2.3) by $\hat{p}_1$, $\hat{p}_2$ and $\hat{Y}$ we get,

$$\hat{Y} = (C + N_1 + N_2 + \hat{Y})\left(1 - \frac{C}{C+N_2}\right)\left(1 - \frac{C}{C+N_1}\right)$$

$$\Rightarrow \quad \hat{Y} = (C + N_1 + N_2 + \hat{Y})\frac{N_1 N_2}{(C+N_2)(C+N_1)}$$

$$\Rightarrow \quad \hat{Y} = \frac{N_1 N_2}{C}$$

Observations: (i) It can be shown [ibid.] that $\hat{Y}$ is unbiased for Y and therefore

$$\hat{N} = \left(C + N_1 + N_2 + \frac{N_1 N_2}{C}\right) \tag{6.2.6}$$

is unbiased for N.

(ii) It can be shown that $\hat{N}$ is a MLE of N [ibid.]

Standard error of $\hat{N}$

$$\hat{N} = C + N_1 + N_2 + \frac{N_1 N_2}{C}, \text{ from (6.2.6).}$$

$$= \frac{(C+N_1)(C+N_2)}{C}.$$

We have $E(C) = NP$ and $\text{Var}(C) = NPQ$,

if C is regarded as the number of successes in a Binomial population of size N with true probability of success P.

We assume that $(C + N_1), (C + N_2)$ and N remain fixed.

Again $P = p_1 p_2$

$$Q = (1 - p_1)(1 - p_2) = q_1 q_2$$

$$\Rightarrow \quad E(C) = N p_1 p_2 \tag{6.2.7}$$

$$\Rightarrow \quad \text{Var}\ (C) = N p_1 q_1 p_2 q_2 \tag{6.2.8}$$

By the Delta method (vide (A.2.21)) of obtaining large sample standard error

$$\text{Var}\ (\hat{N}) \cong \left\{ \frac{\partial}{\partial C} \left[\frac{(C + N_1)(C + N_2)}{C} \right] \right\}^2_{C=E(C)} \text{Var}\ (C)$$

$$= \left. \frac{(C + N_1)^2 (C + N_2)^2}{C^4} \right|_{C=E(C)} N p_1 q_1 p_2 q_2$$

$$= \frac{(C + N_1)^2 (C + N_2)^2}{N^4 p_1^4 p_2^4} \cdot N p_1 q_1 p_2 q_2,$$

$\because (C + N_1)$ and $(C + N_2)$ are assumed to be fixed.

or (6.2.3) $\Rightarrow$ $\text{Var}\ (\hat{N}) \cong \dfrac{N p_1 q_1 p_2 q_2}{p_1^2 p_2^2} = \dfrac{N q_1 q_2}{p_1 p_2}.$

$$\text{The standard error of } \hat{N} \text{ is } = \sqrt{\frac{N q_1 q_2}{p_1 p_2}}. \tag{6.2.9}$$

Precision of $\hat{p}_i$'s, $i = 1, 2$

This is given by their co-efficient of variations.

$$\text{Now, } \hat{p}_1 = \frac{C}{C + N_2} \tag{6.2.10}$$

$$\therefore \quad E(\hat{p}_1) = \frac{1}{C + N_2} E(C), \quad \text{(since } C + N_2 \text{ is assumed to be fixed)}$$

$$= \frac{N p_1 p_2}{C + N_2}, \quad \text{(from (6.2.7))} \tag{6.2.11}$$

$$\text{and} \quad \text{Var}\ (\hat{p}_1) = E(\hat{p}_1^2) - \left[\frac{N p_1 p_2}{C + N_2} \right]^2$$

$$= E\left[\frac{C^2}{(C + N_2)^2} \right] - \left[\frac{N^2 p_1^2 p_2^2}{(C + N_2)^2} \right]$$

$$= \frac{1}{(C + N_2)^2} [N p_1 p_2 q_1 q_2 + N^2 p_1^2 p_2^2] - \frac{N^2 p_1^2 p_2^2}{(C + N_2)^2},$$

from (6.2.8) and $\because (C + N_2)$ is fixed

$$= \frac{N p_1 q_1 p_2 q_2}{(C + N_2)^2} \tag{6.2.12}$$

$$\therefore \text{ Coefficient of variation of } \hat{p}_1 = \frac{\sqrt{N p_1 q_1 p_2 q_2}}{(C + N_2)} \div \left(\frac{N p_1 p_2}{C + N_2} \right)$$

$$= \frac{\sqrt{q_t\left(1 - \frac{C+N_2}{N}\right)}}{\sqrt{Np_1\frac{(C+N_2)}{N}}}, \text{ putting values of } p_2 \text{ and } q_2 \text{ from (6.2.4)}$$

$$= \sqrt{\frac{q_1(N-C-N_2)}{N(C+N_2)p_1}}. \tag{6.2.13}$$

Sometimes a better estimate of the C.V. of $\hat{p}_1$, is obtained, by replacing N by $(N-1)$ in the denominator of (6.2.13), as

$$\left.\begin{aligned} \text{C.V. of } \hat{p}_1 &= \sqrt{\frac{q_1(N-C-N_2)}{(C+N_2)p_1(N-1)}} \\ \text{and similarly,}\quad \text{C.V. of } \hat{p}_2 &= \sqrt{\frac{q_2(N-C-N_1)}{(C+N_1)p_2(N-1)}} \end{aligned}\right\}. \tag{6.2.14}$$

The validity of the method depends on:

(i) Effect of Incomplete investigation of Registrar's.

(ii) Effect of Incompleteness of the coverage of the population.

(iii) The effect of Institutional events i.e. vital events which take place in Institutions as Hospitals, Nursing homes, Dispensaries, Health clinics etc.

(iv) The effect of correlation of events in the list of R and I.

It may be noted that on account of (i), (ii) and (iii) the validity of the method is no doubt vitiated. However, because of (iv), the statistical validity of the method is much vitiated. We, therefore, analyze in details the effect of (iv).

The effect of correlation between *R* and *I*

The first step is to define the correlation. The Registrar and his co-workers will detect some events and miss others. The probability that the Interviewer will detect an event that was missed by R may be different from the probability that he will detect an event that was recorded by R. If these two probabilities are equal then there is complete Independence; but otherwise this is not, in which case the formula given above, for the estimation of the total number of events will be incorrect.

The extent of the error can be investigated:

The situation can be represented in a tabular form as given below,

Class	Group of events	Probability
C	Detected by both	p_1p_{21}
N_1	Detected by Registrar (failed by Investigator)	p_1q_{21}
N_2	Detected by Interviewer (failed by Registrar)	q_1p_{22}
Y	Missed by both	q_1q_{22}

Here, $p_{21}, p_{22}, q_{21}, q_{22}$ are four new conditional probabilities defined as

$$p_{ji} = P_r[\text{Detected by } j/\text{It was detected by } i], j,i = 1,2$$

$$q_{ji} = P_r[\text{Missed by } j/\text{It was missed by } i], j,i = 1,2$$

$$\text{Obviously,}\quad p_{21} + q_{21} = 1$$
$$p_{22} + q_{22} = 1$$

If there is complete independence between the events missed by both R and I then

$$\left.\begin{array}{l} p_{21} = p_{22} = p_2 \\ \text{and}\quad q_{21} = q_{22} = q_2 \end{array}\right\} \tag{6.2.15}$$

When there is dependence, the number of events missed by both (from (6.2.2))

$$\text{viz.}\quad \frac{N_1 N_2}{C} \simeq \frac{(Np_1q_{21})(Nq_1p_{22})}{N(p_1p_{21})} \tag{6.2.16}$$

Since, from the above table, the probability of detection of an individual by both is p_1p_2 and total population is N_1, hence, number of individuals detected by both (i.e. C) $= Np_1p_{21}$, etc.

Whereas if the events are independent

$$\frac{N_1N_2}{C} = N_1q_1q_{22}. \tag{6.2.17}$$

The difference D is

$$D = \frac{Np_1q_{21}q_1p_{22}}{p_1p_{21}} - Nq_1q_{22} = Nq_1\left(\frac{q_{21}p_{22} - P_{21}q_{22}}{p_{21}}\right). \tag{6.2.18}$$

Writing $q_{21} = 1 - p_{21}$ and $q_{22} = 1 - p_{22}$, we get

$$D = N\ q_1\left(\frac{p_{22} - p_{21}p_{22} - p_{21} + p_{21}p_{22}}{p_{21}}\right) = Nq_1\left(\frac{p_{22}}{p_{21}} - 1\right).$$

The Registrar's performance is estimated, as:

$$\text{From (6.2.10)}\quad \hat{p}_1 = \frac{C}{C + N_2} = \frac{Np_1p_{21}}{Np_1p_{21} + Nq_1p_{22}},\ \text{(from table)}$$
$$= \frac{p_1p_{21}}{p_1p_{21} + q_1p_{22}}, \tag{6.2.19}$$

whereas the true value is p_1.

Then the difference is,

$$\left(\frac{p_1p_{21}}{p_1p_{21} + q_1p_{22}} - p_1\right) = \frac{p_1p_{21} - p_1^2p_{21} - p_1q_1p_{22}}{p_1p_{21} + q_1p_{22}}$$
$$= \frac{p_1q_1(p_{21} - p_{22})}{p_1p_{21} + q_1p_{22}}. \tag{6.2.20}$$

Now two cases may arise

Case I: If $p_{21} > p_{22}$, the total number of events is underestimated.

Case II: If $p_{22} > p_{21}$, the total number of events are over-estimated.

Method to Reduce the Correlation

It is important to note that this correlation signifies the heterogeneity in the population viz. the events that fail to be detected by both the agencies do not form a random sample of the events of the whole population. The heterogeneity arises because of the difference in the reporting rates in the different segment of the population.

It therefore follows that the correlation can be minimised by dividing the population into homogeneous groups and estimating the total number of events from each group then by addition the estimate of the grand total is obtained. For the details the reader is referred to Chandrasekharan and Deming (1949).

6.2.2 Other Methods of Estimating Vital Rates from Registration Data

These methods are due to Biswas (1958) and Ghosh (1956). Biswas's approach was to regress U_B, the under-registration of births and U_D, the under registration of deaths, assuming both U_D and U_B are subject to errors.

If

$$U_B = m U_D, \tag{6.2.21}$$

be the regression relation of U_B on U_D on the assumption that $U_D = 0 \Rightarrow U_B = 0$, then by definition the growth rate G given by

$$G = B - D = (B_r + U_B) - (D_r + U_D) = (B_r - D_r) + (U_B - U_D)$$

$$G = G_r + (m-1)U_D, \text{ using (6.2.21)}, \tag{6.7.22}$$

where B_r, D_r represents the registered birth and death rates, respectively and B and D are the true birth and death rates.

Given G, G_r and $\hat{m}$, U_D can be estimated at once which also enables us to estimate U_B. G and G_r are obtainable from census and registration data, respectively. The regression relation was established on the basis of the data of sample census (1952–58) of births and deaths. Ghosh's (1956) method assumes that the primary reporting rates of the Chowkidars (rural reporting agents) were initially suffered from undernumeration. The level of under reporting progressively increased with the additional coverage of the population in the sector covered by the beat of the Chowkidar. This assumption enabled to reconstruct vital rates.

6.3 A Method of Construction of Life Table (Adopted during 1951 census)

As the death registration data available in India are not at all reliable the Census Actuaries had to depend entirely on the census age returns and infant death rates obtained from the proclaimed clan statistics for the construction of life tables. Census age returns are the reported ages of the population as per the conventions of different censuses, e.g. age on last birth day or age as per nearest birth day etc. The data pertaining to a particular Tribal population collected especially for the analysis of mortality situation of infantile age groups, as these were considered more reliable and representative of the population is the absence of reliable registration data on births and deaths, are called Proclaimed Clan Statistics.

The method adopted in different censuses varied only in details. It essentially consisted of calculating survival rates from two successive censuses. The census Actuarial reports describe in details the procedure adopted in the calculation of survival rates. We shall therefore, present a brief outline of the method adopted by 1951 Census Actuary.

The individual age counts of the successive census (say 1941 and 1951) corrected for the disturbances due to migration are the basic data in this analysis. The individual age counts as returned in Indian Censuses are very much distorted due to peaks at ages ending in the units place at either 5 or 0. To estimate this defect in the single year age counts of each census they are first grouped together in quinary age groups (3–7), (8–12), (13–17)...

By Koza-Keiwics's osculatory interpolation formula (used for 1951 census), we obtain the age counts in the central age intervals (5–6), (10–11),...etc.

Denoting the population in the intervals (5–6), (10–11);... by u_5, u_{10}, etc., we have by the above mentioned formula

$$\left.\begin{aligned} u_5 &= 184W_5 + .040W_{10} - .032W_{15} + .008W_{20}\ldots \\ u_{10} &= -.008W_5 + .216W_{10} - .008W_{15} + \ldots \\ \text{and}\quad u_x &= .0066W_{x-10} - .0344W_{x-5} + .2566W_x - .0344W_{x+5} \\ &\quad + .0066W_{x+10} \text{ for } x \geq 5. \end{aligned}\right\} \quad (6.3.1)$$

The population in the 1951 censuses at age $(x+10)$ are the survivors among those aged x in 1941 census when they passed through the intercensal period 1941–51.

$$\text{Therefore,}\quad \frac{u_{x+10} \text{ of 1951 Population}}{u_x \text{ of 1941 Population}} \simeq \frac{L_{x+10}}{L_x} \simeq \frac{l_{x+10+\frac{1}{2}}}{l_{x+\frac{1}{2}}} \simeq 10p_{x+\frac{1}{2}}. \quad (6.3.2)$$

The 10 years survival rate at age $(x+\frac{1}{2})$ i.e. $10p_{x+\frac{1}{2}}$ is the resultant arising out of the operation of successive single year interval rates viz. $p_{x+\frac{1}{2}}, p_{x+1+\frac{1}{2}}, \ldots, p_{x+9+\frac{1}{2}}$ prevailing during the year 1941, 1942, ..., 1950, respectively. If we now suppose that there was a common survival rate for each calendar year (i.e. for $1941 = 1$, $1942 = 2, \ldots, 1950 = 10$) affecting all the single year age specific survival rates of the year then such a scale a_i for the ith year, is given as:

$$a_i = \frac{p_{x+i}^{(i)}}{\left[\prod_{i=1}^{10} p_{x+\frac{1}{2}}^{(i)}\right]^{1/10}}, \quad (6.3.3)$$

where, $p_{x+\frac{1}{2}}^{(i)}$ is the probability of survival between $x+\frac{1}{2}$ to $x+\frac{1}{2}+1$ years of age in the calender year 'i'.

Denoting the average survival rate for a single year at age $(x+\frac{1}{2})$ over the intercensal period by $p_{x+\frac{1}{2}}$, we have,

$$p_{x+\frac{1}{2}} = \left\{\prod_{i=1}^{10} p_{x+\frac{1}{2}}^{(i)}\right\}^{1/10} \quad (6.3.4)$$

We have the 10 year survival rate $p_{x+\frac{1}{2}}$ reproducing the population at age $(x+10+\frac{1}{2})$ in the 1951 census from that at age $(x+\frac{1}{2})$ of the 1941 census given by

$$\begin{aligned} 10p_{x+\frac{1}{2}} &= p_{x+\frac{1}{2}}^{(1)} \cdot p_{x+1+\frac{1}{2}}^{(2)} \cdot \ldots \cdot p_{x+9+\frac{1}{2}}^{(10)} \\ &= (a_1a_2\ldots a_{10})(p_{x+\frac{1}{2}}p_{x+1+\frac{1}{2}}\cdots p_{x+9+\frac{1}{2}}), \text{ since } a_1a_2\ldots a_{10} = 1. \\ &= \prod_{i=0}^{9} p_{x+i+\frac{1}{2}}. \end{aligned} \quad (6.3.5)$$

Thus we have the survival rate

$$10p_{x+\frac{1}{2}} = \frac{l_{x+10+\frac{1}{2}}}{l_{x+\frac{1}{2}}}, \text{see chapter 2,} \tag{6.3.6}$$

reflects the average mortality condition during the intercensal period (1941–51).

Even though for ages below 9 only a few years condition of mortality were effective in reproducing the population of 1951 census from 1941 census figures, the life table based on $p_{x+\frac{1}{2}}$ at $x = 5, 10, 15, \ldots$ does reflect in general the average condition during the intercensal period.

Let us take as the starting point $l_{6+\frac{1}{2}} = 1000$. By successively multiplying this by $10p_{5+\frac{1}{2}}, 10p_{15+\frac{1}{2}}, \ldots$, we get $l_{15+\frac{1}{2}}, l_{25+\frac{1}{2}}, \ldots$, etc. at duennial intervals. By interpolating from the above values we can obtain $l_{10+\frac{1}{2}}$ and using this as a starting point, we can produce a series of terms $l_{20+\frac{1}{2}}, l_{30+\frac{1}{2}}, \ldots$ etc. by successively multiplying $10p_{10+\frac{1}{2}}, 10p_{20+\frac{1}{2}}, \ldots$, etc. These can be joined to the series of terms $l_{5+\frac{1}{2}}, l_{15+\frac{1}{2}}, \ldots$ etc. to produce a complete series of l's in quin-quennial intervals which we shall designate as series A.

Now, starting with $l_{10+\frac{1}{2}} = 1000$, we can obtain decinnial terms $l_{20+\frac{1}{2}}, l_{30+\frac{1}{2}}, \ldots$, etc. by multiplying successively with $10p_{10+\frac{1}{2}}, 10p_{20+\frac{1}{2}}, \ldots$, etc. which on extrapolation gives $l_{5+\frac{1}{2}}$. Using the value of $l_{5+\frac{1}{2}}$ so obtained as the starting point we can obtain $l_{15+\frac{1}{2}}, l_{25+\frac{1}{2}}, \ldots$, etc. by the same method. The two sets of decinnial terms can by joined and then divided by 1000 to obtain a second series B of quin-quennial terms $l_{5+\frac{1}{2}}, l_{10+\frac{1}{2}}, \ldots$, etc. starting with $l_{5+\frac{1}{2}} = 1000$ that we have now to deal with.

The values of $l_{x+\frac{1}{2}}$ for $x = 5$ onwards upto 60 are graduated to yield a smooth curve. The values of $l_{x+\frac{1}{2}}$ for $x > 60$ have not been taken into account for the usual unreliability of age returns at very old ages. The values of P_x for $x > 60$ were obtained from Gompertz curve by the equation

$$\text{Colog } p_x = BC^x, \tag{6.3.7}$$

B and C are constants, being determined empirically by fitting the above curve to values of p_x at $x = 45$, 55 and 60. Knowing p_x we can extend l_x column to any length as necessary.

The values of l_x below the age 5 are obtained by the following procedure:

A function of the form $A + h \cdot X + BC^x$ is fitted to the values of l_x at $x = 5, 6, 7$, already obtained. The values A, H and B expressed in terms of C are used to derive the values of l_1, l_2, l_3 and l_4 and the value of C is obtained by equating $l_1 = 1000$. Further an additional term $\dfrac{m}{20x+1}$ is introduced in the known function $A + Hx + BC^x$ so that $\dfrac{l_1}{l_0} = 1 - q_0$ (q_0 being the infant mortality rate obtained from the registration data). Thus having obtained l_0, l_1 and also the term $\dfrac{m}{20x+1}$, we obtain l_2, l_3 and $l_4, \ldots$, etc. for the function of $A + H_x + BC^x + \dfrac{m}{20x+1}$.

As the term $\dfrac{m}{20x+1}$ rapidly diminishes it practically leaves the previously calculated values $l_5, l_6, \ldots$, etc. remain unaffected and at the same time ensures a smooth junction of the portion of the curve l_0 to l_1 with the portion l_1 to l_5.

(Source—Census of India, Paper No. 12, life tables 1951, Census (1954))

6.4 Sample Registration System (SRS)

The Sample Registration System (SRS) is a large scale Demography survey with the objective to provide continuously more reliable yearly estimates of Birth rate, Death rate and other indices of measures of fertility and mortality at National and State levels. The SRS was initiated first by the office of the Registrar General, India, on a pilot basis in a few selected states of India in 1964–65. However, the survey became fully operational during 1969–70 with approximately 3,700 sample units. By and large the number of sample units was increased periodically. The sample unit in rural areas is a village or a segment of the same, in case the village had a population of 1,500 or more. In Urban areas, the sampling unit is a census enumeration block with population ranging from 750 to 1,000. Presently SRS has 6671 sample units (4436 rural and 2235 urban) in all states and union territories with coverage of 1.1 million households and 6.41 million population. The mode of field investigation consists of continuous enumeration of births and deaths by a resident part time enumerator, generally a teacher and an independent survey every six months by a computer supervisor. The data obtained by this scheme are duly matched. While the unmatched and partially matched events are re-verified in the field and thereafter counts of births and deaths are obtained while avoiding duplication.

The vital rates in respect of three newly formed states viz. Chattisgarh, Jharkhand and Uttaranchal were first published in 1999 in the April 2001 issue of SRS bulletin (2003) and were continued for subsequent years. SRS bulletins, regularly published from the office of the Registrar General, New Delhi, contains the provisional vital rates for the states. Tables 6.1, Tables 6.2, compiled from the SRS bulletin for October '93 provides complete description of the estimated rates.

Table 6.1
Provisional estimates of Birth rate, Death rate, Natural growth rate and Infant mortality rate, 2002

India/States	Birth rate			Death rate			Natural growth rate			Infant mortality rate		
Union territories	Total	Rural	Urban	Total	Rural	Urban	Total	Rural	Urban	Total	Rural	Urban
1	2	3	4	5	6	7	8	9	10	11	12	13
India*	**25.0**	**26.6**	**19.9**	**8.1**	**8.7**	**6.1**	**16.9**	**17.9**	**13.9**	**64**	**69**	**40**
Bigger States												
1. Andhra Pradesh	20.6	21.0	19.2	8.0	8.8	5.4	12.6	12.2	13.7	62	71	35
2. Assam	26.6	27.5	18.3	9.2	9.6	6.0	17.4	18.0	12.3	70	73	38
3. Bihar	30.9	31.8	23.5	7.9	8.2	6.1	23.0	23.7	17.4	61	62	50
4. Gujarat	24.6	26.5	20.5	7.6	8.3	6.3	17.0	18.3	14.2	60	68	37
5. Haryana	26.5	27.5	22.5	7.0	7.2	6.3	19.5	20.4	16.2	62	65	51
6. Karnataka	22.0	23.4	18.8	7.2	7.8	5.7	14.9	15.6	13.0	55	65	25
7. Kerala	16.8	17.0	16.3	6.4	6.4	6.2	10.5	10.6	10.2	10	11	8
8. Madhya Predesh	30.3	32.2	22.7	9.7	10.4	7.2	20.6	21.8	15.4	85	90	56
9. Maharashtra	20.2	20.5	19.8	7.3	8.3	5.6	12.9	12.2	14.1	45	52	34
10. Orissa	23.1	23.6	19.5	9.8	10.3	6.3	13.3	13.3	13.2	87	91	56

Table 6.1: (Contd.)

11. Punjab	20.8	21.7	18.1	7.1	7.4	6.2	13.7	14.4	11.8	51	55	35
12. Rajasthan	30.6	31.9	24.2	7.7	8.0	6.4	22.9	24.0	17.8	78	81	55
13. Tamil Nadu	18.5	19.1	17.5	7.7	8.6	5.9	10.8	10.5	11.5	44	50	32
14. Uttar Pradesh	31.6	32.6	26.8	9.7	10.2	7.3	21.9	22.4	19.5	80	83	58
15. West Bengal	20.3	22.5	14.0	6.6	6.8	6.3	13.7	15.7	7.7	49	52	36
Smaller States												
1. Arunanchal Pradesh	20.2	20.9	12.3	4.8	5.0	2.6	15.4	15.9	9.7	40	42	11
2 Chhatisgarh	25.0	26.5	22.6	8.7	9.7	7.2	16.3	16.8	15.4	73	85	51
3. Goa	14.0	14.4	13.5	8.3	9.0	7.2	5.7	5.4	6.3	19	21	16
4. Jharkhand	26.4	28.5	19.0	7.9	8.6	5.8	18.4	20.0	13.1	58	61	38
5. Himachal Pradesh	20.7	21.1	16.5	7.5	7.7	5.1	13.2	13.3	11.4	58	60	32
6. Jammu& Kashmir	19.2	19.9	16.0	5.7	5.7	5.7	13.4	14.1	10.3	47	49	38
7. Manipur	16.8	17.4	15.3	4.6	4.2	5.8	12.2	13.2	9.5	17	17	21
8. Meghalaya	25.8	27.8	14.6	7.7	8.4	3.9	18.1	19.4	10.7	60	62	41
9. Mizoram	16.9	20.8	12.0	4.8	5.7	3.7	12.1	15.1	8.3	15	16	14
10. Nagaland	NA	NA	11.9	NA	NA	2.5	NA	NA	9.4	NA	NA	16
11. Sikkim	21.9	22.2	13.4	4.9	5.0	4.1	17.0	17.3	9.4	39	39	27
12. Tripura	14.9	15.3	13.3	5.7	5.6	6.2	9.2	9.6	7.0	38	38	34
13. Uttaranchal	17.0	18.1	16.2	6.4	9.0	4.4	10.6	9.0	11.8	44	66	22
Union territories												
1. Andaman& Nicobar Islands	16.8	17.6	14.6	5.5	6.1	3.8	11.3	11.5	10.7	21	25	8
2. Chandigarh	14.6	21.5	13.8	3.4	4.1	3.4	11.2	17.4	10.4	25	30	24
3. Dadra& Nagar Haveli	30.4	31.3	20.4	6.8	7.1	4.0	23.6	24.2	16.3	56	59	16
4. Daman&Diu	22.4	23.5	21.4	6.6	7.4	5.9	15.8	16.2	15.5	39	35	42
5. Delhi	17.2	18.2	17.1	5.1	5.7	5.1	12.1	12.4	12.1	32	31	32
6. Lakshadweep	19.3	20.2	18.3	5.2	5.3	5.1	14.0	14.9	13.2	25	25	25
7. Pondicherry	17.9	18.1	17.7	6.7	8.1	5.7	11.2	10.1	12.0	23	32	17

*: Excludes Nagaland (Rural) due to part-receipt of returns.

NA (Not available) due to part-receipt of returns.

Note: Infant mortality rates for Smaller States and Union territories are for the period 2000-2002.

Table 6.2

Estimates of birth rate, death rate and infant mortality rate for the latest three years, 1999-2001

India/States	Year	Birth rate			Death rate			Infant mortality rate		
Union territories		Total	Rural	Urban	Total	Rural	Urban	Total	Rural	Urban
1	2	3	4	5	6	7	8	9	10	11
India*	1999	26.0	27.6	20.8	8.7	9.4	6.3	70	75	44
	2000	25.8	27.6	20.7	8.5	9.3	6.3	68	74	44
	2001	25.4	27.1	20.3	8.4	9.1	6.3	66	72	42
Bigger States										
1. Andhra Pradesh	1999	21.7	22.0	20.6	8.2	9.0	5.7	66	75	37
	2000	21.3	21.7	20.1	8.2	9.0	5.8	65	74	36
	2001	21.0	21.4	19.7	8.2	9.0	5.6	66	74	40
2. Assam	1999	27.0	28.0	18.9	9.7	10.1	6.2	76	79	36
	2000	26.9	27.9	18.6	9.6	10.0	6.1	75	78	35
	2001	27.0	27.9	18.6	9.6	9.9	6.6	74	77	34
3. Bihar	1999	31.5	32.4	25.2	8.9	9.2	7.1	63	64	55
	2000	31.9	32.8	25.6	8.8	9.1	7.1	62	63	53
	2001	31.2	32.3	23.4	8.2	8.5	6.3	62	63	52

Table 6.2: (Contd.)

4. Gujarat	1999	25.4	27.0	22.0	7.9	8.8	5.9	63	70	45
	2000	25.2	26.8	21.9	7.5	8.3	5.8	62	69	45
	2001	25.0	26.7	21.5	7.8	8.9	5.6	60	68	42
5. Haryana	1999	26.8	27.7	23.3	7.7	8.0	6.4	68	70	58
	2000	26.9	28.0	23.1	7.5	7.9	6.2	67	69	57
	2001	26.8	27.9	22.9	7.6	7.7	7.4	66	68	55
6. Karnataka	1999	22.3	23.7	19.2	7.7	8.7	5.5	58	69	24
	2000	22.0	23.3	19.1	7.8	8.6	5.8	57	68	24
	2001	22.2	23.6	19.6	7.6	8.2	6.4	58	69	26
7. Kerala	1999	18.0	18.1	17.7	6.4	6.5	6.3	14	14	16
	2000	17.9	18.0	17.5	6.4	6.5	6.2	14	14	14
	2001	17.3	17.4	16.7	6.6	6.8	6.2	11	12	9
8. Madhya Predesh	1999	31.1	33.1	23.6	10.4	11.1	7.6	90	96	55
	2000	31.4	33.4	23.5	10.3	11.1	7.5	87	93	54
	2001	31.0	32.9	23.1	10.1	10.8	7.2	86	92	53
9. Maharashtra	1999	21.1	21.6	20.3	7.5	8.7	5.6	48	58	31
	2000	21.0	21.4	20.4	7.5	8.6	5.8	48	56	33
	2001	20.7	21.1	20.2	7.5	8.5	5.9	45	55	28
10. Orrissa	1999	24.1	24.6	20.3	10.7	11.1	7.1	97	100	65
	2000	24.3	24.9	20.1	10.5	11.0	7.0	95	99	66
	2001	23.5	24.0	19.7	10.4	10.8	6.8	91	94	61
11. Punjab	1999	21.5	22.5	18.6	7.4	7.9	6.1	53	57	39
	2000	21.6	22.7	18.6	7.4	7.9	5.9	52	56	38
	2001	21.2	22.1	18.7	7.0	7.2	6.4	52	55	37
12. Rajasthan	1999	31.1	32.5	24.9	8.4	8.9	6.4	81	85	59
	2000	31.4	32.8	25.1	8.5	8.9	6.6	79	82	58
	2001	31.1	32.4	24.8	8.0	8.3	6.2	80	84	57
13. Tamil Nadu	1999	19.3	19.8	18.2	8.0	8.7	6.6	52	58	39
	2000	19.3	20.0	18.1	7.9	8.7	6.5	51	56	38
	2001	19.1	19.6	17.8	7.7	8.5	6.0	49	55	35
14. Uttar Pradesh	1999	32.8	33.9	27.5	10.5	11.1	8.1	84	88	66
	2000	32.8	34.0	27.2	10.3	10.8	8.0	83	87	65
	2001	32.1	33.2	27.0	10.1	10.6	7.8	83	86	62
15. West Bengal	1999	20.7	22.9	14.3	7.1	7.2	6.8	52	55	40
	2000	20.7	23.0	14.1	7.0	7.2	6.7	51	54	37
	2001	20.6	22.9	13.8	7.0	7.2	6.4	51	54	37
Smaller States										
1. Arunachal Pradesh	1999	22.3	23.2	13.5	6.0	6.4	1.9	43	45	10
	2000	22.3	23.1	13.9	6.0	6.3	2.5	44	45	11
	2001	22.2	23.1	13.0	5.7	6.0	2.3	39	41	11
2. Chhatisgarh	1999	26.9	29.3	23.6	9.6	11.3	7.0	78	95	47
	2000	26.7	29.2	22.8	9.6	11.2	7.1	79	95	49
	2001	26.5	29.1	22.5	8.8	10.1	7.0	77	88	58
3. Goa	1999	14.3	14.4	14.1	7.2	7.6	6.6	21	23	17
	2000	14.3	14.3	14.2	7.4	7.9	6.7	23	24	21
	2001	14.2	14.2	14.2	7.7	8.3	6.7	19	20	15
4. Jharkhand	1999	26.3	28.5	19.0	8.9	9.7	6.4	71	76	48
	2000	26.5	28.8	19.4	9.0	9.8	6.5	70	74	48
	2001	26.2	28.3	19.5	8.8	9.7	6.0	62	67	40
5. Himachal Pradesh	1999	23.8	24.3	16.8	7.3	7.5	5.2	62	63	38
	2000	22.1	22.5	16.9	7.2	7.3	5.5	60	62	37
	2001	21.2	21.5	17.1	7.1	7.2	5.3	54	56	32
6. Jammu&Kashmir	1999	20.9	22.0	16.5	6.3	6.2	6.3	52	53	44
	2000	19.7	20.5	16.5	6.2	6.3	5.9	50	51	45
	2001	20.2	21.1	16.4	6.1	6.1	6.1	48	49	39
7. Manipur	1999	18.6	19.4	16.4	5.4	5.0	6.6	25	22	24
	2000	18.3	19.1	16.2	5.6	5.4	6.0	23	23	25
	2001	18.3	19.0	16.1	5.2	4.8	6.1	20	19	23

Table 6.2: (Contd.)

8. Meghalaya	1999	28.7	31.1	15.7	9.1	10.2	3.3	56	59	33
	2000	28.5	31.0	15.3	9.2	10.1	4.6	58	61	32
	2001	28.3	30.7	15.0	9.0	9.9	3.9	56	57	41
9. Mizoram	1999	17.0	18.7	14.8	5.5	6.9	3.7	19	22	14
	2000	16.0	19.2	12.2	5.2	6.2	3.9	21	23	17
	2001	15.7	17.7	13.2	4.4	5.2	3.4	19	23	12
10. Nagaland	1999	NA	NA	11.8	NA	NA	2.3	NA	NA	16
	2000	NA	NA	12.2	NA	NA	3.0	NA	NA	23
	2001	NA	NA	12.4	NA	NA	2.6	NA	NA	13
11. Sikkim	1999	21.6	21.9	14.7	5.8	5.9	3.4	49	50	33
	2000	21.8	22.1	14.8	5.7	5.7	4.0	49	49	36
	2001	21.6	21.8	16.7	5.1	5.2	3.2	42	43	31
12. Tripura	1999	17.0	17.5	14.4	5.4	5.3	5.6	41	42	32
	2000	16.5	17.0	14.0	5.4	5.3	5.6	41	42	32
	2001	16.1	16.6	13.5	5.6	5.6	5.2	39	40	30
13. Uttaranchal	1999	19.6	24.5	16.1	6.5	10.5	3.5	52	75	27
	2000	20.2	24.6	17.1	6.9	10.3	4.5	50	73	26
	2001	18.5	21.1	16.6	7.8	10.0	6.1	48	69	26
Union territories										
1. Andaman& Nicobar Islands	1999	18.1	17.8	19.2	5.5	6.2	3.5	25	30	9
	2000	19.1	19.0	19.3	5.1	5.7	3.4	23	27	10
	2001	16.8	17.8	14.2	4.7	5.0	4.1	18	21	8
2. Chandigarh	1999	17.9	24.5	17.1	3.9	2.6	4.1	28	36	27
	2000	17.5	18.9	17.3	3.9	3.8	3.9	28	38	26
	2001	16.3	20.8	15.7	3.6	2.2	3.7	24	28	23
3. Dadra&Nagar Haveli	1999	34.2	35.2	23.6	6.6	7.0	2.5	56	61	7
	2000	34.9	35.9	24.0	7.8	8.2	3.5	58	62	14
	2001	29.5	30.4	20.0	6.5	6.9	2.9	58	62	9
4. Daman&Diu	1999	26.9	24.9	28.8	7.1	8.9	5.4	35	34	36
	2000	23.7	21.8	25.4	6.6	7.1	6.2	48	38	57
	2001	22.3	22.7	22.0	6.7	7.6	5.9	40	42	35
5. Delhi	1999	20.3	20.7	20.3	4.8	4.5	4.9	31	33	31
	2000	20.3	21.4	20.1	5.1	5.0	5.1	32	32	32
	2001	18.9	23.3	18.3	5.1	5.6	5.0	29	34	28
6. Lakshadweep	1999	25.1	25.1	25.0	4.7	4.3	5.0	32	26	37
	2000	26.1	27.6	24.6	6.0	7.1	4.9	27	25	29
	2001	20.4	22.1	18.7	5.0	4.7	5.2	33	34	33
7. Pondicherry	1999	17.7	18.0	17.5	6.9	7.9	6.1	22	32	15
	2000	17.8	18.4	17.4	6.5	7.2	6.0	23	33	15
	2001	17.9	18.8	17.3	7.0	7.8	6.6	22	31	15

* Excludes Nagaland (Rural) due to part-receipt of returns.

NA (Not available) due to part-receipt of returns.

Note: Infant mortality rate for smaller states/union territories shown against 1999, 2000 and 2001 are for the periods 1997–99, 1998–00 and 1999–2001 respectively.

References

1. Ambannavar Jaipal, P. (1974): Long term perspectives of population growth and labour force in India—A draft, Department of Economics, University of Bombay, Bombay.
2. Biswas (1958); A note on a method of estimating birth and death rates from registration data—Calcutta Statistical Association Bulletin, Vol. 8, Nos. 30–31, Oct. 58, pp. 65–72.

3. Biswas, S. (1974): On a probability model of predicting the labour force of a stable population. The Manpower Journal, Vol. IX, pp. 9–26.

4. Biswas, S. and Berry, V.K. (1975): On the problem of projecting the labour force of a Quasi-stable Population-Demography India, Vol. IV, No. 1, pp. 49–71.

5. Central Statistical Origination (1981): Statistical Pocket Book, Dept. of Statistics, Ministry of Planning, Govt. of India.

6. Chandrasekharan, C. and Deming, W.I. (1949): On a method of estimating birth and death rates and extent of registration-Journal of the American Statistical Association, Vol. 49, pp. 101–115.

7. Dandekar, V.M. and Dandekar, K. (1958): Survey of fertility and mortality in Poona District "Publication No. 27" Gokhale Institute of Politics and Economics, Poona.

8. Davis, Kingsley (1951): The Population of India and Pakistan. Princeton University Press, New Jersey.

9. Government of India Report of the Vital Statistics and Development Committee (1946). Vol. II, Chapter XVI.

10. Government of India Report of the Vital Statistics Committee appointed by the second Health Ministers Conference (1948).

11. Government of India Sample of census of births and deaths (1952–58), Census Paper No. 8.

12. Ghosh, A. (1956): A study of demographic trends in West Bengal-Population Studies, Vol. 9. No. 8.

13. Jain, S.P. (1939): Relationship between Fertility and Economic and Social Status in Punjab, Board of Economic, Govt. of Punjab.

14. Indian Statistical Institute (1958): A study of existing information on Population in relation to development planning in India, Unpublished report.

15. Office of the Registrar General, India—Sample Registration System Bulletin, Volume 37, No. 1, October 2003.

16. Sovani, N.V. and Dandekar, K. (1955): Fertility Survey of Nasik, Kolaba and Satara district (North), Publication No. 81, Gokhale Institute of Politics and Economics, Poona.

Appendix

Some Mathematical and Statistical concepts and results, which are frequently required in the book, are given here (Ommiting the proofs), for the purpose of ready reference.

A.1 Mathematical Preliminaries

A.1.1 Analysis

1. Monotone Functions

A function is said to be monotone in $[a, b]$ if it is either increasing or decreasing in $[a, b]$. A function $f(x)$ is increasing or decreasing in the interval $[a, b]$ according as:

$$f(x_2) \geq f(x_1) \quad \text{or} \quad f(x_2) \leq f(x_1)\ \forall\ x_2 \geq x_1;\ x_1, x_2 \in [a, b]. \tag{A.1.1}$$

It is said to be strictly monotonically increasing if strict inequalities in (A.1.1) holds.

A function $f(x)$ is increasing at a point $x = c$ if

$$f'(c) > 0, \tag{A.1.2}$$

and decreasing at c if

$$f'(c) < 0. \tag{A.1.3}$$

2. Lagrange's Mean Value Theorem

If a function is continuous in $[a, b]$ and differentiable in (a, b) then there exist at least one real number c in the interval $(a, b), a < c < b$, such that

$$f(b) - f(a) = (b - a)f'(c) \tag{A.1.4}$$

or

$$f(a + h) - f(a) = h\, f'(a + \theta h), \tag{A.1.5}$$

where $b = a + h$ and $c = a + \theta h,\ 0 < \theta < 1$.

3. Cauchy's Mean Value Theorem

If two functions f and g are continuous in $[a, b]$ and differentiable in (a, b) then

$$[f(b) - f(a)]g'(c) = [g(b) - g(a)]f'(c), \quad \text{where } c \in [a, b] \tag{A.1.6}$$

or

$$[f(a + h) - f(a)]g'(a + \theta h) = [g(a + h) - g(a)]f'(a + \theta h) \tag{A.1.7}$$

when $b = a + h$ and $c = a + \theta h,\ 0 < \theta < 1$.

4. Taylor Series Expansion

(A) For One Variable: If f be a real valued function on $[a, a + h]$ such that $f^{(n)}$ exists everywhere in $(a, a + h)$ and $f^{(n-1)}$ is continuous in $[a, a + h]$, then

$$f(a+h) = f(a) + h\,f'(a) + (h^2/2!)f''(a) + \cdots + (h^n/n!)f^{(n)}(a) + \cdots, \quad \text{(A.1.8)}$$

this is called Taylor Series, where $f'(a) = \left[\dfrac{df(x)}{dx}\right]_{x=a}$

$$\text{or} \quad f(x) = f(a) + (x-a)f'(a) + \{(x-a)^2/2!\}f''(a) + \cdots, \quad \text{(A.1.9)}$$

where $x = a + h$. This is called Taylor series expansion of $f(x)$ about the point a.

(B) For Two Variables: If $f(x, y)$ is a function of two variables which possesses continuous partial derivatives of order n in the rectangle $[a, a+h; b, b+k]$, then

$$f(x,y) = f(a,b) + \left[(x-a)\frac{\partial}{\partial x} + (y-b)\frac{\partial}{\partial y}\right] f(a,b)$$

$$+ \frac{1}{2!}\left[(x-a)\frac{\partial}{\partial x} + (y-b)\frac{\partial}{\partial y}\right]^2 f(a,b)$$

$$+ \cdots + \frac{1}{n!}\left[(x-a)\frac{\partial}{\partial x} + (y-b)\frac{\partial}{\partial y}\right]^n f(a,b) + \cdots, \quad \text{(A.1.10)}$$

when $x = a + h, y = b + k$. (A.1.10) is the Taylor series expansion of $f(x, y)$ about the point (a, b).

5. Mean Value Theorems of Integral Calculus

(A) First Mean Value Theorem: If a function $f(x)$ is continuous in $[a, b]$ then there exists a number z in $[a, b]$ such that

$$\int_a^b f(x)dx = f(z)(b-a). \quad \text{(A.1.11)}$$

(B) Second Mean Value Theorem: If $\int_a^b f(x)dx$ and $\int_a^b g(x)dx$ both exist and $f(x)$ is monotone on $[a, b]$ then there exists a $z \in [a, b]$ such that

$$\int_a^b f(x)g(x)dx = f(a)\int_a^z g(x)dx + f(b)\int_z^b g(x)dx. \quad \text{(A.1.12)}$$

(C) Generalized Mean Value Theorem: If $f(x)$ and $g(x)$ are both integrable on $[a, b]$ and $g(x)$ keeps the same sign over $[a, b]$, then there exists a number $c \in$ [Min$f$, Max$f$] such that

$$\int_a^b f(x)g(x)dx = c\int_a^b g(x)dx. \quad \text{(A.1.13)}$$

6. Differentiation Under the Sign of Integration

(A) If $f(x, y)$ is defined and continuous on the rectangle $D = [a, b; c, d]$ and if $\dfrac{\partial f(x,y)}{\partial x}$ exists and is continuous on D then

$$\frac{d}{dx}\left\{\int_c^d f(x,y)dy\right\} = \int_c^d \frac{\partial f(x,y)}{\partial x}dy, \ \text{(Leibnitz's Rule)}. \quad \text{(A.1.14)}$$

(B) Further, if $\phi(x)$ and $\Psi(x)$ are both differentiable on $[a, b]$, then

$$\frac{d}{dx}\left\{\int_{\phi(x)}^{\Psi(x)} f(x,y)dy\right\} = \int_{\phi(x)}^{\Psi(x)} \frac{\partial}{\partial x} f(x,y)dy + \Psi'(x)f(x,\Psi(x)) - \phi'(x)f(x,\phi(x)),$$

(General Lebinitz's rule). (A.1.15)

(C) Also $$\int_a^b \left\{\int_c^d f(x,y)dy\right\} dx = \int_c^d \left\{\int_a^b f(x,y)dx\right\} dy,$$

(Change in the order of integration). (A.1.16)

A.1.2 Laplace Transform (LT) (cf. Biswas (1991) Smith (1966))

Definition: Let $f(t)$ be a function of t specified for $t > 0$. Then the LT of $F(t)$ is defined as,

$$L\{F(t)\} = \int_0^\infty e^{-st} f(t)dt = f(s), \tag{A.1.17}$$

where s is real, $|s| < 1$. 'L' is called LT operator. An LT exists if

1. *LT* of Some Elementary Functions

The following Table A.1 present some elementary functions and their LT's Table of Laplace Transform:

Table A.1

S.N.	$f(t)$	$\overline{f}(s)$
1.	1	$1/s$
2.	t^{n-1}	$(n-1)!/s^n,\ n = 1, 2, \ldots$
3.	t^{a-1}	$\Gamma(a)/s^a,\ a > 0$
4.	$\dfrac{1}{\sqrt{\pi t}}$	$\dfrac{1}{\sqrt{s}}$
5.	e^{at}	$1/(s-a)$
6.	$t^{n-1}e^{at}/(n-1)!$	$1/(s-a)^n$
7.	$(at)^n e^{-at}$	$\dfrac{a^n\Gamma(n+1)}{(s+a)^{n+1}}, n > -1$
8.	$\sin \alpha t$	$\alpha/(s^2+\alpha^2)$
9.	$\cos \alpha t$	$s/(s^2+\alpha^2)$
10.	$(e^{bt}\sin \alpha t)/\alpha$	$s/\{(s-b)^2+\alpha^2\}$
11.	$f(t)\begin{cases} = 0, & 0 < t < a \\ = 1, & a < 1 \end{cases}$ (Heaviside unit function)	e^{-at}/s
12.	$\delta(t-a)$ (Dirac δ function located at a)	e^{-at}
13.	$\lambda e^{-\lambda t}, (\lambda > 0)$	$\lambda/(s+\lambda)$
14.	$\lambda^k t^{k-1} e^{-\lambda t}/\Gamma(k)$	$[\lambda/(s+\lambda)]^k$
15.	$(e^{bt} - e^{at})/t$	$\log[(s-a)/(s-b)]$

2. Some Properties of *LT*

(i) $L\{c_1 f_1(t) + c_2 f_2(t)\} = c_1\overline{f}_1(s) + c_2\overline{f}_2(s)$ (A.1.18)
where c_1, c_2 are constants and $f_1(s), f_2(s)$ are LT's of $F_1(t), F_2(t)$.

(ii) $L\{e^{at}f(t)\} = \overline{f}(s-a)$; (first translation or shifting property) (A.1.19)

(iii) $f(t) = \begin{cases} f(t-a), & \text{for } t \geq a \\ 0, & \text{for } t < a, \end{cases}$ (A.1.20)
then $L\{G(t)\} = e^{-as}\overline{f}(s)$,
(second translation or shifting property).

(iv) $L\{f(at)\} = \frac{1}{a}\overline{f}\left(\frac{s}{a}\right)$, (change of scale). (A.1.21)

(v) $L\{f'(t)\} = s\overline{f}(s) - f(0)$, where $F'(t)$ is the derivate of $F(t)$. (A.1.22)

(vi) $L\{f''(t)\} = s^2 f(\overline{s}) - sf(0) - f'(0)$ (A.1.23)

(vii) $L\left\{\int_0^t f(u)du\right\} = \overline{f}(s)/s.$ (A.1.24)

(viii) $L\{t^n f(t)\} = (-1)^n \frac{d^n}{ds^n}\overline{f}(s).$ (A.1.25)

(ix) $L\left\{\frac{f(t)}{t}\right\} = \int_s^\infty \overline{f}(u)du$, provided $\lim\limits_{t\to 0}\frac{f(t)}{t}$ exists. (A.1.26)

(x) (a) $\lim\limits_{t\to\infty} f(t) = \lim\limits_{s\to 0} s\overline{f}(s).$ (A.1.27)

(b) $\lim\limits_{t\to 0} f(t) = \lim\limits_{\to\infty} s\overline{f}(s).$ (A.1.28)

(xi) If $f(t)$ and $g(t)$ are distribution function and density function, respectively, of the r.v. T, then

$$\overline{f}(s) = [\overline{g}(s)]/s. \quad \text{(A.1.29)}$$

(xii) If T_1 and T_2 are two independent r.v.'s with distribution functions $f_1(t)$ and $f_2(t)$ respectively, then the LT of their convolution $f_1 f_2$ (see (A.2.1)) is the product of their LT's i.e.

$$L\{f(t)\} = L\{f_1(t)\}L\{f_2(t). \quad \text{(A.1.30)}$$

(xiii) If $S_n = T_1 + T_2 + \cdots T_n$, n is fixed, where T_i's are i.i.d.r.v.'s then

$$L\{h(S_n)\} = [L\{g(t)\}]^n, \quad \text{(A.1.31)}$$

where $h(S_n)$ is the p.d.f. of S_n and $g(t)$ is that of T_i's.

(xiv) If in (xiii) n is a r.v. with p.g.f. $P(s)$, then

$$L\{h(S_n)\} = P[L\{g(t)\}]. \quad \text{(A.1.32)}$$

3. Inverse Laplace Transform

Definition: If $L\{f(t)\} = \overline{f}(s)$ then $L^{-1}\{\overline{f}(s)\} = f(t)$, (A.1.33)
is called Inverse LT of $f(s)$; L^{-1} is called Inverse Laplace Operator.

The Table A.1. can be used to find the Inverse LT of elementary functions.

A.1.3 The Notations of 'O' and 'o'

To compare the 'Order of magnitude' of a function $f(x)$, as x tends to zero or infinity, with that of some known simple function $g(x) > 0$, we use the following notations

(i) When $\frac{f(x)}{g(x)}$ remains bounded as x tends to its limit, we write $f(x) = O(g(x))$; it means that $f(x)$ is at most of the order $f(x)$. (A.1.34)

(ii) When $\frac{f(x)}{g(x)}$ tends to zero, as x tends to its limit, we write $f(x) = o(g(x))$; it means that $f(x)$ is of a smaller order than $g(x)$. (A.1.35)

(iii) We write $f(x) = o(1)$ to express the fact that $f(x) \to 0$ as $x \to \infty$. (A.1.36)

A.1.4 Bolzano-Weirstrass Theorems

These are two theorems about the limiting points of an infinite set viz.

(i) Every infinite set which is bounded has attest one limiting point.

(ii) For every bounded infinite set there exist a least and a greatest limiting point.

A.2 Statistical Preliminaries (cf. Biswas (1991), Mukhopadhyay (1996)

A.2.1 (a) Convolution of Distribution Functions

If X_1 and X_2 are two independent r.v.'s with distribution functions F_1 and F_2 then the distribution function of $X = X_1 + X_2$ is given by the **convolution** of F_1 and F_2 as

$$F(x) = F_1(x_1) * F_2(x_2) = \int_{-\infty}^{\infty} F_2(x - x_1) dF_1(x_1) = \int_{-\infty}^{\infty} F_1(x - x_2) dF_2(x_2). \quad \text{(A.2.1)}$$

If X_1 and X_2 are continuous r.v.'s, then pdf of X is

$$f(x) = \int_{-\infty}^{\infty} f_2(x - x_1) f_1(x_1) dx_1 = \int_{-\infty}^{\infty} f_1(x - x_2) f_2(x_2) dx_2. \quad \text{(A.2.2)}$$

If X_1 and X_2 are positive r.v.'s then the distribution function of X is

$$F(x) = \int_0^x F_2(x - x_1) dF_1(x_1) = \int_0^x F_1(x - x_2) dF_2(x_2) \quad \text{(A.2.3)}$$

and pdf of X is

$$f(x) = \int_0^x f_2(x - x_1) f_1(x_1) dx_1 = \int_0^x f_1(x - x_2) f_2(x_2) dx_2 \quad \text{(A.2.4)}$$

A.2.2 (b) Some Fundamental Probability Distributions

A. Discrete Distributions

Binomial Distribution: Consider a series of n independent Bernoullian (i.e. with two outcomes viz. success or failure). The probability of x Successes ($x \leq n$) in n trials is given by

$$P(X = x) = \binom{n}{x} p^x (1 - p)^{n-x}$$

where p is the probability of a success n and p are the parameters of the probability distribution (or the probability mass function) $E(X) = np$, Var $(X) = np(1-p)$; Obviously,

$$\text{Var }(X) < E(X).$$

As $p \to 0$ and $n \to \infty$ subject to the condition $np =$ a fixed quantity, say λ then we get the Poisson Distribution given by

$$P(X = x) = e^{-\lambda}\frac{\lambda^x}{x!}; \quad x = 0, 1, 2, \ldots$$

λ is the parameter of the **Poisson Distribution**.

A third distribution which can be still be generated from **Binomial Distribution** is known as **Negative Binomial Distribution**. Assuming that the series of Bernoulli trials is restricted in such a way that the last trial leads to success i.e. that is the first $(n-1)$ trials leads to $(k-1)$ successes and the nth trial leads to kth success then the probability distribution of k number of successes out of n trials is given by

$$P(X = k) = \binom{n-1}{k-1} p^{k-1}(1-p)^{n-k} \cdot p = \binom{n-1}{x-1} p^k (1-p)^{n-k}.$$

If $n = (x + k)$ where $k \geq 0$

$$\Rightarrow \quad P(X = k) = \binom{x+k-1}{k-1} p^k (1-p)^{n-k}$$

$$\text{Now } \binom{x+k-1}{k-1} p^x (1-p)^{n-k} = \sum_{x=0}^{\infty} \binom{-k}{x} (-1)^x (s(1-p))^x$$

$$E(X) = \frac{k(1-p)}{p} \quad \text{and} \quad \text{Var }(X) = \frac{k(1-p)}{p^2}$$

Thus Var $(X) > E(X)$ in case of a **Negative Binomial Distribution**.

Negative Binomial Distribution can also be generated from **Poisson Distribution** by assuming λ, the Poisson parameter without being fixed has a **Gamma distribution*** of the form.

If under **Bernoullian** set-up instead of finding the probability of x successes in n trials if we consider the number of failure trials (X) preceding a success with probability of success as p and probability of failure as $q = (1-p)$ then the probability of $X = x$ failures preceding the first success i.e.

$P(X = x) = q^x p; \quad x = 1, 2, \ldots, \quad 0 \leq p \leq 1$ is called a **Geometric distribution**

which is a discrete distribution in terms of its waiting times taken as integers.

$$E(X) = \sum_{x=0}^{\infty} x q^x p = \frac{q}{p} \quad \text{and} \quad \text{Var }(X) = E(X^2) - [E(X)]^2 = \frac{q}{p^2}$$

which shows that for Geometric distribution Var $(X) > E(X)$.

Proceeding in this way, if we consider the distribution of the number of failures preceding the nth success we get again the negative Binomial Distribution.

B. Absolutely Continuous Distributions

The most fundamental and the earliest discovered distribution is given by a symmetrical, bell shaped distribution is known as Normal distribution or Gaussian distribution which basically arises from the distribution of random errors which occur in the measurement of a physical object when all ascribable cause of variation is eliminated.

The distribution of the random variable X taking a value between $(x, x+dx)$ is given by,

$$f(x) = \frac{1}{\sqrt{2\pi}\sigma} e^{-\frac{1}{2\sigma^2}(x-\mu)^2}, \quad -\infty < x < \infty$$

$$E(X) = \mu \quad \text{and} \quad \text{Var}\,(X) = \sigma^2$$

It can be shown that if n tends to infinity then Binomial distribution converges to Normal distribution. The curve is perfectly symmetrical. The coefficient of Skewness is given by $\beta_1 = \dfrac{\mu_3^2}{\mu_2^3} = 0$ and the coefficient of Kurtosis (which measures the extent of sharpness (or flatness of the curve) given by $\beta_2 = \dfrac{\mu_4}{\mu_2^2} = 3$, shows that the Normal Curve is neither too sharp nor too flat near the median. μ_i, s $(i = 1, 2, 3, \ldots,)$ are moments measured from the mean.

All moments of odd orders are zeros for Normal Distribution.

The recurrence relation between moments of even order is given by

$$\mu_{2n} = \frac{(2n-1)(2n-3)(2n-5)\ldots 1}{2^{n-1}} \cdot 2^{n-1}\mu_2^n = [(2n-1)(2n-3)(2n-5)\ldots 1]\mu_2^n$$

Sometimes a Skew distribution can be made Normal (or Normalized) by using $\log X = Z$.

The distribution of Z is called log Normal distribution.

The probability density of a log Normal distribution is given by

$$f(\log x) = \frac{1}{\sqrt{2\pi}\cdot\sigma x} \exp\left[-\frac{1}{2}(\log x - \mu)^2/\sigma\right], \quad 0 < x < \infty$$

The moment of order j from the origin (i.e. jth raw moment) μ_j' $(j = 1, 2, \ldots)$ is given by

$$\mu_j' = \exp\left(j\mu + \frac{j^2\sigma^2}{2}\right) \Rightarrow \mu_1' = \exp\left(\mu + \frac{\sigma^2}{2}\right)$$

$$\text{and} \quad \mu_2' = \exp\left(2\mu + \frac{2^2\sigma^2}{2}\right) = \exp(2\mu + 2\sigma^2) \Rightarrow \mu_2 = \exp(2\mu + \sigma^2)[\exp(\sigma^2) - 1]$$

$\therefore$ coefficient of variation $= [\exp(\sigma^2) - 1]$ is independent of μ.

Other Absolutely Continuous Distributions

$$f_x(x) = \lambda e^{-\lambda x}; \quad 0 \leq x < \infty$$

where $\lambda > 0$ is the parameter of the distribution.

$$E(X) = \frac{1}{\lambda}, \; \text{Var}\,(X) = \frac{1}{\lambda^2}$$

This distribution has a remarkable property known as the Loss of Memory Property (LMP).

This means that given a system is surviving upto age x, the probability the system will fail immediately after x is independent of x.

Probability of surviving upto age $x = 1 - F(x) = 1 - \lambda \int_0^x e^{-\lambda x} dx = e^{-\lambda x}$

Therefore, the conditional probability of dying immediate after $x = \dfrac{f(x)}{1 - F(x)} = \lambda$ which is independent of x. It may be checked that Geometric distribution has LMP property.

(i) Exponential Distribution, (ii) *Gamma Distribution, (iii) Beta Distribution

An **exponential distribution** is given by

If $X_1, X_2, \ldots, X_n$ are independently and identically distributed as exponential then the sum $Y = X_1 + X_2 + \ldots + X_n$ is distributed as $f_Y(y)$ given by

$$f_Y(y) = \frac{\lambda^n}{\Gamma(n)} e^{-\lambda y} y^{n-1}, \; 0 \le \lambda < \infty, \; n \ge 1$$

which is known as a Gamma distribution with parameters λ and n.

$$E(Y) = \int_0^\infty \frac{\lambda^n}{\Gamma(n)} e^{-\lambda y} y^n dy = \frac{\lambda^n}{\Gamma(n)} \int_0^\infty e^{-z} \left(\frac{z}{\lambda}\right)^n d\left(\frac{z}{\lambda}\right) = \frac{\Gamma(n+1)}{\Gamma(n)} \frac{1}{\lambda} = \frac{n}{\lambda}$$

Similarly, $E(Y^2) = \dfrac{n(n+1)}{\lambda^2} \Rightarrow \text{Var}\,(Y) = \dfrac{n}{\lambda^2}$

If $0 < \lambda < 1 \Rightarrow \text{Var}\,(Y) > \text{Mean}\,(Y)$, a **property analogous to Negative Binomial or Geometric distribution in the family of discrete r.v.s.**

If $\lambda = 1$ then $f_Y(y) = \dfrac{1}{\Gamma(n)} e^{-y} y^{n-1}$ for which $E(Y) = \text{Var}\,(Y) = n$, a property analogous to Poisson r.v.s in discrete family of distributions.

(iii) Beta Distribution: If X and Y are distributed as Gamma with parameters (λ_1, n_1) and (λ_2, n_2) respectively then the distribution of

(a) $Z = \dfrac{X}{X+Y}$ is distributed as Beta distribution of type I given by

$$\frac{1}{B(n_1, n_2)} Z^{n_1-1}(1-Z)^{n_2-1}, \quad n_1, n_2 > 0; \quad 0 < Z < 1$$

and (b) $U = \dfrac{X}{Y}$ is distributed as Beta distribution of type II given by

$$f(u) = \frac{u^{n_1-1}}{(1+u)^{n_1+n_2}} \frac{1}{B(n_1, n_2)}, \quad 0 < u < \infty$$

It may be noted that although the range of Beta I is from (0,1) and that of Beta`II is from $(0, \infty)$ the moments of any order for both the distributions are the same.

For, $\mu'_k = E(Z^k) = E(U^k) = \dfrac{B(k + n_1, n_2 - k)}{B(n_1, n_2)} \; \forall k = 1, 2, 3, \ldots$

A.2.3 (a) Two Results of Expectation

If X_1 and X_2 are two jointly distributed r.v.'s then

$$E[X_1] = E[E(X_1|X_2 = x_2)] \tag{A.2.5}$$

$$\text{Var}[X_1] = E[\text{Var}(X_1|X_2 = x_2)] + \text{Var}[E(X_1|X_2 = x_2)] \tag{A.2.6}$$

A.2.4 (b) Central Limit Theorems

(i) Lindberg Levy Central Limit Theorems: Consider a sequence of identically independently distributed random variables $X_1, X_2 \ldots$ each having a finite mean m and finite standard deviation $\sigma \neq 0$. Let $\{X_n\}$ be a sequence of distribution functions where

$$Z_n = \frac{Y_n - nm}{\sigma\sqrt{n}} \quad \text{and} \quad Y_n = X_1 + X_2 + \cdots + X_n$$

and $F_n(Z)$ is the cumulative distribution function of Z_n, i.e.,

$$P[Z_n \leq Z] = F_n(Z)$$

then $\lim_{n\to\infty} F_n(Z) = \frac{1}{\sqrt{2\pi}} \int_{-\infty}^{Z} \exp(-t^2/2)dt = \Phi(z)$ for $\forall Z$.

(ii) Liapourv's form for the Central Limit Theorem: Let $X_1 X_2 \ldots X_n$ be n independent random variables and $E(X_r) = \mu_r$ and Var $(X_r) = \sigma_r^2$; $r = 1, 2, \ldots, n$. Assume that the third absolute moments ν_r of X_r, about its mean μ_r given by

$$\nu_r^3 = E(|X_r - \mu_r|)^3$$

exists and is finite $\forall r$. If $\nu^3 = \nu_1^3 + \nu_2^3 + \cdots + \nu_r^3$ and if $\lim_{n\to\infty} \frac{\nu}{\Sigma} = 0$ then $\sum_{i=1}^{n} X_i$ is asymptotically normally distributed with mean

$$\Lambda = \mu_1 + \mu_2 + \cdots + \mu_n \quad \text{and} \quad \Sigma^2 = \sigma_1^2 + \sigma_2^2 + \cdots + \sigma_n^2$$

A.2.5 Probability Generating Functions (pgf)

For a discrete r.v. X, its pgf is defined as

$$P_X(s) = P(s) = E[s^X] = \sum_{n=0}^{\infty} s^x p_x, \tag{A.2.7}$$

where $p_x = P[X = x]$ and s is any real quantity with $|s| < 1$.

We can easily see that

$$p_x = \frac{1}{x!}\left[\frac{d^x}{ds^x} P(s)\right]_{s=0}, \quad x = 0, 1, 2, \ldots \tag{A.2.8}$$

$$\text{and} \quad \mu_{(r)} = r\text{th factorial moment of } X = \left[\frac{d^r}{ds^r} P(s)\right]_{s=1}, \quad r = 1, 2, \ldots. \tag{A.2.9}$$

If $X_1, X_2, \ldots, X_N$, (N is fixed), are N independent r.v.'s with pgf's $P_1(s), P_2(s), \ldots, P_N(s)$, then the pgf of $Z = X_1 + X_2 + \cdots + X_N$ is

$$P_Z(s) = P_1(s)P_2(s)\ldots P_N(s). \tag{A.2.10}$$

$$= [P(s)]^N, \quad \text{if } X_i\text{'s are i.i.d. with pgf } P(s). \tag{A.2.11}$$

If N is a r.v. with pgf, $G(s)$, and X_i's are i.i.d. with pgf $P(s)$, then

$$P_Z(s) = G[P(s)]. \tag{A.2.12}$$

The pgf of $Z = X_1 - X_2$ is given by

$$P_Z(s) = P_1(s)P_2\left(\frac{1}{s}\right). \tag{A.2.13}$$

If X_i's are not independent rather jointly distributed, then the pgf of $\widetilde{X} = (X_1, X_2, \ldots, X_n)$, N is fixed, is defined as

$$P(s_1, s_2, \ldots, s_N) = P(\widetilde{s}) = E[s_1^{X_1} s_2^{X_2} \ldots s_N^{X_N}], \quad \text{where} \quad |s_i| < 1. \tag{A.2.14}$$

It is called Multivariate pgf, so $P(s)$ may by called univariate pgf.

We can easily see that

$$E(X_i) = \left[\frac{\partial P(\widetilde{s})}{\partial s_i}\right]_{\widetilde{S}=t}, \tag{A.2.15}$$

$$E[X_i(X_i-1)(X_i-2)\ldots(X_i-k+1)] = \left|\frac{\partial^k}{\partial s^k}P(\widetilde{s})\right| s = 1$$

$$\text{and} \quad E[X_iX_j] = \left[\frac{\partial^2}{\partial s_i \partial s_j}P(\widetilde{s})\right]_{s_j} = s_j = 1;\ i \neq j = 1, 2, \ldots, N. \tag{A.2.16}$$

$$\begin{aligned} \text{Also} \quad p_{x_1, x_2, \ldots, x_N} &= P[X_1 = x_1, X_2 = x_2, \ldots, X_k = x_k] \\ &= \frac{1}{x_1! x_2! \ldots x_k!}\left[\frac{\partial^{x_1+x_2+\cdots+x_{k_2}}}{\partial s_1^{x_1} \partial s_2^{x_2} \ldots \partial s_k^{x_k}}\right] \begin{matrix} s_i = 0 \text{ for } i \leq k \\ = 1 \text{ for } i > k \end{matrix} \end{aligned} \tag{A.2.17}$$

$$= \frac{1}{x_1!}\frac{\partial^{x_1}P(\widetilde{s})}{\partial s_1^{x_1}}\Bigg| \begin{matrix} s_1 = 0 \\ s_i = 1, i \neq 1 \end{matrix} \quad \frac{1}{x_2!}\frac{\partial^{x_2}P(\widetilde{s})}{\partial s_2^{x_2}}\Bigg| \begin{matrix} s_2 = 0 \\ s_i = 1, i \neq 2 \end{matrix}$$

$$\ldots \frac{1}{s_k!}\frac{\partial^{x_k}P(\widetilde{s})}{\partial s_k^{x_k}}\Bigg| \begin{matrix} s_k = 0 \\ s_i = 1, i \neq k \end{matrix}, \quad k = 1, 2, \ldots, N. \tag{A.2.18}$$

A.2.6 Delta Method For Large Sample S.E. of any Function of a Statistic

Let T be a statistic with $E(T) = \mu$. We require the large sample standard error (S.E) of some function of T, say $\Psi(T)$.

Suppose $\Psi(T)$ can be expand in a Taylor series (see (A.1.9) about $T = \mu$, then)

$$\Psi(T) = \Psi(\mu) + (T-\mu)\left(\frac{\partial \Psi}{\partial T}\right)_{T=\mu} + \ldots. \tag{A.2.19}$$

Ignoring terms involving powers higher than 2 of $(T - \mu)$ in (A.2.19) we get

$$E[\Psi(T)] = \Psi(\mu), \tag{A.2.20}$$

$$\text{Var}[\Psi(T)] \cong \text{Var}(T)\left(\frac{\partial \Psi}{\partial T}\right)^2_{T=\mu}. \tag{A.2.21}$$

Delta method has a multivariate analogue. Suppose T_1 and T_2 are two statistics, jointly distributed and $E(T_1) = \mu_1$ and $E(T_2) = \mu_2$. We want sample S.E. of $\Psi(T_1, T_2)$, a function of T_1 and T_2.

Suppose $\Psi(T_1, T_2)$ is capable of Taylor series expansion about μ_1 and μ_2, as (see (A.1.10))

$$\Psi(T_1,T_2) = \Psi(\mu_1,\mu_2) + (T_1-\mu_1)\left(\frac{\partial\Psi}{\partial T_1}\right)_{T_1=\mu_1} + (T_2-\mu_2)\left(\frac{\partial\Psi}{\partial T_2}\right)_{T_2=\mu_2} + \ldots \tag{A.2.22}$$

Following the above argument, we have

$$E[\Psi(T_1,T_2)] \cong \Psi(\mu_1,\mu_2) \tag{A.2.23}$$

$$\text{and}\quad \text{Var}[\Psi(T_1,T_2)] \cong \text{Var}(T_1)\left(\frac{\partial\Psi}{\partial T_1}\right)^2_{T_1=\mu_1} + \text{Var}(T_2)\left(\frac{\partial\Psi}{\partial T_2}\right)^2_{T_2=\mu_2} + 2\text{Cov}(T_1,T_2)\left(\frac{\partial\Psi}{\partial T_1}\cdot\frac{\partial\Psi}{\partial T_2}\right)_{T_1,\mu_1,T_2=\mu_2}. \tag{A.2.24}$$

The Delta method is very useful. For example, using it we can get approximate value of Var $\left(\frac{\overline{T}_1}{\overline{T}_2}\right)$ or Var $(\overline{T}_1, \overline{T}_2)$ etc. where $\overline{T}_1$ and $\overline{T}_2$ are sample means of T_1 and T_2, respectively.

A.2.7 Estimators and Their Properties

Any *statistic* (i.e. a function of sample values) used to estimate a population *parameter*, say θ, is called an *esitmator* of θ. The properties of a good estimator are, (i) Unbiasedness (ii) Consistency (iii) Sufficiency and (iv) Efficiency

Suppose, $T = T(\widetilde{X})$ is statistic, based on a *sample* $\overline{X} = (X_1, X_2, \ldots, X_n)$ of size n, from a *pulation* with *probability density function* (pdf) $f(x;\theta)$, to estimate the parameter θ.

(i) T is *unbiased* for θ, if $E(T) = \theta$. (A.2.25)

(ii) $T = T_n$ (say), is *consistent* for θ if,

$$\text{or}\quad \lim_{n\to\infty} P_r[|T_n - \theta| < \varepsilon] = 1, \tag{A.2.26}$$

where ε is a small positive number.

(iii) T is *sufficient* for θ provided if T_1 is another estimator of θ, then the conditional distribution of T_1 given T is independent of θ. If $f(\widetilde{x};\theta)$ is the joint density function of the sample then the necessary and sufficient condition (NASC) that T is sufficient for θ, is

$$f(\widetilde{s};\theta) = g(t;\theta)h(\widetilde{x}), \text{ (factorization theorem)}, \tag{A.2.27}$$

where $g(t;\theta)$ is function T and θ and h is independent of θ.

(iv) If T_1 and T_2 are two estimators of θ then T_1 is more efficient than T_2 if

$$\text{Var}(T_1) < \text{Var}(T_2). \tag{A.2.28}$$

An unbiased estimator T is called the *best* or *minimum variance unbiased estimator* (MVBU) of θ if $E[(T-\theta)^2]$ is minimum.

The lower bound for the variance of an unbiased estimator T is given by

$$\text{Var}(T) \geq 1/E\left[\frac{\partial}{\partial\theta}\log L\right]^2 = -1/E\left[\frac{\partial^2 \log L}{\partial\theta^2}\right], \quad \text{(Cramer-Rao inequality)}, \tag{A.2.29}$$

where $L\{= f(x_1;\theta)\cdot f(x_2;\theta)\ldots f(x_n;\theta)\}$, is called the *Likelihood* (LH) *function* of the sample $(x_1, x_2, \ldots, x_n)$.

A statistic for which (A.2.29) is an equality, is called minimum variance bound (MVB) estimator or *most efficient* unbiased estimator.

Rao Blackwell Theorem

The theorem is useful in obtaining the minimum variance unbiased estimator (MVUE) of a parameter (which may be a vector) by a process known as "Blackwellization". Before we state the theorem, two fundamental characteristics of an estimator viz. (1) Sufficiency (2) Completeness should be defined.

An estimator is said to be *sufficient* if it provides all the information contained in a sample in respect of the parameter θ.

Let $\hat{\theta}$ and $\widetilde{\theta}$ be two estimators of θ subject to the condition that one is not a function of the other. $\hat{\theta}$ is said to be sufficient if the conditional density function of $\widetilde{\theta}$ given $\hat{\theta}$ is independent of θ. A simple criterion of checking the sufficiency of an estimator is known as *Fisher-Neyman factorization theorem* which runs as follows:

Let $x_1 \ldots x_n$ be a random sample of size n from a population with density function $g(x;\theta)$, $a < x < b$, when a and b does not depend upon θ. Then $\hat{\theta}$ is a sufficient statistic for θ if the joint density function of the sample is capable of being expressed in the form

$$g(x_1 x_3 \ldots x_n : \theta) = g_1(\hat{\theta}, \theta) g_2(x_1 x_2 \ldots x_n) \tag{A.2.30}$$

where $g_1(\hat{\theta},\theta)$ is the density function of $\hat{\theta}$ (the sufficient statistics) involving the parameter θ and $g_2(x_1 x_2 \ldots x_n)$ is the conditional density function of $g(x_1 x_2 \ldots x_n : \theta)$ which does not involve θ, if $\hat{\theta}$ is sufficient. Conversely if the representation (A.2.30) holds then $\hat{\theta}$ is sufficient for θ.

Completeness of a Statistic: A sufficient statistic T is said to be complete if $E_\theta(T) = 0 \Rightarrow T = 0$, almost everywhere, except for a set of points with probability measure zero. A statistic which is unbiased and complete is necessarily unique and is the minimum variance unbiased estimator (MVUE) of its expectation provided MVUE exists.

Rao Blackwell Theorem: Let T be any sufficient statistic for θ (where T and θ may be vector valued) and T_1 be any other statistic

Let $\quad h(t) = E(T_1|T = t)$

Then $\;E(T_1 - g(\theta))^2 \geq E(h(t) - g(\theta))^2$

where $g(\theta)$ is any parametric function of θ.

A Particular Case

If T_1 is unbiased for $g(\theta)$ implying $h(t)$ is also unbiased for $g(\theta)$ then we have Var $(T_1) \geq$ Var $(h(T))$

Blackwellization: The implication of the result is that if we are given an unbiased estimator, say, T_1 of $g(\theta)$ we can improve upon T_1 by taking $E(T_1|T_2 = t) = h(t)$ based on the sufficient statistic T_2. This process of finding an improved estimator using the concept of unbiasedness and then reducing the variance by taking the conditional mean of the unbiased statistic given a sufficient statistic is known as 'Blackwellization' $h(T)$ is better than T_1 but the best if $h(T)$ is complete.

Consistent Asymptotically Normal (CAN) Estimator: If an estimator t_n is such that $\sqrt{N}(t_n - \theta)$ is asymptotically normal $N(0, \tau^2(\theta))$ for some function $\tau^2(\theta)$, then it is called a CAN estimator.

The quantity $I(\theta) = 1/\tau^2(\theta)$ is called Fisherian information about θ contained in the estimator.

A.2.8 Methods of Estimation

(a) Method of Maximum Likelihood (ML) Estimation

This is the most important and most common method of estimation.

A ML estimator (MLE) $\hat{\theta}$ for the parameter θ is the solution of the equation

$$\frac{\partial \log L}{\partial \theta} = 0, \tag{A.2.31}$$

which satisfies
$$\left|\frac{\partial^2 \log L}{\partial^2 \theta}\right|_{\theta=\hat{\theta}} < 0. \tag{A.2.32}$$

A mle is, generally, a biased estimator (though asymptotically unbiased). It is consistent, sufficient (if a sufficient estimator exist), efficient (if an efficient estimator exist) and is asymptotically normally distributed.

If $\hat{\theta}$ is the mle of θ and $g(\theta)$ is a monotone function of θ, then $g(\hat{\theta})$ is the mle of $g(\theta)$. This is called the Invariance property of mle's.

(b) Best Linear Unbiased Estimator (BLUE)

Let $(Y)_{n\times 1}$ be a column vector called observational vector and $(A)_{n\times m}$ be $(n \times m)$ matrix and $(\theta)_{m\times 1}$ be parametric vector. Then

$$(Y)_{n\times 1} = (A)_{n\times m}(\theta)_{m\times 1} + (U)_{n\times 1}$$

$$\left.\begin{aligned} &\text{with} \quad E(Y) = A^{\theta} \Rightarrow E(U) = (0)_{n\times 1} \\ &\text{and} \quad D^2(U) = \sigma^2 I_n \end{aligned}\right] \tag{A.2.32'}$$

where $D^2(U)$ refers to the variance covariance matrix of U and $(b')_{1\times m}(\theta)_{m\times 1}$ is called an estimable parametric function if $\ni$ one $c'Y$

$\ni \; E(c'Y) = (b')_{1\times m}(\theta)_{m\times 1}$

$\Rightarrow c'A\theta = b'\theta$ holds for $\forall 0$

$\Rightarrow A'c = b$

Under the set up of (A.3.1), $c'Y$ a linear function of the observations is known as the *best linear unbiased estimator (BLUE)* which satisfies the conditions

$$\left.\begin{aligned} E(c'Y) &= b'\theta \\ \text{Var }(c'Y) &\leq \text{Var }(d'Y) \end{aligned}\right] \tag{A.2.33'}$$

where $d'Y$ stands for any other linear function subject to $c'A = b'$. The set up (A.3.1) is called *Gauss Markov set up.*

Gauss-Markov Linear Model

Let $Y_1, Y_2, \ldots, Y_n$ be n independent r.v.'s with

$$E[Y_i] = \mu_i \quad \text{and} \quad \text{Var}(Y_i) = \sigma^2, i = 1, 2, \ldots, n. \tag{A.2.33}$$

Let μ_i, be expressible as a linear function of k unknown parameters $\theta_j, j = 1, 2, \ldots, k$, with known coefficients as

$$E[Y_i] = \mu_i = a_{i1}\theta_1 + a_{i2}\theta_2 + \cdots + a_{ik}\theta_k; \; i = 1, 2, \ldots, n. \tag{A.2.34}$$

The same model can be written in the matrix form as

$$Y_{n\times 1} = A_{n\times k}\; \theta_{k\times 1} + e_{n\times 1}, \tag{A.2.35}$$

where $Y = (Y_1, Y_2, \ldots, Y_n)'$, $e = (e_1, e_2, \ldots, e_n)'$, $A = [a_{ij}]$ and $\theta = (\theta_1, \theta_2, \ldots, \theta_k)'$ with $E(e_i) = 0$ and Var $(e_i) = \sigma^2 \Rightarrow E(e) = O_{n\times 1}$ and $\text{Var}(e) = \sigma^2 I_{n\times 1}$, where O is a null matrix and I is a unit matrix.

The BLUE of θ is given by

$$\hat{\theta} = (A'A)^{-1}A'Y \quad \text{provided } (A'A)^{-1} \text{ exists.} \tag{A.2.36}$$

A.2.9 Non-Parametric Estimation

If we only assume that a sample is drawn from a general class of populations (not a particular population) e.g. class of continuous (or discrete) distributions and the problem is to draw inference about the population-characteristics on the basis of this sample, then it is called *non-parametric estimation* (for detail see Seigal (1966)).

A.3 Matrices (cf. Biswas (1997))

A.3.1 Characteristic Equation, Vectors and Roots

For the square matrix $A = A_{n\times n}$, the matrix $A - \lambda I_n$, where I_n is a unit matrix of order n and λ is a scalar, is called *Characteristic* (ch.) *matrix* of A, its determinant is called *Ch. Determinant* or *Ch.* Function of A. The equation

$$f(\lambda) = |A - \lambda I_n| = O_{n\times 1}. \tag{A.3.1}$$

is called the *Ch Equation* of A and its roots are called *Ch. Roots* or *Latent Roots* or *Eigen Values* of A. If $\lambda_1, \lambda_2, \ldots, \lambda_n$ are the ch. roots of A then,

$$|A| = \lambda_1\lambda_2 \ldots \lambda_n \tag{A.3.2}$$

If k is a scalar, ch. Roots of kA are k times that of A. Characteristic roots of A are also characteristic roots of A'. But the characteristic vectors of A' are not the same as that of A unless $A = s'$. The characteristic vectors of A' are called left characteristic vectors as opposed the characteristic vector A is called the right characteristic vector.

If λ is a ch. Root of A and

$$AX = \lambda X, \tag{A.3.3}$$

then X is a *ch. vector* of A corresponding to the root λ.

Any set of ch. vectors $X_1, X_2, \ldots, X_k$, corresponding to a set of distinct ch. roots $\lambda_1, \lambda_2, \ldots, \lambda_k$ of A is linearly independent.

The ch. roots of a real symmetric matrix are all real and that of skew symmetric matrix are either imaginary or zero.

Every square matrix satisfies its own ch. equation i.e. if $f(\lambda) = 0$ is the ch. equation of A then $f(A) = 0$. The Result is known as Cayley Hamilton Theorem.

If the ch. roots of $A_{n\times n}$ are $\lambda_1, \lambda_2, \ldots, \lambda_n$ and $\phi(A)$ is a rational function of A then the ch. roots of $\phi(A)$ are $\phi(\lambda_1), \phi(\lambda_2), \ldots, \phi(\lambda_n)$ and

$$|\phi(A)| = \phi(\lambda_1)\phi(\lambda_2)\ldots\phi(\lambda_n) \tag{A.3.4}$$

A.3.2 Leslie Matrix

A Leslie matrix L is a special non-negative matrix (i.e. whose elements are non-negative) used in population dynamics for *Population Projection.* It is a square matrix of the following form:

$$L = \begin{bmatrix} a_1 & a_2 & a_3 & \ldots & a_{n-1} & a_n \\ b_1 & 0 & 0 & \ldots & 0 & 0 \\ 0 & b_2 & 0 & \ldots & 0 & 0 \\ \vdots & \vdots & \vdots & & & \vdots \\ 0 & 0 & 0 & \ldots & b_{n-1} & 0 \end{bmatrix}. \tag{A.3.5}$$

subject to $0 \le a_i < 1$ and $0 < b_i \le 1, i = 1, 2, \ldots, n$.

The ch. equation of this matrix has only one real root λ_1 which is positive and simple and all the elements of the corresponding ch. vector X_1 are positive. If λ_i is any other complex ch. root of L then $|\lambda_i| \le \lambda_1$.

A.3.3 Quadratic Form (Q.F.)

Definition: $Q = X'_{1\times n}A_{n\times n}X_{n\times 1}$, where $X'_{1\times n} = (x_1, x_2, \ldots, x_n)$, (A.3.6)

is called a *quadratic form* if $A = [a_{ij}]_{n\times n}$ is a symmetric matrix i.e. $a_{ij} = a_{ji}$.

If we use the linear transformation $X = PY$, where $Y' = (y_1, y_2, \ldots, y_n)$, then $Q = Y'P'APY$, where P is a non singular matrix i.e. determinant $P \ne 0$.

If $P'AP = I_n$ and rank of A is n i.e. $\rho(A) = n$, then, $Q = \sum_{i=1}^{n} y_i^2$, (A.3.7)

is called *canonical form* of Q.

If $Q > 0$, for all x_i's $\ne 0$, then Q is called *positive definite* Q.F. and A is called a *Positive definite matrix.* On the other hand if

$$P'AP = -I_n \quad \text{then} \quad Q = -\sum_{i=1}^{n} y_i^2, \tag{A.3.8}$$

is called *negative definite* Q.F.

If $\rho(A) = r < n$, then it can be shown that

either $$Q = \sum_{i=1}^{n} y_i^2 \quad \text{or} \quad Q = -\sum_{i=1}^{n} y_i^2 \tag{A.3.9}$$

then Q is, respectively, called positive semi-definite or negative semi-definite. The characteristic roots of a positive definite, negative definite, positively semi definite and negative semi definite Matrices are > 0, < 0, ≥ 0, ≤ 0 respectively (cf. Scarborough (1966))

A.4 Numerical Integration

Numerical integration is the process of evaluating a definite integral, say $\int_a^b f(x)dx$, from a set of numerical values of the integrand $f(x)$.

Let the interval (a, b) be divided into n equidistant subinterval of length h by the points $x_1, x_2, \ldots, x_n$, such that

$$a = x_0,\ b = x_n \quad \text{and} \quad h = (x_n - x_0)/n \tag{A.4.1}$$

The following are some commonly used numerical integration formulas:

1. Trapezoidal Rule (for any +ve integer n):

$$\int_a^b f(x)dx \cong \int_{x_0}^{x_0+nh_0} y\,dx = h\left[\frac{1}{2}y_0 + y_1 + y_2 + \cdots + y_{n-1} + \frac{1}{2}y_n\right]. \tag{A.4.2}$$

2. Simpson's $\frac{1}{3}$rd Rule (for n even):

$$\int_a^b f(x)dx \cong \int_{x_0}^{x_0+nh_0} y\,dx = \frac{h}{3}[y_0 + 4\{y_1 + y_3 + \cdots + y_{n-1}\}$$
$$+ 2\{y_2 + y_4 + \cdots + y_{n-2}\} + y_n], \text{ where } y = f(x). \tag{A.4.3}$$

3. Simpson's $\frac{3}{8}$rd Rule:

$$\int_a^b f(x)dx = \int_{x_0}^{x_0+nh} = \frac{3h}{8}[y_0 + 3(y_1 + y_2 + y_4 + y_5$$
$$+ \cdots + y_{n-2} + y_{n-1} + 2(y_3 + y_6 + y_9 + \cdots + y_{n-3}) + y_n]$$

where n is a multiple of 3.

4. Euler-Maclaurin's formula:

$$\frac{1}{h}\int_a^{a+nh} f(x)dx = \frac{1}{2}f(x) + f(a+h) + \cdots + f(a+(n-1)h)$$
$$+ \frac{1}{2}f(a+nh) - \frac{h}{12}[f'''(a+nh) - f'''(a)] + \frac{h^3}{720}[f^{(v)}(a+h) - f^{(v)}(a)]$$

A.5 Survival Function and Hazard Function

Failure (or death) is inherent in any machine (or living being). The time to failure (or death) T is often called 'Life Time'; it is a r.v. The distribution $F(t)$ of T is called failure time or life time distribution, it gives probability of death on or before time t, i.e.

$$F(t) = P[T \leq t]. \tag{A.5.1}$$

Survival function $S(t)$ (or *Reliability* $R(t)$) is defined as

$$S(t) = P[T \geq t] = 1 - F(t) = R(t). \tag{A.5.2}$$

The *hazard function* (or failure rate or mortality rate or instantaneous force of mortality), $h(t)$, is defined as,

$$h(t) = \frac{f(t)}{1 - F(t)} = \frac{f(t)}{S(t)} = \frac{f(t)}{R(t)}. \tag{A.5.3}$$

Here $h(t)dt$ is the conditional probability of dying (or failure) in the interval t to $t + dt$, on the condition that it survived up to time t.

From (A.5.3),

$$-\log[1 - F(t)] = \int_0^t h(u)du$$

or
$$1 - F(t) = S(t) = \exp\left[-\int_0^t h(u)du\right] \tag{A.5.4}$$

Further, *mean survival time* (or mean life time) is,

$$E(T) = \int_0^\infty t\, f(t)dt = -\int_0^\infty t\, dS(t), \text{ from (A.5.2)}$$

$$= \int_0^\infty S(t)dt. \tag{A.5.5}$$

A.6 Mathematical Models

'The principle of mathematical modeling has now become so widely accepted in the Physical, Biological and Social Sciences that it scarcely seems necessary to present a case for it', Golomb (1971).

In mathematical modeling same mathematical equation or inequalities are setup among observed variables of a phenomenon in order to explain and give quantitative measures to the characteristics of the phenomenon.

In order that a chosen model should serve a useful purpose it should yield an adequate idealized description of the phenomenon in terms of the mathematical concepts used. If it is verified that a proposed model has a sufficiently accurate and permanent agreement with observed facts, it may be used for deduction of the characteristics of the phenomenon, for prediction about future behavior of the phenomenon and for practical decision-makings.

The mathematical models can be broadly divided into two groups: (a) Deterministic and (b) Stochastic.

(a) A model is *deterministic* if all the variables (observable as well as unobservable) and the relationship among them are deterministic and randomness has no role to play.

(b) In *stochastic* models, at least some of the variables involved are regarded as r.v.'s the probability distributions of which are more or less known.

References

1. Biswas, S. (1991): Topics in Statistical Methodology, Wiley Eastern Ltd., New Delhi.

2. Biswas, S. (1997): A Text Book of Matrix Algebra, New Age International Publishers (P) Ltd., New Delhi.

3. Biswas, S. (2003): Statistics of Quality Control Sampling Inspection and Reliability, New Central Book Agency (P) Ltd., Kolkata.

4. Golomb, S.W. (1971): Mathematical Models: Uses and Limitations, IEEE Trans. on Reliability, R-20, pp. 130–131.

5. Malic, S.C. and Arora, S. (1992): Mathematical Analysis, New Age International Publishers (P) Ltd., Calcutta.

6. Mukhopadhyay, P. (1996): Mathematical Statistics, New Central Book Agency (P) Ltd., Calcutta.

7. Scarborough, J.B. (1966): Numerical Mathematical Analysis, Oxford & IBH Publishing Co. Pvt. Ltd., New Delhi.

8. Siegel, S. (1966): Nonparametric Statistics for Behavioural Sciences. McGraw Hill Pvt. (Ltd.), New Delhi.

9. Smith, M.G. (1966): Laplace Transform Theory, D. Van Nostrand Co. Ltd., London.

INDEX